Cell Biology Monographs

Volume 7

Springer-Verlag
Wien GmbH

Peter Böck, Robert Kramar, and Margit Pavelka

Peroxisomes and Related Particles in Animal Tissues

Springer-Verlag
Wien GmbH

Prof. Dr. PETER BÖCK

Anatomisches Institut
Technische Universität München, Federal Republic of Germany

Doz. Dr. ROBERT KRAMAR

Institut für Medizinische Chemie
Universität Wien, Austria

Dr. MARGIT PAVELKA

Institut für Mikromorphologie und Elektronenmikroskopie
Universität Wien, Austria

ISBN 978-3-7091-2057-6   ISBN 978-3-7091-2055-2 (eBook)
DOI 10.1007/978-3-7091-2055-2

© 1980 by Springer-Verlag Wien
Originally published by Springer Vienna in 1980.

With 60 Figures

Library of Congress Cataloging in Publication Data. Böck, Peter. Peroxisomes and related particles in animal tissues. (Cell biology monographs; v. 7.) Bibliography: p. Includes index. 1. Microbodies. 2. Tissues—Analysis. I. Kramar, Robert, 1934—joint author. II. Pavelka, Margit, 1945—joint author. III. Title. IV. Series. [DNLM: 1. Organoids. 2. Microbodies. W1 CE128H v. 7/QH603.M35 B669p.] QH603.M35B63. 591.87'34. 80-19480

ISSN 0172-4665

Dedicated to Prof. Dr. L. Stockinger
on the occasion
of his 60th birthday

Preface

In modern scientific investigation the fields of biochemistry, molecular biology, and morphology comprise an indivisible area of study. The present book results from the cooperation of a biochemist and morphologists: the revision and unified treatment of available data is the primary object of our work. A comprehensive review of all the available literature is therefore beyond the scope of this volume. It is intended to be a manual to be used in the laboratory, with convenient guidelines for practical work.

Plant microbodies have been treated by B. Gerhardt in Volume 5 of this series. The discovery of fatty acid β-oxidation in animal peroxisomes has proved once more that plant and animal microbodies are members of the same family of organelles. It provided new insights into the physiological meaning of these particles; our understanding of these "classical" cell organelles is undergoing continual alteration and development.

Vienna, July 1980

Peter Böck
Robert Kramar
Margit Pavelka

Acknowledgements

We wish to express our gratitude to Prof. Dr. D. H. FAHIMI and Dr. P. KALMBACH (Heidelberg) for kindly providing Figure 14, to Prof. Dr. KARIN GORGAS (Heidelberg) for allowing the reproduction of Figure 43, to Profs. Dr. L. STOCKINGER and Dr. E. KAISER for helpful criticism, and to all our colleagues in our respective institutes. We are especially grateful to Drs. H. GOLDENBERG and M. HÜTTINGER for continuous discussion, to Mrs. JUTTA SELBMANN for typing the references, and to Mr. P. KAMPFER and Mr. H. WAGNER for carefully drawing some of the figures.

We are indebted to Dr. W. SCHWABL and Dr. E. UNGERSBÄCK, Springer-Verlag, Vienna, for their patient cooperation.

One of us (R. K.) was supported by "Fonds zur Förderung der wissenschaftlichen Forschung", Projects 2368 and 3294.

Contents

Abbreviations

AIA	allylisopropylacetamide
APAD	acetylpyridine analogue of NAD
CPs	catalase positive particles
DAB	3,3'-diaminobenzidine
DEHP	Di (2-ethylhexyl)phthalate
DHAP	dihydroxyacetone phosphate
DNA	desoxyribonucleic acid
DNB	5,5'-dithiobis (2-nitrobenzoic acid)
EDTA	ethylenediaminetetraacetic acid
E-face	convex fracture of microbody membrane
FAD	flavin adenine dinucleotide
FMN	flavin mononucleotide
IU	international unit of enzymatic activity
L	light mitochondrial fraction
LDH	lactate dehydrogenase
NAD	nicotinamide adenine dinucleotide
NADP	nicotinamide adenine dinucleotide phosphate
NBT	nitro blue tetrazolium
OD	optical density
P-face	concave fracture of microbody membrane
PVP	polyvinylpyrrolidone
RNA	ribonucleic acid
rpm	revolutions per minute
RSA	relative specific activity
SE	standard error
SD	standard deviation
SDS	sodium dodecyl sulfate
TCA	trichloroacetic acid
TNBT	tetranitro blue tetrazolium

I. Introduction

I.A. Catalase and Hydrogen Peroxide–Producing Oxidases in Subcellular Particles

Microbodies were first described in the cytoplasm of mouse proximal kidney tubules (RHODIN 1954, 1958) and in rat hepatocytes (ROUILLER and BERNHARD 1956). Far from being recognized as cell organelles *sui generis,* they were considered to be precursors of mitochondria (ROUILLER and BERNHARD 1956).

The recognition of microbodies as biochemically defined organelles was closely related to the development and improvement of cell fractionation methods. It started with a differential centrifugation study on rat liver tissue (DE DUVE *et al.* 1955). This work led to the concept that lysosomes are organelles that house a group of hydrolytic enzymes of marked latency. Urate oxidase was also studied, and DE DUVE *et al.* (1955) suspected that urate oxidase (which is essentially insoluble and shows a distribution different from mitochondrial and lysosomal markers) might be "associated with a special group of granules resembling large microsomes rather than small mitochondria and differing from lysosomes by a greater uniformity in sedimentation properties". However, the alternative possibility of firm binding of urate oxidase to lysosomes has not been excluded.

THOMSON and KLIPPEL (1957) introduced catalase as a second marker enzyme. Using rate sedimentation in a sucrose gradient, they were able to demonstrate that in mouse liver catalase and urate oxidase are bound to particles of identical size distribution.

D-Amino acid oxidase has been shown to sediment with the lighter mitochondrial fraction (PAIGEN 1954); actually its distribution pattern resembles that of catalase and urate oxidase (BAUDHUIN *et al.* 1964) (Fig. 1). The final conclusion that catalase, urate oxidase, and D-amino acid oxidase are located in distinct particles different from lysosomes, was drawn from results obtained by isopycnic gradient centrifugation. DE DUVE *et al.* (1960) and BEAUFAY *et al.* (1964) systematically varied density gradients of glycogen (plus a constant proportion of sucrose) and of sucrose (with water or D_2O as solvents). Starting with a large-particle (premicrosomal) fraction of rat liver tissue, these authors demonstrated that the distribution of the three enzymes mentioned differs from the distribution of mitochondrial and lysosomal markers. In sucrose gradients (under optimal conditions) the separation of these enzymes from cytochrome oxidase, the mitochondrial marker, was nearly perfect. The dissociation of catalase, urate oxidase, and D-amino acid oxidase from lysosomal markers (acid hydrolases) became particularly obvious in glycogen gradients (Fig. 2, two histograms are shown for comparison). The extent of separation was measured in terms of area displacement (percentage of total activity of an enzyme not overlapping with the second one considered). These findings led BEAUFAY et al. (1964) to the conclusion "that urate oxidase, catalase, and D-amino acid oxidase are associated together and ... with a single group of particles ... probably identical with the so-

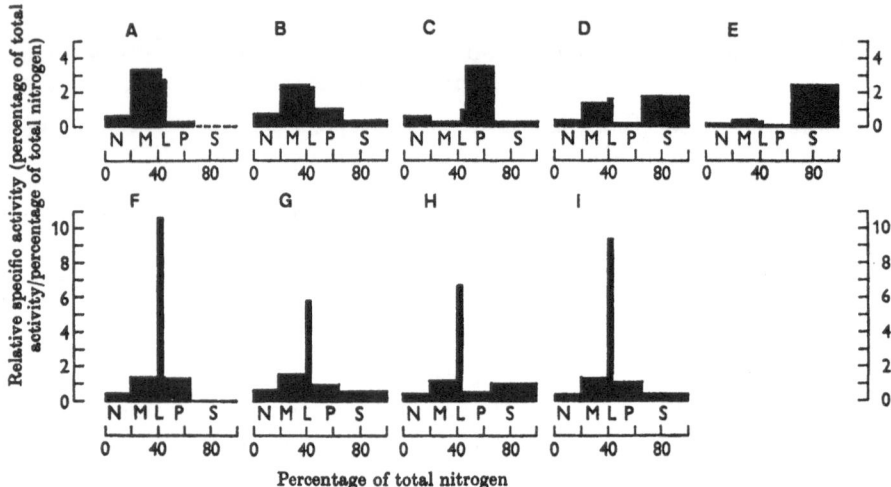

Fig. 1 A–I. Distribution pattern of enzymes as histogram ("Duvogram"). Five-fraction centrifugation scheme (APPELMANS et al., 1955). Fractions: N nuclear; M heavy mitochondrial; L light mitochondrial; P microsomal; S soluble (from left to right, in order of their isolation). A Cytochrome oxidase. B Monoamine oxidase. C Glucose-6-phosphatase. D Aspartate aminotransferase. E Alanine aminotransferase. F Urate oxidase. G D-Amino acid oxidase. H Catalase. I Acid phosphatase. From: BAUDHUIN et al. 1964.

called 'microbodies' ... The only link between the three enzymes so far identified as constituents of the new particles appears to be hydrogen peroxide, which is formed by the oxidases and destroyed by catalase" (see also BAUDHUIN and BEAUFAY 1963, DE DUVE et al. 1960).

The addition of microbodies to the group of biochemically defined particles (i.e., mitochondria, lysosomes, and microsomes) was accomplished by indirect evidence, via exclusionis; when it turned out that pericanalicular dense bodies of the liver are the morphological correlate of lysosomes, microbodies became appropriate candidates for the particles carrying catalase and the above mentioned oxidases. The unequivocal proof that microbodies are identical with these particles came from BAUDHUIN et al. (1965 a), who studied liver cell fractions from sucrose or glycogen-sucrose gradients (BEAUFAY et al. 1964). Electron microscopy of osmium tetroxide-fixed preparations revealed the predominance of microbodies in specimens rich in catalase, urate oxidase, and D-amino acid oxidase. When preparations of analogous fractions from the

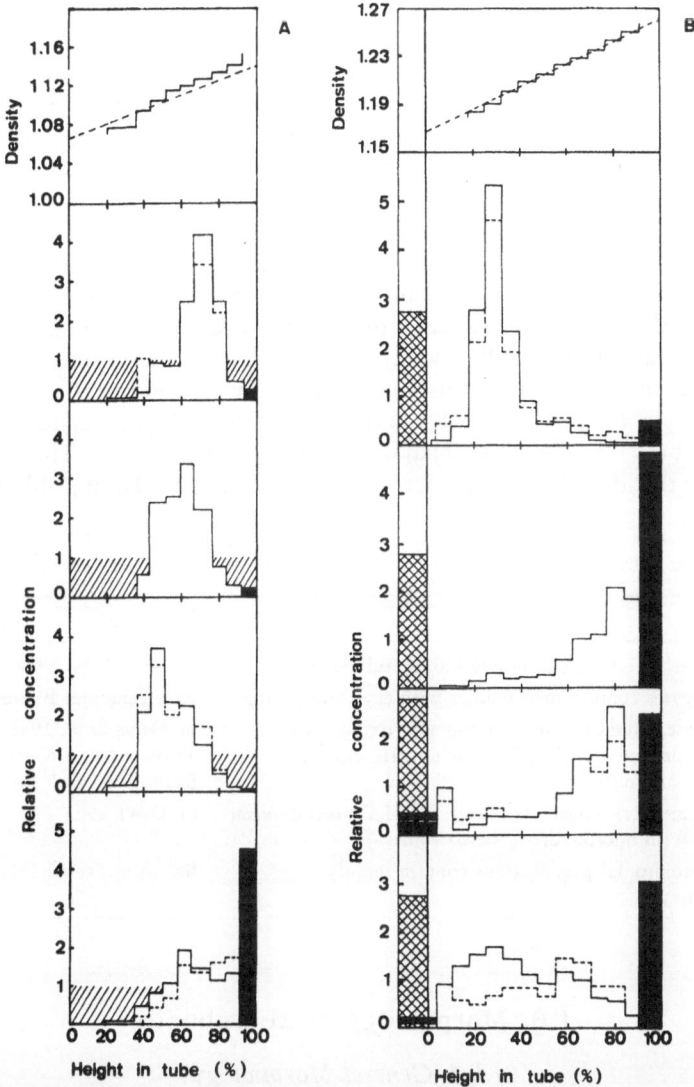

Fig. 2. Centrifugation of a "large-particle" (mitochondrial) fraction (prepared from rat liver as illustrated in Fig. 27) in two different gradients. *A* Linear glycogen-sucrose gradient: density range 1,068–1,142; light end (top), 0.5 M sucrose; heavy end, 0.5 M sucrose plus 30.6 g glycogen/100 g water. *B* Linear sucrose gradient; density range 1,170–1,260; light end (top), 1.3 M sucrose; heavy end, 2.0 M sucrose. Graphs (from top down) demonstrate the following: *1* densities measured in the fractions (———) and in the starting gradient (--------); *2* cytochrome oxidase (———) and protein (-------); *3* urate oxidase; *4* catalase (———) and D-amino acid oxidase (-------); *5* acid phosphatase (———) and acid deoxyribonuclease (-------). *Relative concentration:* concentration of the component in the fraction divided by its concentration if it were evenly distributed throughout the gradient. *Height in tube* (%): the percentage height of the fraction in a cylindrical tube in practice corresponds to its percentage volume (see Section II.A.1.a.). *Black bars:* enzyme activities found in the fractions out of the gradient (material not entering the gradient or being pelleted). *Shaded area* (below the ordinate 1): this area corresponds to 100% of the relative amount. Comparison with the distribution pattern after centrifugation shows the rearrangement during separation. From: BEAUFAY *et al.* 1964.

liver of rats injected with Triton WR-1339 4 days before examination [this procedure lowers the density of lysosomes and therefore facilitates their separation from microbodies by isopycnic density gradient centrifugation (cf., WATTI-AUX et al. 1963)] were studied, even more clear-cut morphological results were obtained. The recognition of L-α-hydroxy acid oxidase as another enzyme that is exclusively located in microbodies (BAUDHUIN et al. 1965 b) fully confirmed these results [LEIGHTON et al. 1968; in this study particles were prepared on a large-scale zonal rotor (see Section II.A.1.b.) from the liver of Triton WR-1339-treated animals]. These experiments revealed the full congruence of the microbody as an organelle, which had been discovered merely by morphological methods, with an enzymologically defined cell compartment (BEAUFAY et al. 1964, DE DUVE 1960). DE DUVE (1965) coined the name "peroxisome" for this functional entity (see Section I.C.). The discoveries leading to the identification of peroxisomes are summarized in Table 1.

Table 1. *Discoveries leading to the recognition of microbodies as "peroxisomes"*

Discovery	Reference
"Microbodies" are found in mouse kidney tubule cells	RHODIN 1954, 1958
Rat hepatocytes contain microbodies with crystalline cores	ROUILLER and BERNHARD 1956
Urate oxidase, catalase, and D-amino acid oxidase are possibly connected with cell particles different from lysosomes and mitochondria	DE DUVE et al. 1955 THOMSON and KLIPFEL 1957 BAUDHUIN et al. 1964
Are these particles, which are centers of H_2O_2 metabolism, identical with microbodies ("peroxisomes")?	DE DUVE 1965
Purified peroxisomal preparations contain mainly microbodies	BAUDHUIN et al. 1965 a

I.B. Morphology of Microbodies

I.B.1. General Morphology

Spherical cell organelles with diameters up to light microscopic dimensions (0.5 μm in hepatocytes, 3.0 μm in epithelial cells of proximal kidney tubules) that are bounded by a single trilaminar membrane and contain matrix material slightly more dense than the cytoplasm have been termed "microbodies". These organelles were first described in proximal tubules of mouse kidney (RHODIN 1954) and have been identified as regular constituents of hepatocytes (GÄNSLER and ROUILLER 1956, ROUILLER and BERNHARD 1956), where they have been thought to be progenitors of mitochondria. The matrix material is homogeneous to finely granular. The limiting unit membrane is 6–8 nm thick. No halo is observed between matrix and limiting membrane. A close spatial relationship between smooth-surfaced endoplasmic reticulum and microbodies has been noted (NOVIKOFF and SHIN 1964). Continuities between microbodies and smooth endoplasmic reticulum have frequently been described. From these observations it has been postulated that microbodies do not exist as

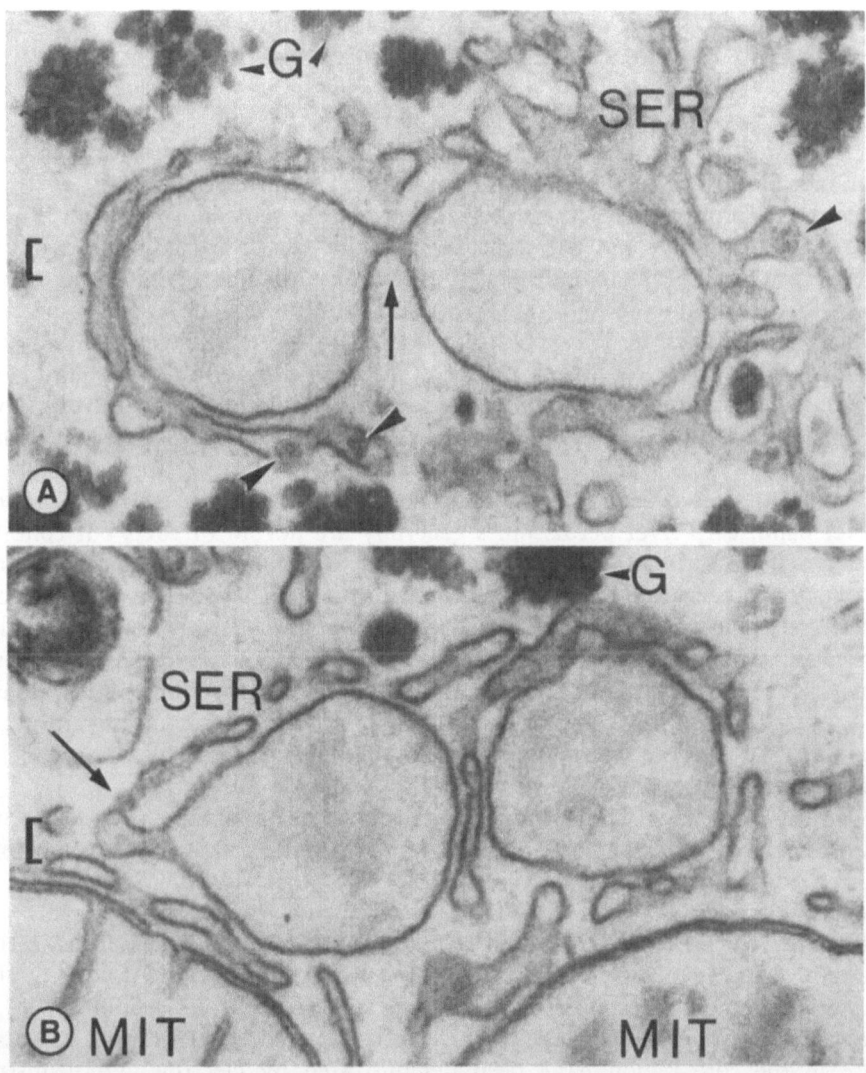

Fig. 3 A, B. Membrane continuities between peroxisomes (A) and between peroxisomes and smooth endoplasmic reticulum (B) are indicated by *arrows*. Tubules and cisternae of smooth endoplasmic reticulum (*SER*) contain very low density lipoproteins (*arrowheads* in part A), which may be of functional significance. *Bars at lefthand margins:* approximate section thickness. G Glycogen; *MIT* mitochondria. Glutaraldehyde/prereduced osmium fixation, mouse hepatocytes. ×100,000.

individual free organelles but appear to be reversible dilatations of smooth endoplasmic reticulum (REDDY and SVOBODA 1973 b) (Fig. 3 A, B). Morphological examination of isolated fractions obtained from rat liver tissue definitely has shown that microbodies are identical to peroxisomes (BAUDHUIN et al. 1965 a). Direct evidence has been obtained by the cytochemical demonstration of catalase via its peroxidatic activities (FAHIMI 1969).

"Microbody" is a morphological, merely descriptive term. The term "peroxisome" includes a biochemical concept (see Section I.C.).

I.B.2. Nucleoids

Mammalian liver and kidney tissue has been the main source of material for morphological and biochemical studies of peroxisomes. In most of these cases the microbody matrix displays a central or eccentric density which may exhibit a crystalline pattern. The terms "core" or "nucleoid" are used for these structural peculiarities. The presence of nucleoids in microbodies has been found to coincide with the occurrence of urate oxidase in the respective tissue (HAYASHI et al. 1976 a, HUANG and BEEVERS 1973, LATA et al. 1977, LEIGHTON et al. 1969, SHNITKA 1966); therefore the term "uricosome" was proposed (AFZELIUS 1965). However, several exceptions to this rule have been noted. In rat kidney, for instance, no urate oxidase activity is found (BAUDHUIN et al. 1965 b), although various types of cores are seen in peroxisomes of this tissue (HRUBAN and RECHCIGL 1969) (see below). On the other hand, carp liver peroxisomes generally are devoid of cores but contain urate oxidase activity (KRAMAR et al. 1974). As a matter of fact, the morphological observation of crystalline cores or densities has lost its significance. However, it must be emphasized that urate oxidase is firmly attached to the cores when nucleoids and the enzyme are present in microbodies. This led to the confusion of crystalline cores of microbodies with crystalline urate oxidase (HRUBAN and SWIFT 1964). In peroxisomes that lack a core but contain urate oxidase, the enzyme seems to be bound to the membrane of the organelle (carp liver, GOLDENBERG 1977 a, b).

Recent biochemical studies have shown that other peroxisomal oxidases are also associated with the core, namely D-amino acid oxidase and L-α-hydroxy acid oxidase (HAYASHI et al. 1971, 1973, 1976 a). In addition to oxidases, a density-lowering material rich in cholesterol, the "light compartment", is associated with the core (HAYASHI et al. 1976 a). Catalase is dissolved in the matrix. This organization prompts the speculation that the aim of cores might be to obtain high concentrations of oxidases homogeneously distributed within low concentrations of catalase. Significantly lowered concentrations of catalase in the core region have been demonstrated by electron cytochemistry (FAHIMI 1969) (Fig. 4 A–C). It has been pointed out by BAUDHUIN et al. (1965 a) that microbody cores must also contain proteins other than urate oxidase.

The term "core" is used by morphologists to designate regions of higher electron density within the microbody matrix. Densities with a highly organized structure are called "crystalloid". The classification scheme proposed by HRUBAN and RECHCIGL (1969) is presented in Table 2.

The crystalloid of peroxisomes is generally formed by tightly packed tubules, called "primary tubules". The primary tubules may be isolated or may be arranged in polytubular crystalloids. The diameter of primary tubules ranges from 4.5 to 5 nm, and the wall is 5 nm thick. Secondary tubules are composed of 6, 10, or 12 primary tubules in a circular arrangement. Secondary tubules with a 1 : 10 pattern build up coarsely polytubular crystalloids,

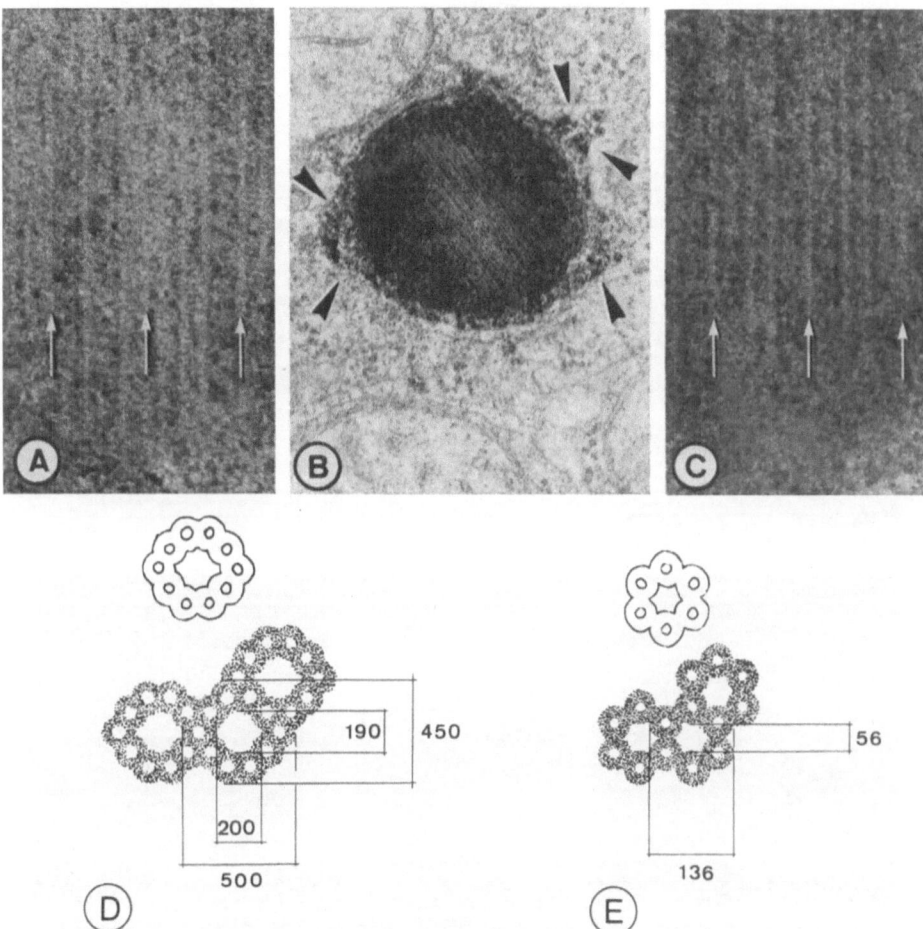

Fig. 4 A–E. Rat liver peroxisomes stained for peroxidatic activity of catalase. *A* High magnification of the central region of the crystalline core seen in part B. ×160,000. *B* Moderate magnification. Part of the reaction product is seen outside the organelle, mostly attached to the surfaces of adjacent lamellae of endoplasmic reticulum (*arrowheads*). This phenomenon is due to diffusion of catalase and/or reaction product. No staining of ribosomes more distant from the peroxisome is observed. Mitochondria also do not react under these conditions. Peroxidatic activity of catalase is low in the region of the nucleoid. ×30,000. *C* Structural details are accentuated by linear image superimposition photography. Two varieties of light lines are clearly demonstrated: thicker ones and thinner ones. There is either one light line accompanied by two thinner lines at each side (*arrows*), or one thin line accompanied by two thicker and lighter lines (not indicated). This picture fully corresponds to results obtained after negative staining of isolated cores (cf., TSUKADA *et al.* 1966). ×160,000. *D* and *E* Crystalline cores as found in liver peroxisomes of the rat, mouse, cat, and rabbit (part *D*, so-called 1 : 10 pattern), or in guinea pig peroxisomes (part *E*, so-called 1 : 6 pattern). Dimensions are given in angstroms (0.1 nm) for rat microbody nucleoids (in part *D*) and guinea pig microbody nucleoids (in part *E*) according to the work of TSUKADA *et al.* (1966, 1971).

8 Introduction

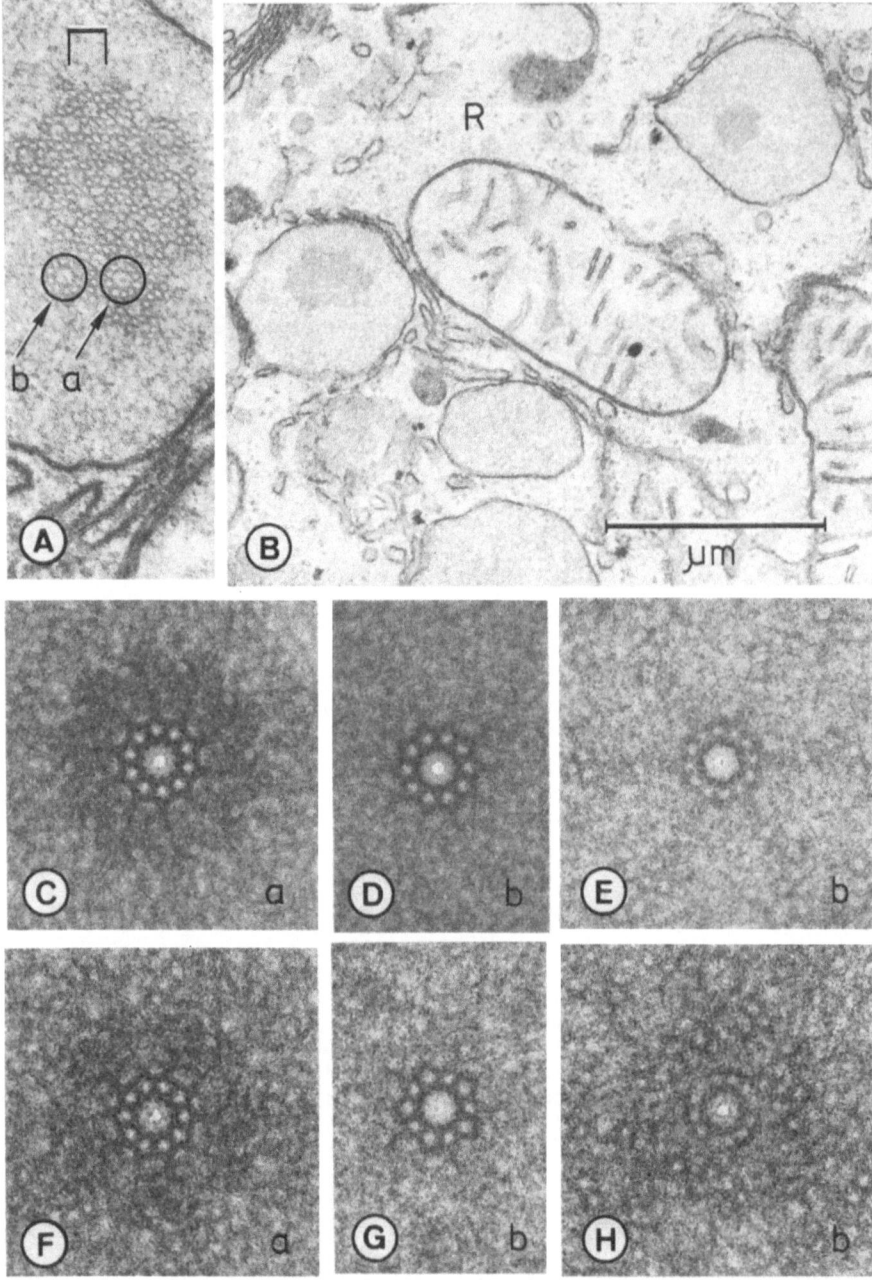

Fig. 5 A–H. Mouse liver peroxisomes. *A* Detail of the nucleoid seen at the left margin in part *B*. *Bar:* 100 nm. *a, b:* These encircled regions of the crystalloid were used to accentuate structural details by image superimposition photography in *C–H*. ×60,000. *B* Low-power electron micrograph showing four microbody profiles, two with and two without a crystalline core. ×30,000. *C* and *D* Photographs taken at rotation number 10. *F* and *G* The corresponding regions demonstrated at rotation number 5. *E* Photograph taken at rotation number 12. *H* The identical region at rotation number 6.

secondary tubules with a 1 : 6 or 1 : 12 pattern form finely polytubular crystalloids (Figs. 5 and 6). Linear crystalloids are composed of tubules arranged in single rows.

A comprehensive and detailed description of microbody crystalloids in various species has been given by HRUBAN and RECHCIGL (1967, 1969). It is

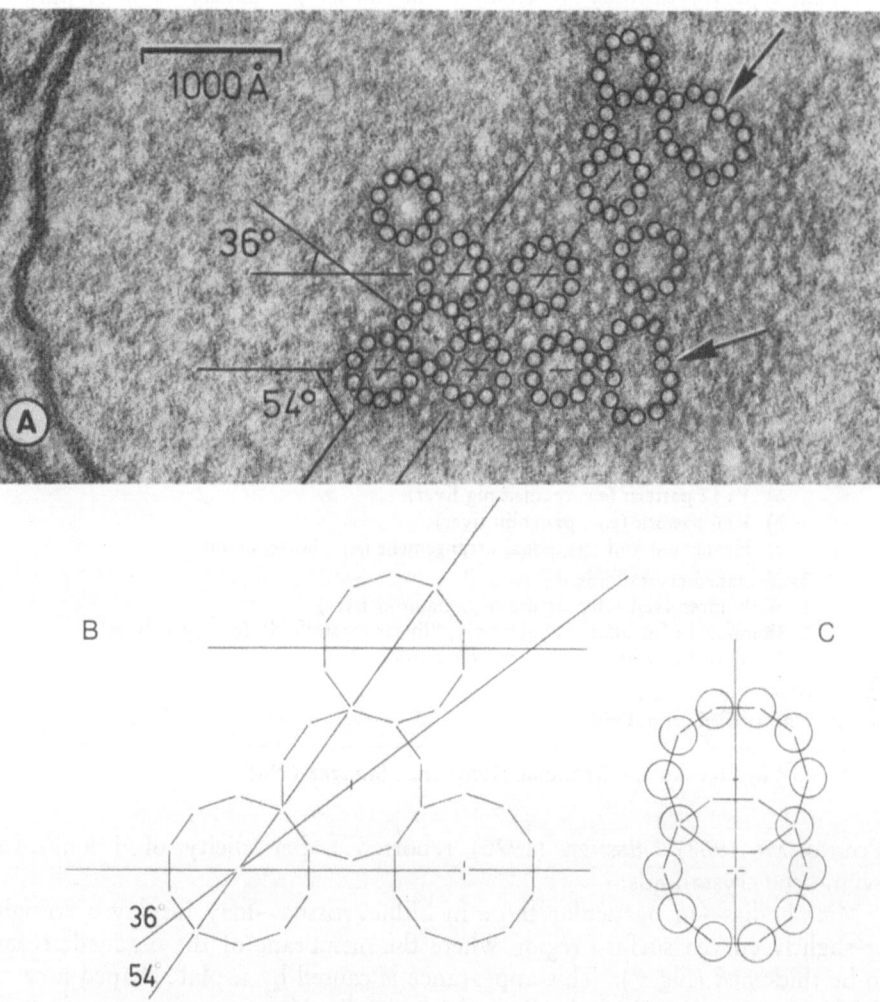

Fig. 6 A–C. Tubules in crystalline core from mouse liver peroxisome. *A* High magnification of a cross-sectioned core. The nucleoid is composed of primary tubules with an outer diameter of about 10 nm. Cylinders are formed by 10 of these tubules, as indicated; they are called secondary tubules. Secondary tubules either remain isolated or adhere to other secondary tubules to form more complex structures. In the latter case, secondary tubules share 2 of the 10 subunits. The triangles that are then defined by the centers of the secondary tubules are indicated (and schematically illustrated in *B*). The decagonal array of two of the secondary tubules may also be interlaced (*arrows*) (schematically illustrated in *C*). Fragments of decagonal arrays of primary tubules occur but are not indicated.

only briefly noted here that coarsely polytubular crystalloids with secondary tubules of the 1 : 10 pattern are usually found in hepatic microbodies of the rat, dog, cow, cat, and pig. In mouse and hamster hepatocytes crystalloids of the linear type are generally seen (although crystalloids of the compact type may also be observed; Fig. 6). Finely polytubular crystalloids are found in guinea pig hepatic peroxisomes.

Avian-type crystalloids are rhomboidal, showing a substructure of parallel dense lines (100 nm periodicity) that are about 35 nm thick (SHNITKA and

Table 2. *Proposed classification of cores (or nucleoids)*

I. Coreless (or anucleoid) microbodies

II. Noncrystalloid cores with
   A. Low electron density (also called "density")
   B. High electron density (also called "nucleoid")

III. Crystalloids (cores with recognizable regular substructure)
   A. Isolated tubules
      1. Primary (*e.g.*, kidney of mud puppy, tobacco roots)
      2. Secondary (*e.g.*, cotton rat)
   B. Compact polytubular crystalloids
      1. Coarsely polytubular crystalloid, 1 : 10 pattern (*e.g.*, rat liver)
      2. Finely polytubular crystalloid
         a) 1 : 12 pattern (*e.g.*, guinea pig liver)
         b) 1 : 6 pattern (*e.g.*, pangolin liver)
         c) Hexagonal and tetragonal arrangement (*e.g.*, horse, plants)
   C. Band-shaped crystalloids
      1. With unresolved substructure (*e.g.*, hamster liver)
      2. Composed of primary tubules only, "linear crystalloid" (*e.g.*, goat liver)
      3. Composed of primary and secondary tubules
   D. Avian type
   E. Unclassified crystalloids

Modified by HRUBAN and RECHCIGL (1969) from SHNITKA (1966).

YOUNGMAN 1968). ESSNER (1970) reported a periodicity of 115 nm for avian-type crystalloids.

Microbodies—in particular those in kidney tissue—may display a straight or slightly curved surface region where the membrane of the organelle seems to be thickened (Fig. 7). This appearance is caused by a plate-shaped area of condensed matrix, called the "marginal plate". Marginal plates are located in the periphery of microbodies; they are separated from the limiting membrane by a narrow electrolucent space. The plate is only slightly thicker than the limiting membrane and is very electron-dense. It gives an uneven edge to otherwise round microbodies. In obliquely or tangentially sectioned marginal plates periodicities of 10, 20, or 30 nm can be identified (rat renal proximal tubules, BARRETT and HEIDGER 1975). The presence of one or more marginal plates within a peroxisome does not exclude the presence of nucleoids. Marginal plates react when incubated in the cytochemical catalase medium (whereas crystalline cores are not reactive; *cf.*, Figs. 4 B and 31 B). A short incubation time reveals that the enzyme activity is higher in marginal plates than in the matrix material (BARRETT and HEIDGER 1975).

Circular profiles are frequently seen in rat renal microbodies. These profiles are formed by 6-nm granules separated by a 4-nm space. Series of circular profiles make up cylinders (these are also called "tubular profiles"). The

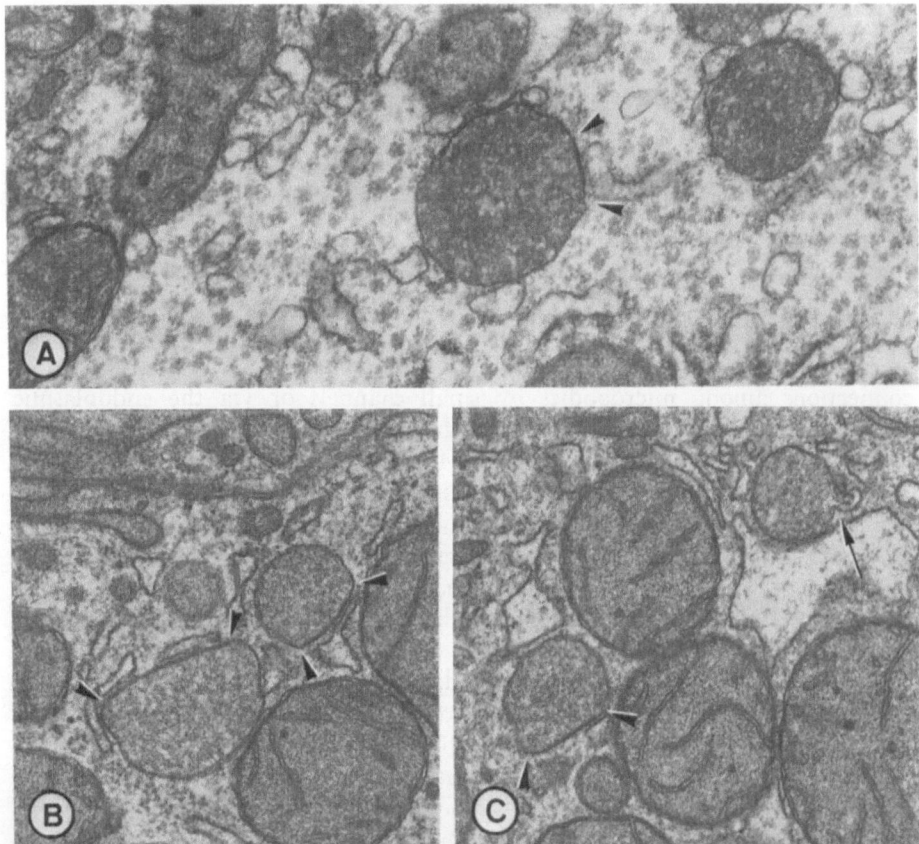

Fig. 7 A–C. Peroxisomes in human liver (*A*) and in human kidney proximal tubules (*B*, *C*). *Arrowheads:* marginal plates. *Arrow* (in part *C*): connection between microbody and smooth endoplasmic reticulum. Glutaraldehyde/osmium tetroxide. *A* ×40,000; *B*, *C* ×30,000.

diameter of cylinders or tubular profiles ranges from 85 to 140 nm. Occasionally, duplication of tubular profiles is seen: the cylinders are concentrically arranged and separated by a 15-nm space. Tubular profiles are preferentially located near the limiting membrane of the peroxisome, frequently causing deformation of the spherical organelles (Figs. 13 and 31 B). Like marginal plates, tubular or circular profiles react with the histochemical catalase medium (BARRETT and HEIDGER 1975, REDDY *et al.* 1976 b). Transformation of microbody matrix protein(s) to tubular structures with a diameter of approximately 110 nm has been observed in hepatic peroxisomes of hypophysectomized rats after prolonged treatment with clofibrate (REDDY and SVOBODA 1973 a) (see Section III.G.1.a.). The presence of crystalline cores and tubular structures concomitantly in the matrix has been frequently observed. Transitional forms between microbodies with large tubular inclusions and the so-called Φ-bodies (see Section III.F.) have been identified (REDDY and SVOBODA 1973 a). Comparable tubular profiles were seen in microbodies of LEYDIG cell tumors of rat testis. These structures stain with the histochemical catalase method (REDDY and SVOBODA 1972 c, d, REDDY et al. 1973 a), as do tubular profiles induced by clofibrate in hepatocytes of hyperblastic nodules of rats fed 2-acetylaminofluorene (TSUKADA et al. 1975 a).

I.B.3. Continuities Between Microbodies and Between Microbodies and Endoplasmic Reticulum

Microbodies are often very close to the endoplasmic reticulum. Cisternae of the endoplasmic reticulum are wrapped around the organelles and direct continuities can be seen between microbodies and the endoplasmic reticulum, as well as between microbodies (Figs. 8 and 9). Biochemical studies of catalase turnover in liver peroxisomes have indicated that hepatic microbodies behave as a single organelle (POOLE et al. 1970). Thus, morphological observations of connections among microbodies by small channels or via the endoplasmic reticulum are of particular interest. Such morphological evidence has already been published (NOVIKOFF and SHIN 1964, REDDY and SVOBODA 1971 a).

In view of the small diameters of these connections (less than 40 nm), their identification obviously depends upon a favorably oriented section plane. The smaller the organelles are, the greater is the likelihood that these delicate channels will be observable in single sections. The use of a tilting stage in analyzing relatively thick sections has been recommended to overcome this difficulty (NOVIKOFF and NOVIKOFF 1973 a, NOVIKOFF et al. 1973 a). NOVIKOFF et al. claim that many slender continuities between coreless peroxisomes (microperoxisomes; see Section VI.) and endoplasmic reticulum exist, besides a few larger connections 40 nm in diameter. NOVIKOFF and coworkers used a goniometer to demonstrate these slender connections by specimen tilting. The finding that has been interpreted as slender connections characteristically appears in extreme angles of view (+ 30° or — 30°) at opposite sides on the microbody circumference. Either the limiting mem-

---

Fig. 8 A–D. Continuities between smooth endoplasmic reticulum (*SER*) and microbodies. *A* and *B* In rat hepatocytes relatively wide continuities between peroxisomes and the smooth endoplasmic reticulum are seen (*arrowheads*). The electron density of the matrix material continuously decreases in the transient region between peroxisome and endoplasmic reticulum. Conventional fixation with glutaraldehyde/osmium tetroxide followed by uranyl acetate/lead citrate staining of thin sections. ×70,000. *C* and *D* In mouse hepatocytes connections between microbodies are indicated *b* and *c;* continuities between peroxisomes and smooth endoplasmic reticulum may be assumed in regions indicated *a* and *d.* The approximate section thickness is indicated in *D* (*bar within the peroxisome near the right-hand margin*) to demonstrate that the entire channel *c*, which connects two peroxisomes, is included in the section. *RER* Rough endoplasmic reticulum. Fixation with glutaraldehyde and postfixation with prereduced osmium tetroxide (KARNOVSKY 1971), followed by lead citrate staining of the sections. ×55,000

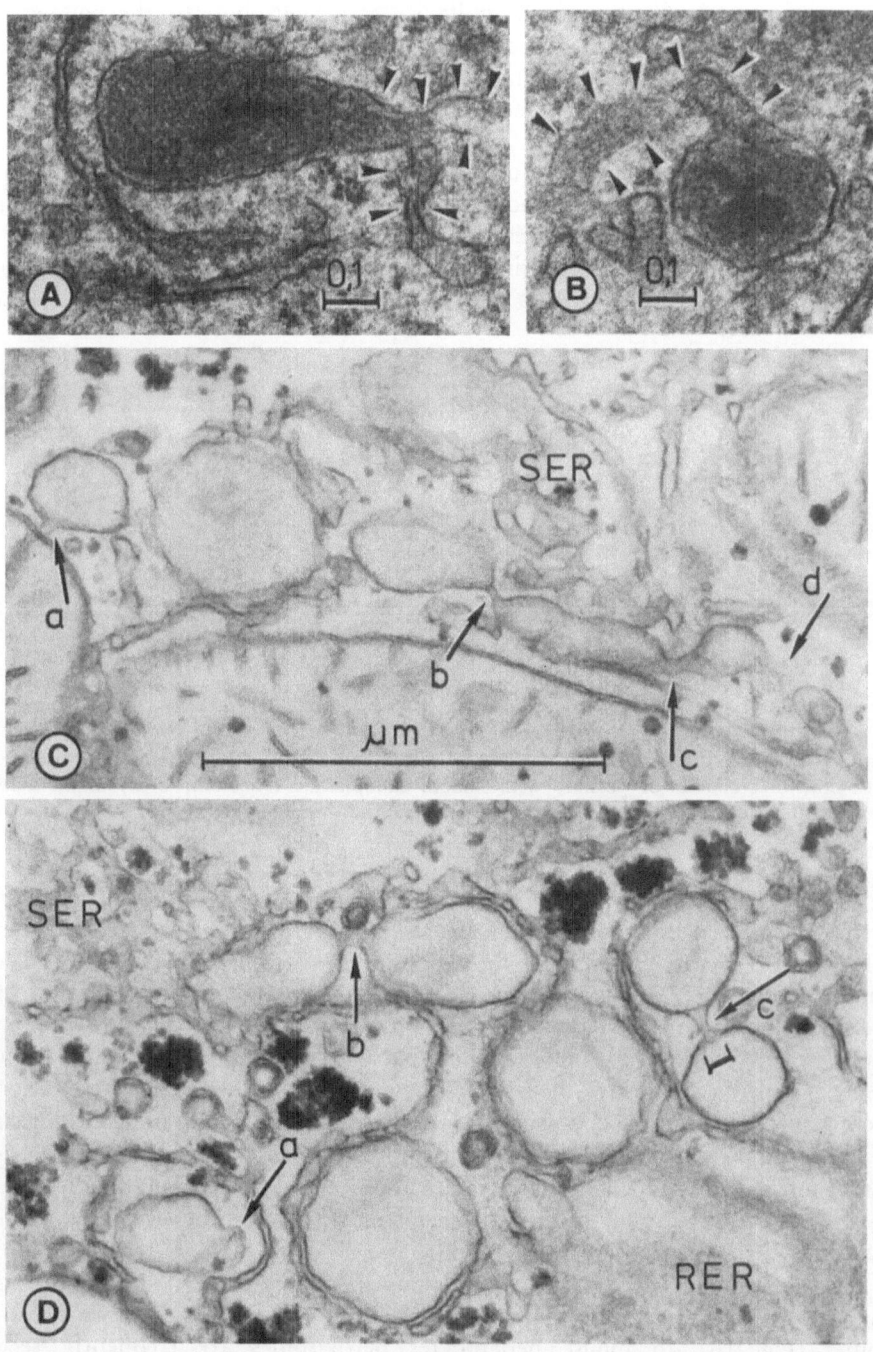

Fig. 8.

brane of the peroxisome or the membrane of the endoplasmic reticulum appears in perpendicular view, while its continuation onto the connected organelle becomes obscure. We feel that multiple slender connections betweeen peroxisomes and endoplasmic reticulum have not been demonstrated convincingly by this method. It seems rather that specimen tilting has produced a misleading superimposition of membranes that suggests membrane continuities. A tilting stage should be used to ascertain whether possible membrane continuities appear separated under a certain angle of view, but not to obtain an angle of view under which membranes seem to be continuous.

Serial sectioning more convincingly demonstrates connections between microbodies and between microbodies and endoplasmic reticulum (Fig. 10). These

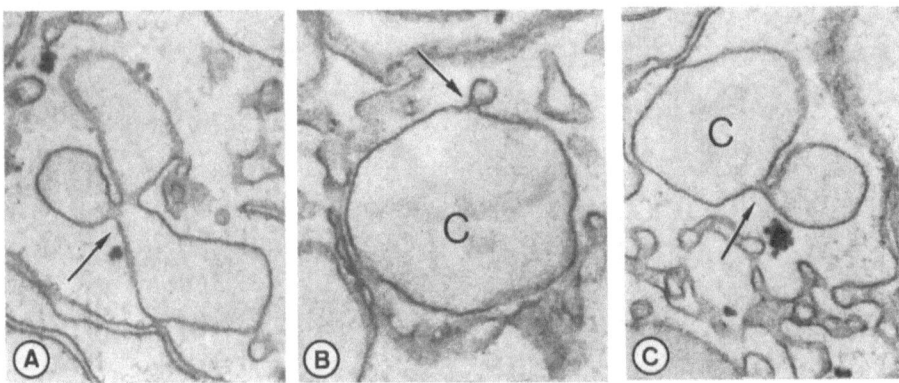

Fig. 9 A–C. Mouse hepatic peroxisomes do not exist as individual entities. *Arrows*, continuities with other microbodies or possibly with smooth endoplasmic reticulum; *C* crystalline core. Fixed with glutaraldehyde, postfixed with prereduced osmium, and stained with uranyl acetate and lead citrate. × 40,000.

narrow channels (with diameters smaller than the section thickness) are seen better in material postfixed with KARNOVSKY's ferrocyanide-reduced osmium solution (1971) than after conventional osmium tetroxide fixation (RUSSELL and BURGUET 1977) (Fig. 8). The electron density and granular texture of the microbody matrix is reduced after fixation with the ferrocyanide-osmium mixture, and nucleoids of the organelles appear less prominent. As cytoplasmic electron density is also low, membranes are clearly discerned. This preparation method has been shown to be suitable for the study of membrane continuities (Figs. 9 and 10). Connections are easily discerned between microbodies, between microbodies and endoplasmic reticulum, and between microbodies and particles of the size of microperoxisomes (for the definition of microperoxisomes see Chapter VI.).

Serial sectioning has shown that in mouse hepatocytes every microbody is connected with the endoplasmic reticulum, either directly or via other microbodies (Fig. 10). In mouse hepatocytes peroxisomes are frequently seen clustered in circumscribed regions (DAEMS 1966); this clustering facilitates the study of serial sections. Morphological results support the biochemical findings, which lead to the assumption that microbodies behave as a single or-

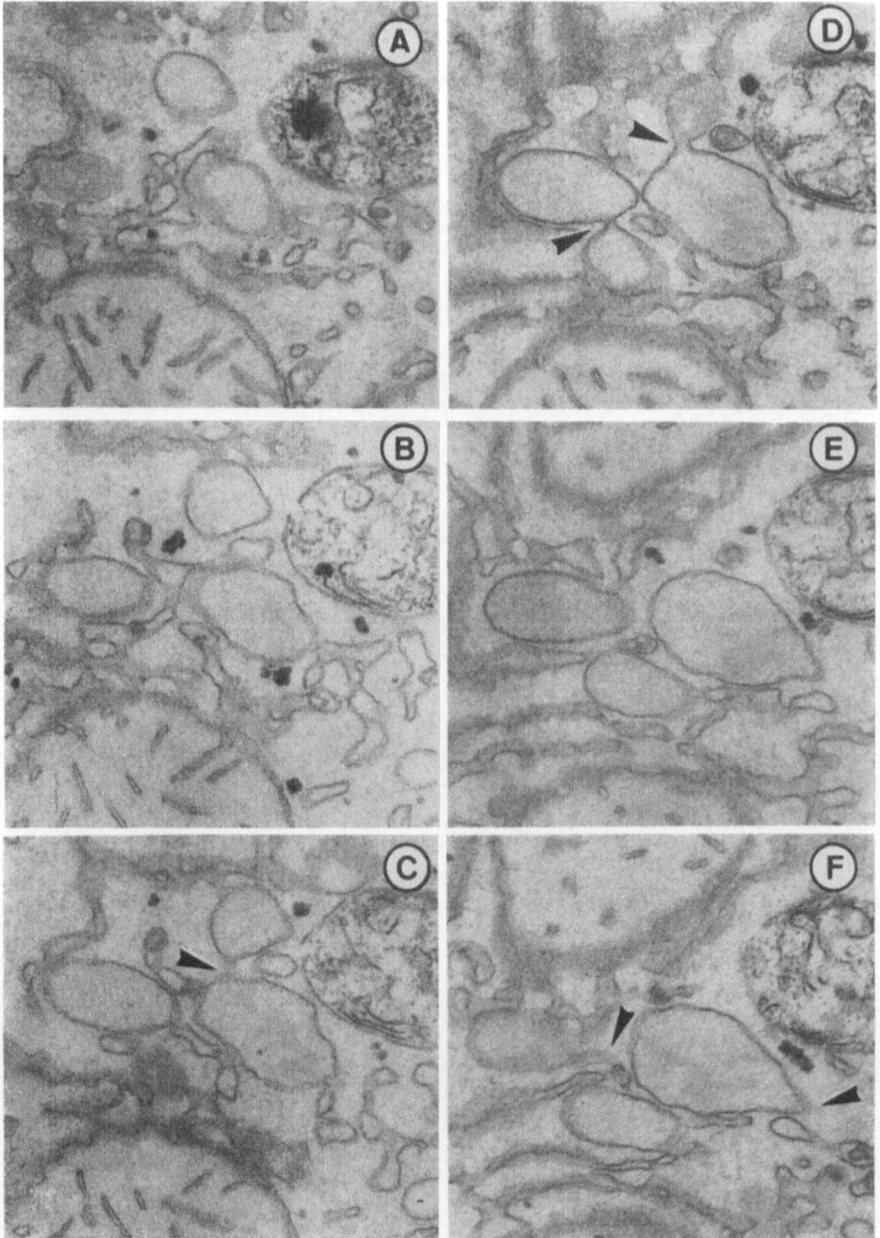

Fig. 10 A–F.

Fig. 10 A–J. Mouse liver peroxisomes. *A–I* Serial sectioning. *Arrowheads:* membrane continuities between microbodies and between microbodies and smooth endoplasmic reticulum. Fixed with 2.5% glutaraldehyde in cacodylate buffer and postfixed with prereduced osmium tetroxide (KARNOVSKY 1971). ×30,000. *J* Reconstruction of peroxisomal profiles: *D, F, H, I* indicate essential profiles that can be seen in the corresponding figures.

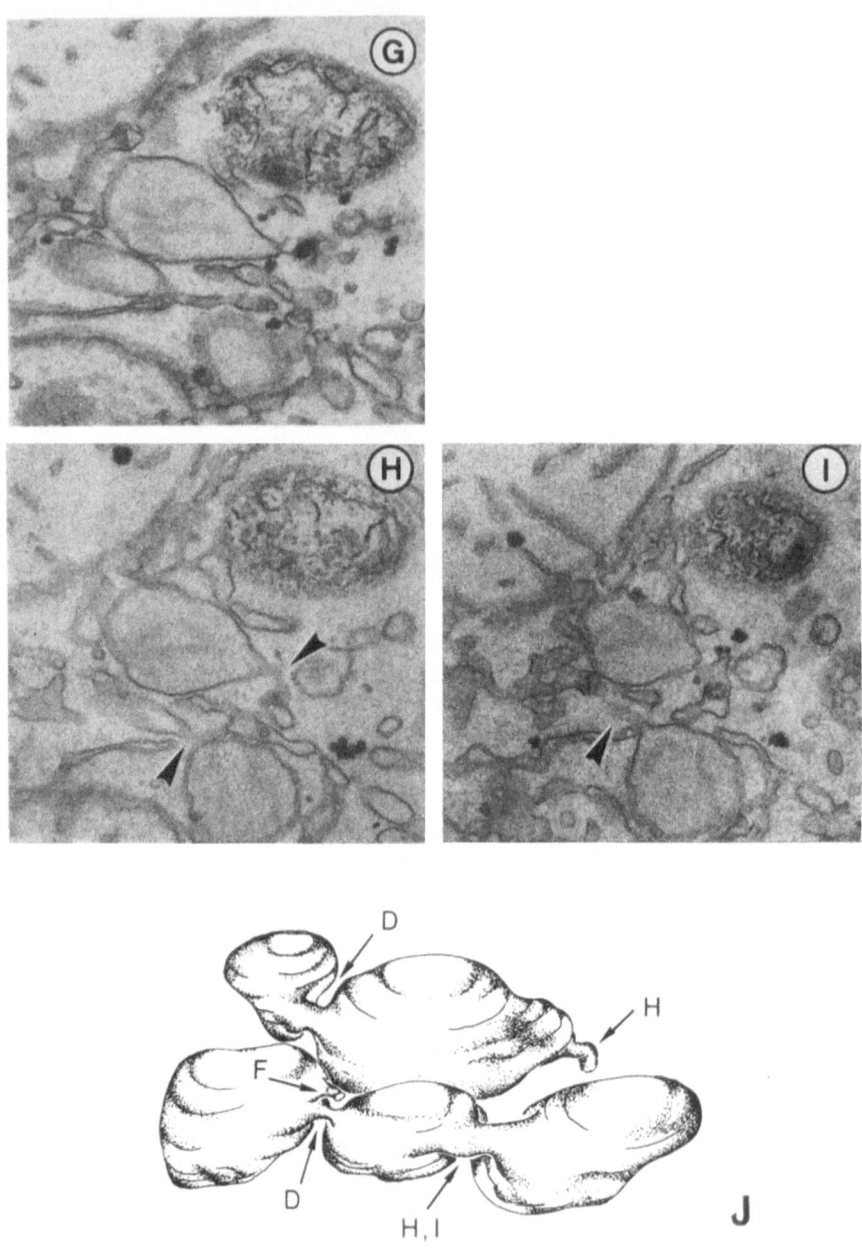

Fig. 10 G–J.

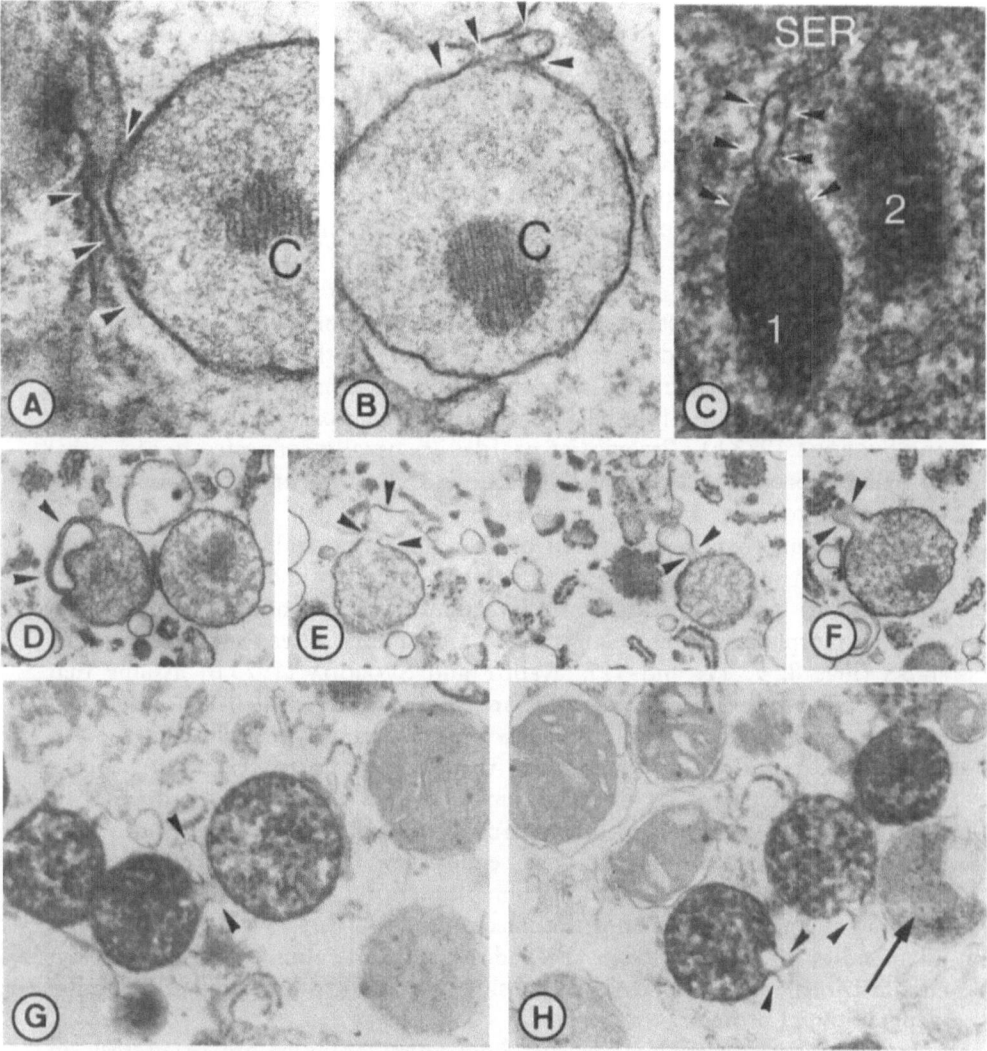

Fig. 11 A–H. Continuities between smooth endoplasmic reticulum (*SER*) and peroxisomes or microperoxisomes. *A, B* Rat liver peroxisomes. The endoplasmic reticulum and peroxisomes are connected by axilla-like continuities (*arrowheads*). Glutaraldehyde/osmium tetroxide fixation; sections stained with uranyl acetate and lead citrate. *A* ×80,000; *B* ×65,000. *C* Microperoxisomes from carp kidney epithelium. The continuity between one of the microperoxisomes (*1*) and smooth endoplasmic reticulum is clearly seen (*arrowheads*). Although the endoplasmic reticulum is free of reaction product, no morphological boundary is seen between its lumen and that of the peroxisome. The reaction product shows coarse granularity. Microperoxisome 2 is cut tangentially. Specimen fixed with glutaraldehyde, postfixed with osmium tetroxide, and briefly stained with uranyl acetate and lead citrate. ×50,000. *D–F* Pellet of a crude mitochondrial fraction prepared from rat liver homogenate by differential centrifugation. *Arrowheads:* several profiles that suggest the presence of small appendices of endoplasmic reticulum that adhere to isolated peroxisomes. Glutaraldehyde/osmium tetroxide fixation, thin section stained with uranyl acetate and lead citrate. ×25,000. *G, H* Mitochondrial fraction from rat liver homogenate prepared by differential centrifugation. *Arrowheads:* finger-like processes of several peroxisomes. The limiting membranes of these processes are continuous with the peroxisomal membranes. The lumina of the processes remain free of reaction product, which homogeneously fills the peroxisomes. *Arrow:* lysosome. Glutaraldehyde/osmium tetroxide fixation and histochemical identification of peroxidatic activities of catalase; brief staining of thin sections with lead citrate. ×30,000.

ganelle (DE DUVE 1973). DE DUVE concluded that microbodies represent protrusions or specialized segments of smooth endoplasmic reticulum. Similar
conclusions were drawn by REDDY and SVOBODA (1973 b, c), who studied
single sections after having applied a drug (clofibrate) that stimulates the
proliferation of peroxisomes. Numerous continuities among microbodies and
between microbody profiles and endoplasmic reticulum were seen. These
profiles have been interpreted as accumulations of microbody proteins in
dilated segments of the endoplasmic reticulum and it has been suggested that
this matrix material exists as a circulating cellular pool.

Membranes of the endoplasmic reticulum close to the connected microbody
(connecting stalk) are free of ribosomes. It seems that the loss of ribosomes
might be a prerequisite for the accumulation of peroxisomal proteins (REDDY
and SVOBODA 1973 b). Peroxisome profiles are usually connected to each other
by a single continuity; occasionally multiple continuities are seen. Connections
between microbodies are also free of ribosomes.

It is evident that in the region of the connecting stalk a boundary must
exist between the peroxisome content (which stains by the cytochemical catalase method) and the endoplasmic reticulum (which is not reactive). However, no morphological correlate to such a boundary is seen in the connecting
stalks (Figs. 8, 9, 11, and 12). REDDY and SVOBODA (1973 b) assume that
microbodies in electron micrographs are "simply the profiles of accumulated
electron-opaque peroxisomal proteins, which constantly circulate in the ER
channels *in vivo.*" This view raises the question as to the nature of the
mechanism that causes the focal accumulations of peroxisomal proteins (the
assumption that catalase and other peroxisomal proteins are redistributed and
interchanged via channels of the endoplasmic reticulum requires free
diffusion of these substances within the entire endoplasmic reticulum).
Nothing is known concerning this mechanism. It is, however, feasible that
microbodies are connected to the endoplasmic reticulum only in a certain
circumscribed region, which is responsible for the synthesis and distribution
of peroxisomal proteins.

In hepatocytes very low density lipoprotein granules are often seen in
cisternae of the endoplasmic reticulum wrapped around the peroxisomes, or in
segments of the endoplasmic reticulum that are connected to microbodies
(Fig. 3). This morphological observation might be of functional significance.

Peroxisomes with pseudopod-like or finger-like extensions can also be seen
when isolated organelles are studied (Fig. 11 D–H). Membrane continuities
among peroxisomes or between microbodies and endoplasmic reticulum may
be responsible for this particular shape. Continuities between microbodies
and endoplasmic reticulum have also been observed in freeze-fracture replicas
(REDDY et al. 1974).

Fig. 12 A–C. Peroxisomes in mouse kidney epithelium. Glutaraldehyde fixation followed by
cytochemical demonstration of catalase and postfixation with prereduced osmium tetroxide.
A possible continuity between smooth endoplasmic reticulum (*SER*) and a peroxisome.
B, C Obvious continuities between these compartments, and elongated peroxisomes with
various degrees of constriction (*arrowheads*). ×30,000.

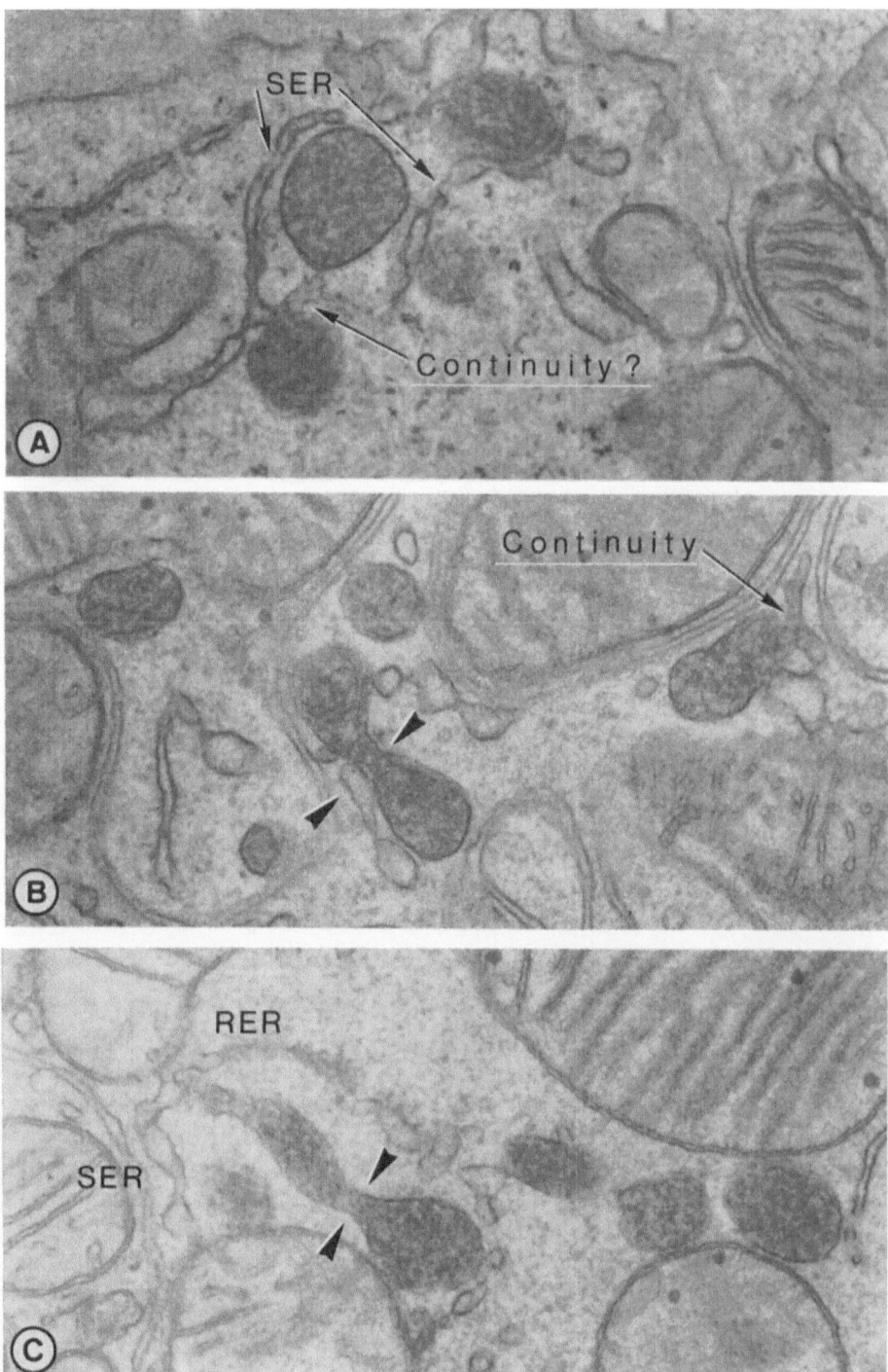

Fig. 12.

Connections between peroxisomes and endoplasmic reticulum have been studied in normal and simfibrate-treated mouse hepatocytes by HIRAI and OGAWA (1975), who payed particular attention to loop- or hook-shaped profiles of smooth endoplasmic reticulum, which often have been seen to be continuous with peroxisomes or their connecting stalks. It has been suggested that these loops are not derived from peroxisomes, but "rather that they are formed from a part of the rER cisterna before peroxisomes are newly generated" (TSUKADA et al. 1968). HIRAI and OGAWA (1975) assume that this loop-shaped segment of smooth endoplasmic reticulum is a small, flattened

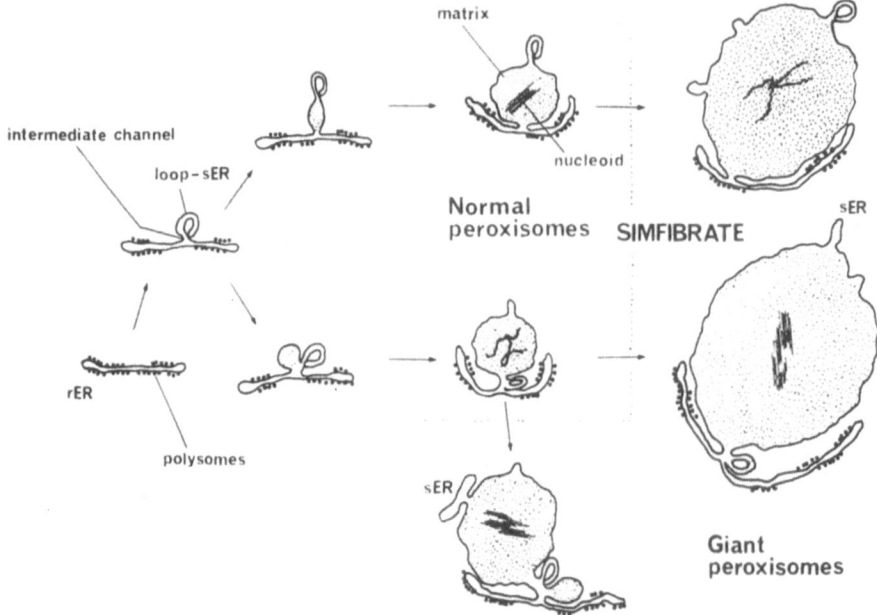

Fig. 13. Loop-shaped profiles of smooth endoplasmic reticulum (sER) and their relationship to outgrowing normal hepatic (mouse) peroxisomes, as well as to simfibrate-induced "megalo-peroxisomes". From HIRAI and OGAWA (1975): Acta Histochem. Cytochem. 8, 27.

cisterna that forms a bowl. It may play a role in determining the site of peroxisome formation from the rough endoplasmic reticulum. The findings of HIRAI and OGAWA (1975) are summarized in Fig. 13. Similar observations were made by HRUBAN et al. (1974 a), who studied rat hepatic microbodies after treatment with acetylsalicylic acid, clofibrate, and dimethrin. Loops or finger-like segments of smooth endoplasmic reticulum have been interpreted as sections through "gastruloid cisternae", which correspond to the bowl-shaped cisternae of HIRAI and OGAWA (1975). The significance of connections between endoplasmic reticulum and peroxisomes is also considered in Section III.E.

Recently two types of interrelationships between microperoxisomes (see Chapter VI) and smooth endoplasmic reticulum have been discerned in lutein cells (GULYAS and YUAN 1977). The first type is represented by narrow, continuous channels, as usually described; the second type is not a continuity of the lumina of microperoxisomes and smooth endoplasmic reticulum, but merely a peculiar spatial relationship between these organelles. An extended, blunt-ended lingula originates from the surface of the peroxisome and is inserted into an invagination of the smooth endoplasmic reticulum. The lumen of this lingula is continuous with that of the peroxisome, but not with the endoplasmic reticulum. In addition, thread-like structures between the membranes of peroxisomes and smooth endoplasmic reticulum have been described where the two organelles are in close proximity (GULYAS and YUAN 1977). The latter observation has been reported also for rat hepatic peroxisomes (KARTENBECK and FRANKE 1974).

Table 3. *Thickness of microbody membranes*

Microbody membrane (nm)	Lysosome membrane (nm)	References	Tissues
6–8		COSSEL 1964	Human liver
6	9	MAUNSBACH 1966	Rat kidney, proximal tubules
6		ERICSSON and TRUMP 1966	Rat kidney
6.5–8		TISHER et al. 1968	Rhesus monkey kidney
6–8	10	YOKOTA and FAHIMI 1978	Rat liver

I.B.4. Microbody Membrane

Peroxisomes are limited by a single membrane of the usual trilaminar structure. This membrane is thinner than the membrane of lysosomes, resembling that of the endoplasmic reticulum. Earlier observations and results of a recent study are summarized in Table 3. These values differ markedly from those reported by WEDEL and BERGER (1975). These authors identified two populations of microbodies in rat liver: (a) larger organelles with a mean diameter of 410 nm and a membrane, 11 nm thick, and (b) smaller organelles, called "Ω particles", with a mean diameter of 260 nm and a membrane, 13 nm thick.

Diffusion of the reaction product is not observed when the cytochemical catalase staining is performed immediately after tissue fixation. Storage of glutaraldehyde-fixed tissue in buffer solutions (FAHIMI 1973, 1974) or in various solvents or detergents causes morphological destruction of the peroxisome membrane and leakage of peroxidatic activity (YOKOTA and FAHIMI 1978). Diffusion of catalase from microbodies into surrounding tissue has been shown directly by cytochemistry and indirectly by examination of the catalase activity in the storage solutions. Diffusion is marked after treatment with Triton X-100, deoxycholate, ethanol, aceton, *t*-butanol, and propanol, whereas PVP does not alter the peroxisome membrane. Glycerol causes leakage of catalase, even in glutaraldehyde-fixed tissues. Discontinuities or disappearance of the membranes are morphological correlates to these findings (YOKOTA and FAHIMI 1978). Since glutaraldehyde-treated catalase remains (at least partially) soluble, detergents cause the release of peroxidatic activities in glutaraldehyde-fixed tissues, as well as in subcellular fractions (BAUDHUIN 1969 a, FUJIWARA 1964). It is interesting to note that a given concentration of digitonin does not similarly affect the limiting membranes of different microbodies. It seems that there exists a certain heterogeneity of peroxisomes with respect to membrane susceptibility to digitonin.

Freeze-fracture replicas of liver microbodies have been studied by REDDY et al. (1974 b); peroxisomes of normal animals as well as organelles from animals treated with hypolipidemic agents were investigated. The convex fracture of the microbody membrane (E-face) is nearly devoid of particles or displays few aggregated particles, whereas the concave fracture of the microbody membrane (P-face) displays numerous particles that are frequently grouped or arranged in an indistinct network.

The angular shape of peroxisomes in rat kidney proximal tubules has also been used to identify these organelles in freeze-fracture replicas. Their particular shape is caused by tubular inclusion bodies. Numerous particles are seen in the P-face of the peroxisome membrane, whereas the E-face displays only a few particles (Fig. 14). Prolonged etching reveals a crystalline lattice of the E-face (periodicity of 9.2 ± 0.8 nm) which is similar to that of purified catalase (KALMBACH and FAHIMI 1978).

I.B.5. Biogenesis and Life History of Peroxisomes

Microbodies are generally believed to arise as dilatations of the endoplasmic reticulum (ESSNER 1967, NOVIKOFF and SHIN 1964, REDDY and SVOBODA 1971 a, 1973 c). Dilatations of the endoplasmic reticulum may either bud off to become free organelles, or remain connected with the endoplasmic reticulum by delicate channels. Stimulated by the work of NOVIKOFF and associates (1972 a, 1973 a, b, c) and that of REDDY and SVOBODA (1971 a, 1973 c), morphologists are now inclined to think that all microbodies are continuous with the endoplasmic reticulum. This view agrees with biochemical studies on catalase synthesis (HIGASHI and PETERS 1963 a, b; see however below and III.1.). The peak of radioactive labeling is found first in microsomes derived from the rough endoplasmic reticulum and thereafter gradually shifts to the mitochondrial fraction.

More detailed studies of the incorporation of newly synthesized, labeled catalase into microbodies of different sizes have shown that particle size is no criterion for particle age, i.e., microbodies do not continuously grow larger (POOLE et al. 1970). It is possible that peroxisomes continuously or inter-

---

Fig. 14 *A, B*. Peroxisomes in the P_3 segment (pars recta) of the rat nephron. *A* Cytochemical demonstration of catalase (DAB method) shows electron-dense reaction product within the organelles. Microbodies are of angular shape and often display rod-like protrusions. ×40,500. *B* In freeze-etch preparations the E-fracture face of the peroxisomal membrane shows a crystalline pattern where tubular inclusions underlie the membrane. The outline of cross-fractured peroxisomes corresponds to the shape seen in sectioned specimens (*P* in part *A*). Peroxisomes are wrapped with fenestrated cisternae of endoplasmic reticulum (*ER*). *EF* E-fracture face; *PF* P-fracture face; *arrow*, direction of shadowing. ×39,500. Courtsey of P. KALMBACH and H. D. FAHIMI, Heidelberg.

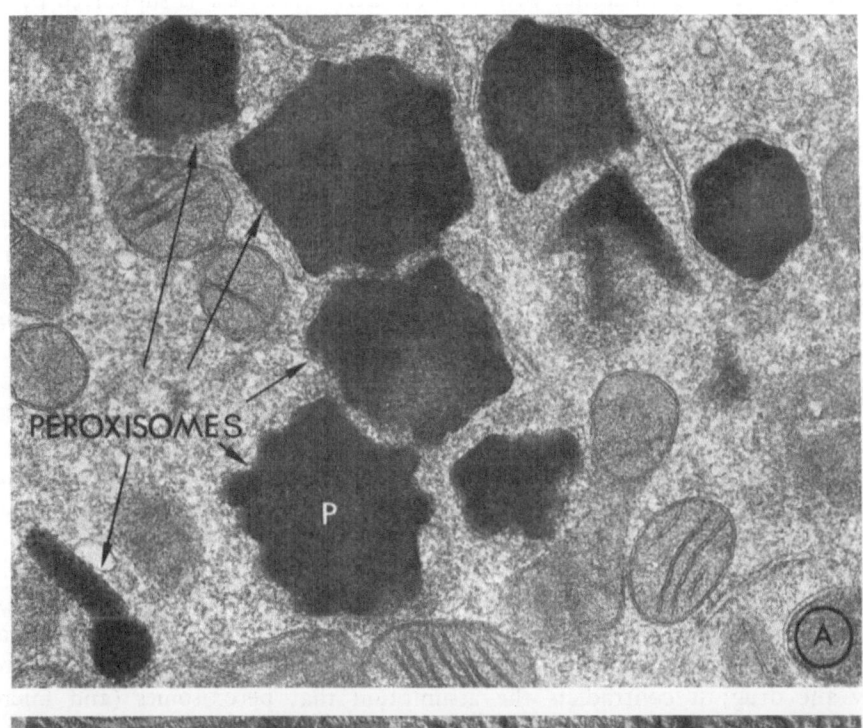

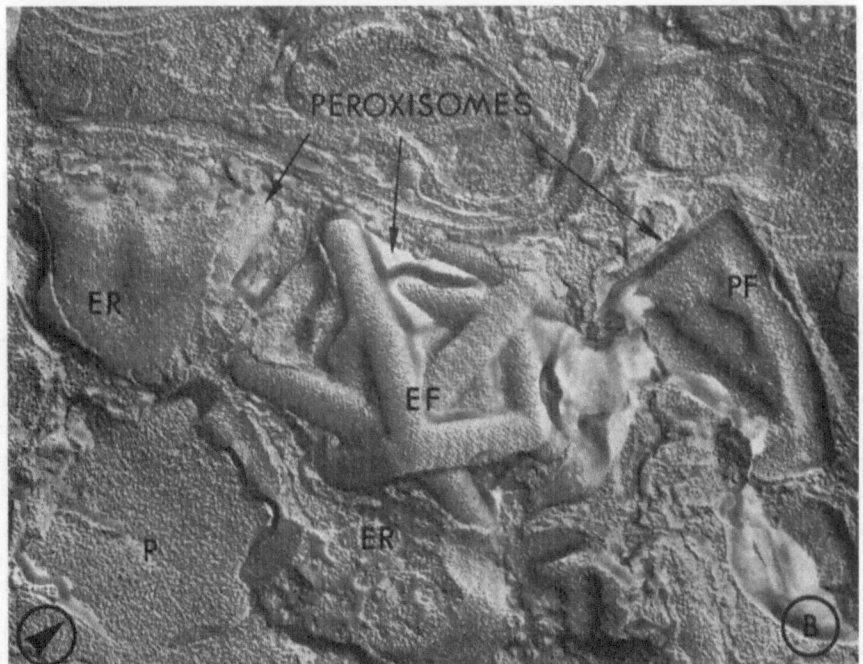

Fig. 14.

mittently exchange material with one another. This view is supported by the morphological observation that microbodies are gathered in clusters in which the individual particles are connected to each other by delicate channels and are also connected to the endoplasmic reticulum (see Section I.B.3.). Alternatively, it may be assumed that peroxisomes continuously interchange material by fusional and/or fissional processes occurring at the level of the organelles (Legg and Wood 1970 a).

Catalase-positive particles devoid of nucleoids and generally smaller than regular peroxisomes have been defined as "microperoxisomes" (Novikoff and Novikoff 1972, 1973). It has been suggested that microperoxisomes represent progenitors of peroxisomes, at least in tissues in which microperoxisomes and peroxisomes are found concomitantly, e.g., in liver tissue (Novikoff et al. 1973 a). Morphometric studies revealed that in rat hepatocytes about 13% of the peroxisomes are indeed devoid of a nucleoid and therefore are microperoxisomes. This population of microperoxisomes remains unaltered during nafenopin-induced expansion of the peroxisome compartment. Thus, a progenitor-mature organelle relationship between microperoxisomes and peroxisomes is unlikely (Stäubli et al. 1977). Preferential incorporation of ^{3}H-arginine into larger peroxisomes has been observed during nafenopin-induced proliferation of microbodies (Stäubli et al. 1977). The gradual shift of radioactivity to larger particles during the experiment supports the assumption that the growth of a particular class of peroxisomes is stimulated by the drug; it contradicts the assumption that peroxisomes (and microperoxisomes) belong to a common metabolic compartment.

It is not known whether peroxisomal proteins are synthesized by bound ribosomes and delivered into cisternae of the endoplasmic reticulum and from there into peroxisomes, or whether the proteins are released from bound or free ribosomes into the cytoplasm to enter the peroxisomes from outside through their membranes (Masters and Holmes 1977). The catalase molecule, a tetramer, is assembled within the peroxisomes, where heme is also built in (Lazarow and de Duve 1973 a, b). The distribution of the apomonomer within peroxisomes of different sizes cannot be distinguished from the distribution of resident catalase. Significant amounts of newly synthesized apomonomer are found in the supernatant before being recovered in the peroxisomal fraction. These findings imply either transport of apomonomer to peroxisomes through the plasmatic matrix of the cell, or unusual vulnerability of certain segments of those regions of the endoplasmic reticulum where newly formed apomonomer is channeled to peroxisomes. Electron microscopic findings of Yokota and Nagata (1974 b) with immunocytochemical methods favor the former possibility.

Cytochemical evidence of catalase activity in the cytoplasm between rough endoplasmic reticulum and adjacent peroxisomes has been obtained with DAB methods (Fahimi and Venkatachalam 1970, Legg and Wood 1970 a, 1972, Rigatuso et al. 1970). However, these observations demonstrate diffusion artifacts rather than transport of newly synthesized catalase, because the apomonomer outside the peroxisome is not enzymatically active (de Duve 1973). Recent critical analysis of the histochemical method established that ribosomal catalase staining is a diffusion artifact (FAHIMI 1973, NOVIKOFF et al. 1972 b).

It must be reiterated that only ribosomes that are near stained microbodies react positively (BÖCK 1972 a). On the other hand, it can be excluded that microbodies are attached to defined segments of the endoplasmic reticulum that synthesize catalase. A preferred occurrence of diffusion artifacts in studies on regenerating or proliferating microbodies might indicate enhanced lability of microbody membranes under these conditions (GOLDENBERG et al. 1975 a).

The assumption that peroxisomes form a single cellular compartment of continually varying shape, which is connected with a special region of the

Table 4. *Morphometric and kinetic parameters of cellular autophagy (rat hepatocytes)*

Component	(1) Fractional volume of the components in cytoplasm $V_{cyto, comp}/V_{cyto}$ $\times 10^{-2}$	(2) Fractional volume of the components in AV compartment $V_{AV, comp}/V_{AV}$ $\times 10^{-2}$	(3) Fractional volume of the AVs in cytoplasm $V_{AV, comp}/V_{cyto}$ $\times 10^{-4}$	(4) Segregated fraction of the components $V_{AV, comp}/V_{cyto, comp}$ $\times 10^{-4}$	(5) Lifetime of the AVs $t_{1/2AV, comp}/\ln 2$ (min)	(6) Rate of autophagic breakdown $1/t_{cyto, comp}$ (day^{-1})
All	100.0	100.0	2.11 ± 0.25	2.11	13.0	0.023
Mitochondria	20.0 ± 0.6	37.4 ± 3.8 *	0.79 ± 0.12	3.95	14.6	0.039
Microbodies	1.4 ± 0.2	4.3 ± 0.7 *	0.09 ± 0.02	6.42	10.4	0.089
ERGS	55.4 ± 1.4	38.3 ± 4.6 *	0.81 ± 0.14	1.47	16.6	0.013
Glycogen	21.4 ± 1.2	20.0 ± 5.5	0.42 ± 0.09	1.95	26.0	0.011

From: PFEIFFER et al. 1978.

The morphometric parameters of rat liver parenchyma at time zero, i.e., immediately after the administration of insulin, are shown in columns 1–4. Each value represents the mean ± the standard error of the mean of 16 animals. Except for glycogen, the fractional volumes in the AV compartment, shown in column 2, are different from the corresponding values in the cytoplasm, shown in column 1. The asterisks in column 2 indicate that by Student's t-test the differences are statistically highly significant ($P < 0.001$). In columns 5 and 6, the kinetic parameters calculated from the decay of the AVs after insulin administration are shown. The values for the average lifetime of the AVs (column 5) were obtained by dividing the values for the half-life by ln 2. AV autophagic vacuole.

endoplasmic reticulum, is in agreement with both biochemical and morphological observations (DE DUVE 1973).

Destruction of peroxisomes is a random process. All peroxisomal proteins seem to have the same kinetics, which indicates that peroxisomes are destroyed as wholes (POOLE et al. 1969).

Microbodies are normally degraded in autophagic vacuoles. The kinetics of this degradation mechanism have been analyzed morphologically in hepatocytes of male rats (PFEIFFER et al. 1978). Autophagic vacuoles remove different amounts of different cytoplasmic components per time unit; they remove, per day, 2.3% of the whole cytoplasm, 3.9% of the mitochondria, 8.9% of the microbodies, and 1.1% of the glycogen. Table 4 is a summary of the morphometric data on microbodies in relation to other cytoplasmic components of hepatocytes; the kinetics of the cellular breakdown of these components by autophagocytosis are also shown. The fractional volumes of different components within the compartment "autophagic vacuole" $V_{AV, comp}/V_{AV}$; column 2) are different from the fractional volumes within the compartment of free cytoplasm ($V_{cyto, comp}/V_{cyto}$; column 1). Therefore,

Table 5. *The segregated fractions (segregated/nonsegregated material) of the whole cytoplasm and of its components calculated as mean values of the diurnal cycle (rat kidney tubules)*

Basis of calculation	Component	Nonsegregated	Segregated	Segregated fraction $\times 10^{-4}$	95% confidence interval
Area	Cytoplasm	76.6×10^4 μm²	155.9 μm²	2.1	(1.6–2.6)
	ERGS	45.4×10^4 μm²	70.6 μm²	1.6	(1.2–2.0)
	Mitochondria	17.3×10^4 μm²	75.9 μm²	4.4	(3.3–5.5)
	Microbodies	0.8×10^4 μm²	9.4 μm²	11.7	(3.1–20.3)
Number of sectioned profiles	Mitochondria	31.8×10^4 Mean area 0.55 (SEM 0.04) μm²	289 Mean area 0.26 (SEM 0.01) μm²	9.1	(6.7–11.5)
	Microbodies	2.4×10^4 Mean area 0.34 (SEM 0.08) μm²	26 Mean area 0.36 (SEM 0.03) μm²	10.8	(5.5–16.1)

From: PFEIFFER and SCHELLER 1975.

segregation of different cytoplasmic components into autophagosomes seems to encompass selection and discrimination. The high segregated fraction of microbodies ($V_{AV, comp}/V_{cyto, comp}$; column 4) may reflect the long lifetime of those autophagic vacuoles that contain microbodies. However, this assumption is in contrast to the estimated lifetimes of various autophagic vacuoles (column 5), which is the shortest for autophagic vacuoles containing microbodies.

The relative rate of microbody degradation (column 6) agrees well with biochemical observations (POOLE *et al.* 1969), when compared with that of mitochondria. Absolute rates of microbody turnover, however, fit less well with observed autophagic degradation rates. Microbody protein degradation of 20–30% per day (DOYLE and TWETO 1975, POOLE *et al.* 1969) contrasts with about 10% autophagic degradation per day, which has been calculated from morphological observations (PFEIFFER *et al.* 1978). It seems that autophagocytosis is responsible for the degradation of about 50% of microbody material (in rat liver).

Morphometric data of cellular autophagy in rat kidney tubules have been given by PFEIFFER and SCHELLER (1975). Autophagic vacuoles containing microbodies are less frequently seen than vacuoles containing mitochondria or other cytoplasmic constituents. Marked diurnal variation in the total number of autophagic vacuoles is observed, with a maximum during the day and a minimum during the night. The number of autophagic vacuoles containing microbodies varies correspondingly, being low during the inactive phase of the animal. Once more it has been found that cellular autophagy is not a random process. Different cytoplasmic components are selected, and the segregated fraction of microbodies is relatively higher than the fractional volume of microbodies in the cytoplasm. The calculated lifetime of micro-

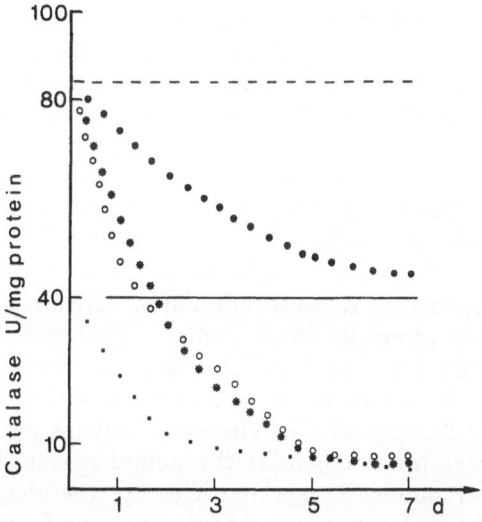

Fig. 15. Effect of ethyl-chlorophenoxyisobutyrate (*CPIB*, clofibrate) and allylisopropyl-acetamide (*AIA*) on rat liver catalase activity. Day 0 represents the beginning of the experiment after groups 1–4 had received CPIB for 4 weeks. Symbols denote: ———— normal rat; -------- CPIB control; ■■■ normal + AIA; ●●●●● CPIB off; ○○○○○ CPIB off + AIA; ***** CPIB continued + AIA. From: Svoboda and Reddy (1972): Am. J. Pathol. 67, 543.

bodies is approximately 8 days. The findings of Pfeiffer and Scheller (1975) are compatible with the assumption that kidney microbodies are destroyed to a major extent—if not exclusively—by autophagocytosis. More details are found in Table 5.

The origin and removal of microbodies have been studied in the synchronously growing cells of an insect fat body (*Calpodes ethlius*; Locke and McMahon 1971). In addition to the final destruction of microbodies in autophagic vacuoles the stages of *atrophy* have been described as follows: (a) reduction of the microbody core, (b) appearance of myelin figures (which may be present along with reduced cores), and, finally, (c) shrinking and disappearance of the remaining vesicles. The effect described as atrophy by Locke and McMahon (1971) may correspond to the breakdown mechanism suggested by Moody and Reddy (1976 a) in that microbody matrix material is retracted within the endoplasmic reticulum. No peroxisomes could be observed in autophagic vacuoles of human hepatocytes (STERNLIEB and QUINTANA 1977).

The mechanism of removal and degradation of microbodies in rats previously given clofibrate seems to be different from that in normal hepatocytes (SVOBODA and REDDY 1972). Biochemical catalase determinations after an initial treatment with clofibrate for 4 weeks show a gradual decrease of catalase activities to normal values during the first week after clofibrate administration is stopped. This decrease to normal levels is reached within 2 days when AIA is given during this time (Fig. 15). Morphological observations during this period indicate that the size and number of autophagosomes remain unaltered. After clofibrate administration is stopped, there is a marked decrease in the electron density of the matrix material in several microbodies, whereas other microbodies remain normal. Fragmentation of the limiting membrane is also observed and frequently only the crystalline cores remain. SVOBODA and REDDY (1972) suggest that whereas a minor portion of the microbodies is removed by autophagic vacuoles, the major portion undergoes gradual dissolution. Membranes of microbodies become ruptured, leaving the crystalline cores in an electrolucent remnant. It has been suggested that this gradual dissolution does not require the participation of any other cell organelles. Reconstitution after nafenopin-induced proliferation of peroxisomes seems to work similarly (STÄUBLI et al. 1977). The mechanism, however, has not been elucidated in more detail. Extruded nucleoids have also been seen (as in case of the so-called Φ-bodies, HANKER et al. 1977 a, b, see Section III.F.). The entire cellular mechanism to remove drug-induced proliferated microbodies (by autophagocytosis and dissolution) does not work at random. Larger particles are preferentially selected for degradation, which leads to an increased number of smaller peroxisomes (STÄUBLI et al. 1977).

I.C. Peroxisome Concept of DE DUVE

The concomitant occurrence of hydrogen peroxide-producing enzymes (urate, D-amino acid, and L-α-hydroxy acid oxidase) and catalase in identical particles led to the concept of a cell organelle concerned with hydrogen peroxide formation and consumption, the peroxisome (BEAUFAY et al. 1964, DE DUVE 1965, DE DUVE and BAUDHUIN 1966, DE DUVE et al. 1960). These particles are morphologically identical to microbodies.

Early studies of KEILIN and HARTREE (1936 b, 1945) had proved that catalase is not only able to decompose hydrogen peroxide (catalatic action) but also catalyzes the coupled oxidation of hydrogen donors such as lower alcohols (peroxidatic action of catalase, CHANCE 1950, THEORELL 1948). Either reaction starts with the formation of a catalase-hydrogen peroxide intermediate called compound I (CHANCE 1947) (Fig. 16). Other intermediates (compounds II and III) derived from compound I are essentially inactive in the catalatic reaction and probably absent in the native tissue (OSHINO et al. 1973 a). In liver tissue the high catalase content is concentrated in peroxisomes: OSHINO et al. (1975 a, b), using dual-wavelength spectroscopy of the

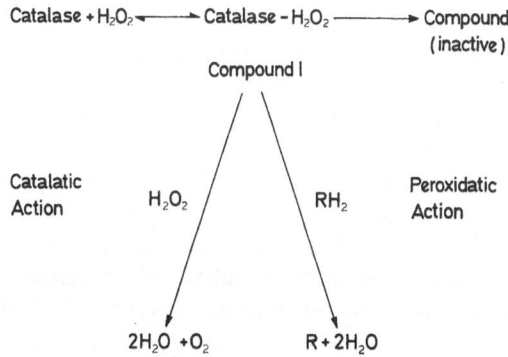

Fig. 16. Formation of catalase-hydrogen peroxide compounds and catalatic and peroxidatic reactions of compound I. RH_2 corresponds to $R'H_2$ in Fig. 17 and stands for hydrogen donors for the peroxidatic reaction, such as alcohols or formic acid. From HALLIWELL 1974.

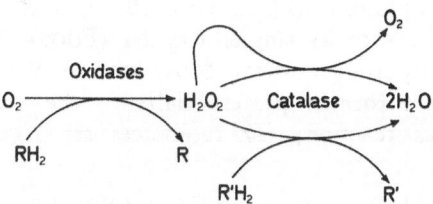

Fig. 17. Schematic representation of the peroxisome concept. RH_2 stands for substrates of the peroxisomal oxidases (urate, D-amino acids, L-α-hydroxy acids). R designates the oxidized substrates. $R'H_2$ stands for hydrogen donors for the peroxidatic reaction of catalase. From DE DUVE and BAUDHUIN 1966.

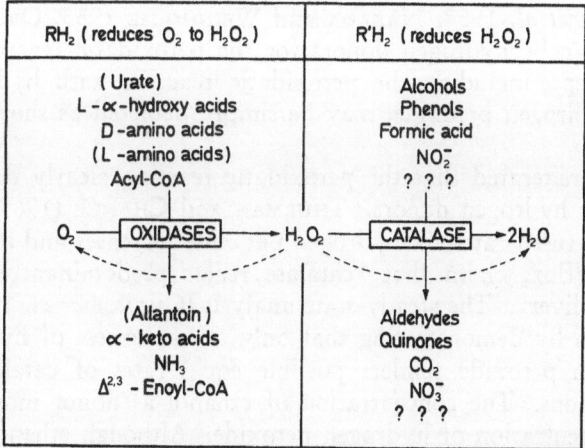

Fig. 18. The two groups of oxidoreductase reactions catalyzed by peroxisomes (scheme slightly adapted from DE DUVE and BAUDHUIN 1966). Right side: hydrogen peroxide consuming reactions ($R'H_2 : H_2O_2$ oxidoreductases). Left side: hydrogen peroxide delivering reactions ($RH_2 : O_2$ oxidoreductases).

catalase-cyanide complex, found 4.8×10^{-9} mol/g tissue (*i.e.*, 1.92×10^{-8} mol heme/g; see also HIGASHI and PETERS 1963 a, OSHINO *et al.* 1973 b). Therefore, the steady-state concentration of hydrogen peroxide becomes very small by formation of compound I: from the generation rate of hydrogen peroxide in perfused rat liver OSHINO *et al.* (1973 b) have assessed it to be on the order of 10^{-9} M; when the tissue is supplemented with hydrogen peroxide-generating substrates (urate or glycolate) it rises to at most 10^{-7} M. For this reason the turnover rate of the catalatic reaction *in vivo* is low. This is in contrast to the extremely high turnover number of catalase *in vitro* (about 10^5 sec^{-1}, the maximum value is reported to be 3.8×10^7 sec^{-1}; OGURA 1955).

Accordingly, DE DUVE and BAUDHUIN (1966) pointed out that catalatic decomposition of hydrogen peroxide becomes important only when it is present in relatively high concentrations. In the absence of suitable hydrogen donors, the catalatic decomposition of hydrogen peroxide plays the role of a safety valve against an excess amount. Although hydrogen peroxide *per se* hardly seems to be cytotoxic (HALLIWELL 1974), it is capable of producing agressive compounds, such as singlet oxygen (FOOTE *et al.* 1968), or free radicals, such as the hydroxyl radical (NORMAN and WEST 1969). Presumably hydroxyl radicals are formed preferentially in the presence of superoxide anions (Haber-Weiss reaction; for references see KOPPENOL and BUTLER 1977):

$$H_2O_2 + O_2^- \longrightarrow OH^{\cdot} + OH^- + O_2$$

Therefore, the maintainance of low hydrogen peroxide levels, as discussed above, is an important function of catalase.

If hydrogen donors other than hydrogen peroxide are available, they compete for compound I. In this case the peroxidatic reaction of catalase is favored. In addition to alcohols, nitrite (HEPPEL and PORTERFIELD 1949) and formate (AEBI *et al.* 1957, NAKADA and WEINHOUSE 1953, ORO and RAPPOPORT 1959) can be hydrogen donors for the peroxidatic reaction. The peroxisome concept, including the peroxidatic reaction with hydrogen donors other than hydrogen peroxide, may be simply depicted as shown in Figs. 17 and 18.

It must be reiterated that the peroxidatic reaction clearly depends on the availability of hydrogen donors. THURMAN and CHANCE (1969), who found that in the perfused rat liver hydrogen peroxide increases and azide decreases the oxygen efflux, claim that "catalase reacts predominantly catalatically *in vivo* in rat liver". The steady-state analysis (CHANCE *et al.* 1952) confirms this conclusion by demonstrating that only a vast excess of hydrogen donor over hydrogen peroxide renders possible equal rates of catalatic and peroxidatic reactions. The concentration of ethanol as donor must be 3×10^4 times the concentration of hydrogen peroxide. Although ethanol (KREBS and PERKIN 1970) and formate (arising in minute quantities from glycine via glyoxylate; NAKADA and WEINHOUSE 1953) seem to be obligate components of mammalian blood, their physiological concentrations are normally too low by far for significant peroxidatic reactions.

Therefore, one must concede that there is no confirmed hypothesis as to the physiological significance of the peroxidatic reaction in animal cells. Even in the case of ethanol ingestion the alcohol dehydrogenase-independent oxidation of ethanol by liver catalase seems to be less important than the microsomal ethanol-oxidizing system, which does not require catalase (TESCHKE et al. 1974, 1975 a, b, 1976). Measurements in situ have shown that catalase contributes 10⁰/o of the total ethanol oxidation in rat liver (OSHINO et al. 1975 b).

In conclusion, if the peroxidatic reaction of catalase plays a significant role at all, its physiological hydrogen donor is as yet unknown.

I.D. Role of Microbodies as Peroxisomes

I.D.1. Enzymes in Animal Peroxisomes

Catalase-containing particles have been detected in numerous animal tissues other than liver or kidney. This raises the question as to whether such particles always contain a set of enzymes compatible with the peroxisome concept. Unfortunately, a considerable body of enzyme data has been accumulated only for liver tissue (especially rat liver tissue), while other tissues have been studied far less intensively. Therefore, the discussion will be restricted to liver tissue. The enzymes known to be present in liver peroxisomes are listed in Table 6.

At first it seems that only some of the enzymes that accompany oxidases and catalase can be interpreted according to the peroxisome concept: glyoxylate reductase in combination with L-α-hydroxy acid oxidase forms an intraperoxisomal NADH-oxidizing system (the net reaction is NADH oxidation and hydrogen peroxide formation; VANDOR and TOLBERT 1970):

$$\text{L-lactate} + O_2 \rightarrow \text{pyruvate} + H_2O_2$$
$$\text{(or glycolate)} \qquad \text{(or gloxylate)}$$

$$\text{pyruvate} + \text{NADH} + H^+ \rightarrow \text{L-lactate} + \text{NAD}^+$$
$$\text{(or glyoxylate)} \qquad\qquad \text{(or glycolate)}$$

It must be mentioned, however, that the capacity of liver peroxisomes to reduce glyoxylate has been doubted. NADH-glyoxylate reductase, which sediments with the particles, may be released from them by salt treatment (SUZUKI et al. 1973). McGROARTY et al. (1974), who identified this activity as being due to lactate dehydrogenase (LDH_5, M_4 isoenzyme), could not exclude the possibility that this LDH might have been occluded only in the peroxisomal fraction from the cytoplasm.

It is tempting to speculate that the (antimycin-insensitive) NADH-cytochrome c reductase activity of the peroxisome membrane (DONALDSON et al. 1972) also takes part in the oxidation of (extraperoxisomal?) NADH, although an appropriate electron acceptor remains to be identified. NADH-cytochrome b_5 reductase in combination with cytochrome b_5 is thought to be responsible for the antimycin-insensitive NADH-cytochrome c activity of endoplasmic reticulum and Golgi membranes (STRITTMATTER 1963). This

Table 6. *Enzymes of vertebrate liver peroxisomes and their intraperoxisomal location*

EC	Recommended names	Comments	References
I. NAD(P)-linked dehydrogenases			
1.1.1.8	Glycerol-3-phosphate dehydrogenase (NAD)		GEE et al. (1974)
1.1.1.27	NADH-glyoxylate reductase; M_4^- Isozyme of lactate dehydrogenase (LDH 5)?	Also reduces pyruvate and hydroxypyruvate to L-lactate and L-glycerate [a]	McGROARTY et al. (1974), VANDOR and TOLBERT (1970)
1.1.1.42	Isocitrate dehydrogenase (NADP)	MA	LEIGHTON et al. (1968, 1969)
1.1.1.101	Acyl-dihydroxyacetone phosphate-NADPH oxidoreductase		HAJRA et al. (1979)
1.2.1.37	Xanthine dehydrogenase [b]		SCOTT et al. (1969)
II. Enzymes of hydrogen peroxide metabolism			
1.1.3.1 (1.1.3.15)	Glycolate oxidase (L-α-Hydroxy acid oxidase)	MAC, also oxidizes other α-hydroxy acids (e.g., L-lactate and L-α-hydroxyisocaproate)	BAUDHUIN et al. (1965 b), HAYASHI et al. (1971, 1973), McGROARTY et al. (1974)
1.4.3.3	D-Amino acid oxidase	MAC	BAUDHUIN et al. (1964, 1965 a), BEAUFAY et al. (1964), HAYASHI et al. (1971, 1973)
1.7.3.3	Urate oxidase	C, not present in avian and primate liver	BAUDHUIN et al. (1965 a), BEAUFAY et al. (1964)
1.11.1.6	Catalase	MA	BAUDHUIN et al. (1965 a), BEAUFAY et al. (1964)
III. Transferases			
2.3.1.7	Carnitine acetyltransferase	MA, induced by clofibrate	MARKWELL et al. (1973), MOODY and REDDY (1974), GOLDENBERG et al. (1976)
2.3.1.x	Carnitine octanoyltransferase	MA, palmitoyltransferase lacking in peroxisomes	MARKWELL et al. (1973, 1976)
2.3.1.42	Dihydroxyacetone phosphate acyltransferase		JONES and HAJRA (1977), HAJRA et al. (1979)
2.6.1.44 (2.6.1.51?)	Glyoxylate aminotransferase [c]	MA, preferred NH_2-donors are L-leucine, L-alanine and L-phenylalanine	HSIEH and TOLBERT (1976) (cf., NOGUCHI and TAKADA 1978 a, b), NOGUCHI et al. (1979 b)
IV. Other			
1.6.2.2?	NADH-cytochrome c reductase (antimycin insensitive)	ME	DONALDSON et al. (1972)

Table 6 (continued)

EC	Recommended names	Comments	References
3.5.2.5	Allantoinase [d]		SCOTT et al. (1969)
—	Palmitoyl-CoA oxidizing system ("β-oxidation")	Cyanide insensitive, carnitine independent, NAD dependent, H_2O_2 producing, oxidizes preferentially long-chain acids, induced by clofibrate	LAZAROW (1978), LAZAROW and DE DUVE (1976), THINÈS-SEMPOUX and BORMANN (1978)
6.2.1.3	Acyl-CoA synthetase	Activates preferentially C_{10}–C_{18} acids (like the microsomal enzyme)	SHINDO and HASHIMOTO (1978)

The table generally refers to rat liver tissue. Enzymes that have not yet been detected in rat liver, or have been proved to be absent from it, are indicated by footnotes.

Abbreviations: ME peroxisomal membranes; MA peroxisomal matrix, "soluble enzyme"; C crystalline core; MAC matrix or weakly bound to the core.

[a] The L-glycerate formed is the "unnatural" isomer, which on the other hand can be reoxidized by L-α-hydroxy acid oxidase. Specific D-glycerate dehydrogenase is absent from rat liver and kidney peroxisomes.

[b] Found in chicken liver and kidney. In rat, frog, and carp liver EC 1.2.1.37 is found only in the soluble cytoplasmic fraction (GOLDENBERG 1977 a, REID et al. 1956, SCOTT et al. 1969).

[c] Since it has a similar amino acid specificity this enzyme might be identical to the peroxisomal serine-pyruvate aminotransferase (EC 2.6.1.51), the "pyruvate(glyoxylate)amino-transferase", purified from rat liver and found to share properties with the mitochondrial alanine-glyoxylate aminotransferase isoenzyme 1 (NOGUCHI and TAKADA 1978 b, NOGUCHI et al. 1979 b). In human liver alanine-glyoxylate aminotransferase (EC 2.6.1.44) was shown to be a peroxisomal enzyme which presumably is missing in mitochondria (NOGUCHI and TAKADA 1979). A comparable activity (serine-pyruvate aminotransferase, EC 2.6.1.51) was localized in the peroxisomes (but not in the mitochondria) of human liver. With L-serine as amino group donor the purified enzyme seems to be specific for pyruvate (NOGUCHI and TAKADA 1978 a).

[d] Found in frog liver. In carp liver it is probably exclusively soluble as is allantoicase (GOLDENBERG 1977 a, b). In mackerel, yellow mackerel and prawn liver and in mantis club hepatopankreas both enzymes were allocated to the peroxisomes (NOGUCHI et al. 1979 a).

enzyme, which projects its hydrophilic part into the cytoplasm, is capable of using cytoplasmic NADH for fatty acid desaturation (OSHINO and OMURA 1973, OSHINO and SATO 1971, OSHINO et al. 1971, STRITTMATTER et al. 1974).

The peroxisomal system(s) that oxidate(s) NADH should be of particular importance in processing the NADH that is produced in the peroxisomes themselves; this is the case with fatty acid β-oxidation, which formerly was thought to occur only in specialized plant microbodies (glyoxysomes; COOPER and BEEVERS 1969) but recently was also detected in rat liver peroxisomes (LAZAROW and DE DUVE 1976). Furthermore, the acyl-CoA dehydrogenase step proceeds with molecular oxygen as acceptor. The reaction delivers hydrogen peroxide and therefore may be connected with catalase:

$$\text{fatty acyl-CoA} + \text{flavoprotein} \rightarrow \Delta^{2,3}\text{-enoyl-CoA} + \text{flavoprotein} \cdot H_2$$

$$\text{flavoprotein} \cdot H_2 + O_2 \rightarrow \text{flavoprotein} + H_2O_2$$

OSUMI and HASHIMOTO (1978 b) were able to purify this acyl-CoA oxidase from the liver of rats treated with di(2-ethylhexyl)phthalate (DEHP) and to prove the formation of hydrogen peroxide and enoyl-CoA from palmitoyl-CoA.

Acyl-CoA oxidase which seems to be the rate limiting enzyme of the peroxisomal β-oxidation system (OSUMI and HASHIMOTO 1979 a, INESTROSA et al. 1979) is most active toward activated fatty acids with a chain length of 12–18 carbon atoms; the maximum activity is found with C_{16} substrates [purified enzyme: $1.45 \, \mu mol/(min \cdot mg)$]. C_{20} and C_{22} substrates are still quite reactive (about 50 and 25%, respectively, of the C_{16} value), whereas C_4 and C_6 substrates are hardly oxidized at all. In liver peroxisomes from untreated rats the peak of β-oxidation and acyl-CoA oxidase activities occurs with C_{12} substrates [$19 \, nmol/(min \cdot mg)$; HRYB and HOGG 1979].

In sharp contrast to these findings, in liver mitochondria the first step of β-oxidation (acyl-CoA dehydrogenase) is performed by several enzymes of different chain length specifity (e.g., in pig liver mitochondria; CRANE et al. 1956, GREEN et al. 1954, HAUGE et al. 1956). Short-chain fatty acyl-CoA's are dehydrogenated preferentially (HRYB and HOGG 1979).

The peroxisomal β-oxidation system differs from its mitochondrial counterpart also in some other respects. In rat liver the mitochondrial 3-hydroxyacyl-CoA dehydrogenase is active with acetoacetyl pantheteine as well as with acetoacetyl-CoA (type I enzyme). The other isozyme found mainly in the peroxisomes only catalyzes the reduction of acetoacetyl-CoA (type II enzyme). It is induced by di-(2-ethylhexyl)phthalate to a much higher extent than type I enzyme (OSUMI and HASHIMOTO 1979 b). The peroxisomal enzyme is copurified with enoyl hydratase (bifunctional enzyme?; OSUMI and HASHIMOTO 1979 c).

The fatty acyl-CoA oxidase of human liver peroxisomes acts maximally on C_{12}–C_{18} saturated fatty acyl-CoA's and on oleyl-CoA (BRONFMAN et al. 1979).

The data cited above seem to establish a role for peroxisomes in the chain shortening of saturated long-chain fatty acids. This conclusion, however, has been questioned by OSMUNDSEN et al. (1979), who found the maximum activity in liver peroxisomes from clofibrate-fed rats with myristoyl-CoA (C_{14}, saturated), whereas the maximum for monounsaturated fatty acids was determined at C_{18}. The activity with C_{18}, C_{20}, and C_{22} unsaturated acyl-CoA's is about 10 times as high as with the respective saturated CoA derivatives. The products of peroxisomal degradation presumably are further metabolized by the mitochondria.

The peroxisomal β-oxidation of palmitoyl-CoA and its shorter homologues is inhibited by high CoA concentrations whereas the degradation of long monounsaturated acyl-CoA's (gadeoyl-CoA, erucoyl-CoA) is hardly influenced by CoA. That means that peroxisomal β-oxidation is optimally active and works as a security valve for fatty acids when the cell is provided abundantly with fatty acids so that most cytosolic CoA is bound to those. On the other hand, at high CoA concentrations (poor fatty acid supply) the peroxisomes are supposed to oxidize essentially only the very long fatty acids (Osmundsen and Neat 1979).

Table 7. *Total activity and recovery of marker enzymes and of 1-¹⁴C-palmitoyl-CoA oxidation after differential centrifugation of rat liver homogenates (see section II.A.1.b)*

	Control ($n = 4$)		Clofibrate ($n = 5$)	
	Total activity (units/g liver)	Recovery after fractionation (%)	Total activity (units/g liver)	Recovery after fractionation (%)
Protein (mg/g)	185.0 ± 9.5	94.3	199.9 ± 10.5	100.3
Glutamate dehydrogenase	216.6 ± 9.7	96.5	301.4 ± 2.8 a	95.3
Carnitine palmitoyltransferase	1.21 ± 0.02	94.0	3.97 ± 0.16 a	107.0
Catalase	69.4 ± 2.5	98.1	85.4 ± 4.5 a	91.5
Mitochondrial palmitoyl-CoA oxidation	0.226 ± 0.029	97.4	0.532 ± 0.092 a	93.1
Peroxisomal palmitoyl-CoA oxidation ($v = 1.67$)	0.099 ± 0.016	107.8	0.582 ± 0.098 a	85.0
Peroxisomal palmitoyl-CoA oxidation ($v = \infty$)	0.672 ± 0.043	86.6	5.28 ± 1.11 a	74.0

From Mannaerts et al. 1979.

Mitochondrial oxidation measured in the presence and absence of bovine serum albumin (v = substrate : albumin ratio). One catalase unit corresponds to the amount that causes the destruction of 90% of the substrate in 1 minute in a volume of 50 ml. Palmitoyl-CoA oxidation rates represent the formation of total acid-soluble oxidation products and are expressed as micromoles of fatty acid oxidized per minute. Glutamate dehydrogenase and carnitine palmitoyltransferase activities are expressed in international units.

[a] Results are significantly different from controls, $p < 0.025$.

Three peroxisomal β-oxidation enzymes (acyl-CoA dehydrogenase, 3-hydroxyacyl-CoA dehydrogenase, and 3-ketothiolase) have been localized in the matrix of the particles rather than in the membrane (unpublished data from our laboratory). Osumi and Hashimoto (1979 a) found enoyl-CoA hydratase, 3-hydroxyacyl-CoA dehydrogenase, and 3-ketoacyl-CoA thiolase only in peroxisomes of the liver of rats that had been treated with a hypolipidemic drug (DEHP; see Section III.G.1.h.), whereas the peroxisomes of control animals were essentially devoid of these enzymes. Therefore, Osumi and Hashimoto raise the question as to whether peroxisomal fatty acid oxidation is significant in the normal liver at all.

In liver homogenates of control and clofibrate-fed rats Mannaerts et al. (1979) found activities of the peroxisomal (cyanide-insensitive) palmitoyl-

CoA oxidation comparable to those of the mitochondria. In the absence of serum albumin (under which condition the mitochondrial process is severely inhibited), the peroxisomal oxidation is even severalfold higher (Table 7). A quite different view comes from the assessment of the peroxisomal β-oxidation in isolated hepatocytes by measurement of the hydrogen peroxide formed during the acyl-CoA dehydrogenase reaction. MANNAERTS et al. (1979) report "that the contribution of the peroxisomes to fatty acid oxidation was less than 10% both in cells from control and clofibrate-treated animals".

Carnitine acetyltransferase (MARKWELL et al. 1973) can be seen as closely related to β-oxidation. An excess of acetyl groups formed in peroxisomes by this pathway can be stored as acetyl carnitine.

The indirect oxidation of extraperoxisomal NADH via α-hydroxy acid oxidase has been discussed by DE DUVE and BAUDHUIN (1966); it depends on the free permeability of the peroxisomal membrane for small α-hydroxy acids and keto acids (e.g., lactate and pyruvate).

The peroxisomal membrane seems to be permeable for NADH (THINÈS-SEMPOUX and BORMANN 1978). Otherwise shuttle systems for hydrogen should exist. Cytoplasmic and peroxisomal glycerol phosphate dehydrogenase (GEE et al. 1974) might represent such a shuttle system, consisting of the NADH-dependent redox couple glycerol-3-phosphate—dihydroxyacetone phosphate (DHAP).

On the other hand, glycerol-3-phosphate dehydrogenase should be able to keep up the peroxisomal level of DHAP, which in the whole cell is much lower than that of glycerol-3-phosphate. JONES and HAJRA (1977) speculated that high dihydroxyacetone phosphate concentrations might favor the acylation of this compound by peroxisomal DHAP acyltransferase, an enzyme formerly thought to be localized in mitochondria (HAJRA 1968 a). JONES and HAJRA suspect that also alkyl-DHAP synthase and acyl/alkyl-DHAP-NADPH oxidoreductase—enzymes which finally yield the O-alkyl analogues of glycerolipids—are located in peroxisomes. This has been proved recently by density gradient centrifugation (HAJRA et al. 1979). Thus, the so-called DHAP pathway of phosphatide biosynthesis demonstrated in liver (HAJRA 1968 a, b, HAJRA and AGRANOFF 1968 a, b) and in other tissues (LA BELLE and HAJRA 1972) might be confined to peroxisomes; if so, it would represent a peroxisomal anabolic chain of phosphatide biosynthesis in addition to the alternative glycerol-3- phosphate pathway, which is located in the smooth endoplasmic reticulum.

Fig. 19. Some metabolic reactions in which peroxisomal enzymes are directly involved. In two of the hydrogen peroxide-producing reactions (urate oxidase and hydroxyacid oxidase) the flavines are omitted. *Broken arrows:* intermediate steps are omitted; *acyl-Tf.* dihydroxyacetone phosphate acyltransferase; *CAT* carnitine acetyltransferase; *DHAP* dihydroxyacetone phosphate; *DHAP Red.* acyl/alkyl-dihydroxyacetone phosphate-NADPH-oxidoreductase; *GLAT* glyoxylate aminotransferase; *G 3 P* glycerol-3-phosphate; *GPDH* glycerol-3-phosphate dehydrogenase (NAD); *HAO* L-α-hydroxyacid oxidase; *ICDH* isocitrate dehydrogenase (NADP); *UO* urate oxidase; *Glu* glutamate; *α-KG* α-ketoglutarate.

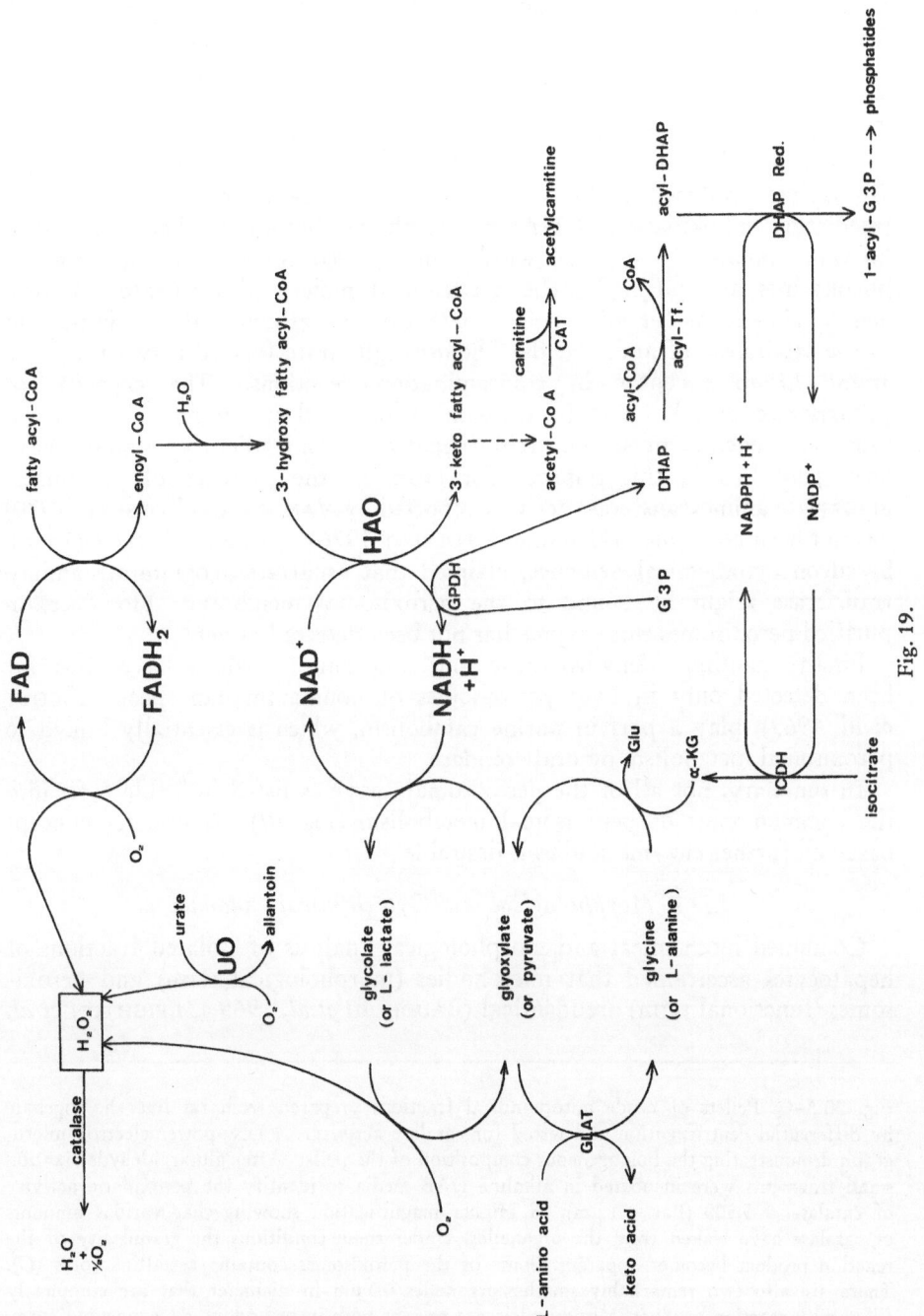

Fig. 19.

In liver peroxisomes some L-amino acids are able to undergo transamination with glyoxylate (glyoxylate aminotransferase; HSIEH and TOLBERT 1976). The most active amino group donors are leucine, phenylalanine, glutamine, and serine (more than 25% of the activity that is found with leucine): The catabolism of amino acids seems to be connected to peroxisomal reactions. The amino group acceptor glyoxylate can arise along with alanine from the degradation of an amino acid other than glycine, namely hydroxyproline. Glyoxylate produced in this way may alternatively serve as a redox couple (mediating the oxidation of NADH; see above), become further degraded or be transaminated to glycine, which can be used for biosyntheses (e.g., of porphyrines and purines). The peroxisomal moiety of isocitrate dehydrogenase (LEIGHTON et al. 1968, 1969) can be assumed to be important in the catabolism of amino acids: The α-oxoglutarate formed may serve as an amino group acceptor in transamination reactions. The capacity of peroxisomes to take part in transaminations and in amino acid metabolism in general must be reinvestigated. Contradictory results have been published in this respect: for example, the presence of glutamate-glyoxylate aminotransferase activity reported by VANDOR and TOLBERT (1970) has not been confirmed (HSIEH and TOLBERT 1976). LEE and TORACK (1968), based on cytochemical evidence, claimed that aspartate-oxoglutarate aminotransferase might be bound to the peroxisomal membrane. However, in purified peroxisomes this enzyme has not been detected as yet.

Finaly, xanthine dehydrogenase and allantoinase, which have thus far been detected only in liver peroxisomes of nonmammalian species (SCOTT et al. 1969), play a part in purine catabolism, which is essentially linked to peroxisomal metabolism by urate oxidase.

In summary, not all of the peroxisomal enzymes listed in Table 6 fit into the common view of peroxisomal metabolism (Fig. 19). A broader concept based on further enzyme studies is desirable.

I.D.2. Morphological and Cytochemical Evidence

Combined biochemical and morphological analysis of isolated fractions of hepatocytes ascertained that microbodies (morphological term) and peroxisomes (functional term) are identical (BAUDHUIN et al. 1965 a, LEIGHTON et al.

Fig. 20 A–C. Pellets of crude mitochondrial fractions prepared from rat liver homogenate by differential centrifugation and tested for catalase activity. A Low-power electron micrograph demonstrating the homogeneous composition of the pellet. After glutaraldehyde fixation small fragments were incubated in alkaline DAB media to identify the peroxidatic activity of catalase. ×5,500 (Bar = 1 µm). B Higher magnification showing that various amounts of catalase have leaked from the organelles. Under these conditions the granularity of the reaction product becomes apparent. Some of the peroxisomes contain crystalline cores (C). There are also two remarkably smaller organelles 0.3 µm in diameter that are completely filled with reaction product. Mitochondria are present both in orthodox and condensed form. Brief staining with lead citrate. ×28,000. C Preparation similar to that demonstrated in parts A and B. All peroxisomes have lost a major portion of their catalase content, most probably as a result of osmotic shock. Remaining peroxidatic activities are attached to the inner surfaces of the limiting membranes. ×28,000.

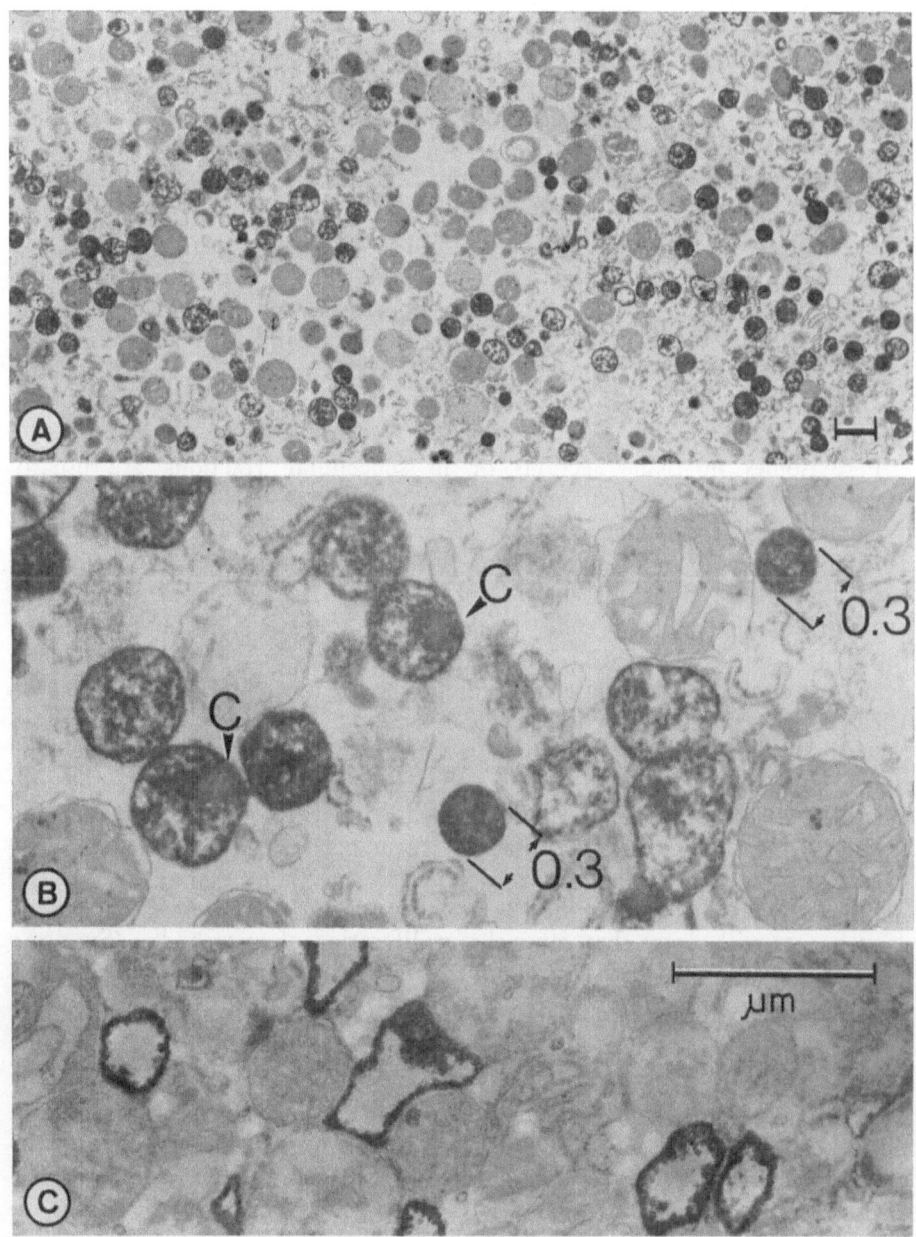

Fig. 20.

1968). Furthermore, the cytochemical 3,3'diaminobenzidine tetrahydrochloride (DAB) method for determining the peroxidatic activities of catalase (FAHIMI 1969) has been successfully used to stain peroxisomes in various tissues as well as in isolated fractions (Figs. 20 and 21). Catalase in mitochondrial fractions has been clearly demonstrated to be restricted to microbodies.

The control of isolated cell fractions by means of cytochemical methods has become more important in studies of isolated microperoxisomes (BÖCK et al. 1975, GOLDENBERG et al. 1975 b, 1978 a, KRAMAR et al. 1974). Whereas hepatic or kidney peroxisomes are clearly defined by size, shape, matrix density, and nucleoids, the morphology of isolated microperoxisomes is less distinct; the organelles, which lack a nucleoid, vary in shape and size. The often oval or worm-like microperoxisomes become spherical when they are isolated (see section VI.B.3.). This causes a certain dilution and rearrangement of matrix material, and cytochemical catalase staining remains the only method of identifying these particles in the various fractions (Figs. 20 and 21).

Peroxisomes in tissue sections or in subcellular fractions are usually identified by means of the cytochemical catalase reaction. Other methods of demonstrating peroxisomal enzymes, described in section II.B., may be used as alternatively to, or in combination with, catalase staining. However, the DAB reaction is the most simple and reliable. Incubation media used to demonstrate L-α-hydroxy acid oxidase stain peroxisomes in rat kidney (SHNITKA and TALIBI 1971); slightly modified media have been used to demonstrate glycolate oxidase in cucumber cotyledons (BURKE and TRELEASE 1975) and in green algae (GRUBER and FREDERICK 1977), as well as in *Hydra* (HAND 1976). D-amino acid oxidase has been demonstrated at the fine structural level in peroxisomes of stickleback kidney cells (VEENHUIS and WENDELAAR BONGA 1977). Malate synthase has been identified with cytochemical methods in glyoxysomes of cucumber and sunflower seedlings (TRELEASE et al. 1974).

Other evidence that microbodies are identical to peroxisomes came from experiments with drugs that cause the proliferation of microbodies (SVOBODA and AZARNOFF 1971) and concomitantly increase the levels of peroxisomal enzymes (see section III.G.).

I.E. Catalase-Positive Particles in Plants

Recently a comprehensive monograph on plant microbodies was issued in this series (GERHARDT 1978). Obviously, there is no demand for a review of plant microbodies here. Only some basic features that are necessary for comparison with metabolic pathways in animal peroxisomes will be discussed.

I.E.1. Higher Plants: Glyoxysomes and Leaf Peroxisomes

From an evolutionary point of view, certain plant microbodies stand closer to the "ancestral peroxisome" (the hypothetical precursor of plant and animal peroxisomes; DE DUVE 1969), which comprehends all capacities of various forms of peroxisomes (Fig. 22). Important metabolic pathways running in plant peroxisomes are missing in animal microbodies, while all principal activities of animal peroxisomes have been detected in plant peroxisomes as well.

Three principal types of peroxisomes can be distinguished in higher plants. All these particles contain catalase and hydrogen peroxide-producing oxidases (α-hydroxy acid oxidase and urate oxidase); thus they represent true peroxisomal systems (HUANG and BEEVERS 1971). Only the first two types of plant microbodies display metabolic activities beyond production and disposal of hydrogen peroxide.

1. Glyoxysomes from the endosperm of germinating seeds (standard object: castor bean, *Ricinus communis*) are characterized by the enzymes of the gly-

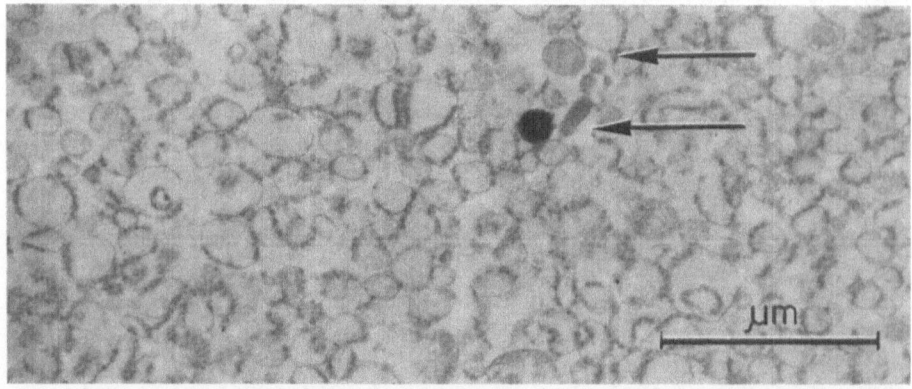

Fig. 21. Pellet from a microsomal fraction prepared by differential centrifugation (Harderian gland of the rat). The cytochemical catalase reaction shows that not all the particles that contain matrix material are microperoxisomes (*arrows*). ×30,000.

oxylate cycle (BEEVERS 1969, BREIDENBACH and BEEVERS 1967). The glyoxylate cycle enables seedlings to transform stored fat into carbohydrates: acetyl-CoA, the product of fatty acid β-oxidation, is converted into C_4-dicarboxylic acids necessary for gluconeogenesis (BEEVERS 1961) (Fig. 22). The enzymes of β-oxidation are bound to the glyoxysomal membrane as well as malate synthase, citrate synthase, and antimycin A-insensitive NADH-cytochrome c reductase (BIEGLMAYER *et al.* 1973, HUANG and BEEVERS 1973), so that the membrane forms a functional unit of production and consumption of acetyl-CoA.

2. Leaf peroxisomes from the mesophyll of green leaves (standard object: spinach, *Spinacea oleracea*) are able to dispose of glycolate, which is delivered by chloroplasts during photosynthesis (McGROARTY and TOLBERT 1973, TOLBERT and YAMAZAKI 1969, TOLBERT *et al.* 1968). Glycolate is oxidized to glyoxylate by glycolate oxidase; in the chloroplasts glyoxylate may be changed back into glycolate by NADP-linked glyoxylate reductase ("photorespiration"; GIBBS 1969). This pathway dissipates photosynthetic NADPH. On the other hand, glyoxylate may be transaminated to glycine, which, when

taken up by mitochondria, gives serine. This amino acid—again, inside the peroxisomes—is capable of transamination with glyoxylate and forms hydroxypyruvate, which finally is reduced by NADH and hydroxypyruvate reductase to form glycerate, a precursor of carbohydrates.

A summary of the reactions taking this course in glyoxysomes and leaf peroxisomes is given in Fig. 22. As this figure summarizes the reactions in different types of plant microbodies, it essentially corresponds to a scheme of the "ancestral peroxisome".

3. Besides the rather specialized glyoxysomes and leaf peroxisomes, particles similar to animal peroxisomes have been found in most plant tissues and in microorganisms; these are the so-called "unspecialized peroxisomes" (TOLBERT and YAMAZAKI 1969). Thus far, these particles have been found to contain only the "peroxisomal" set of enzymes (catalase and hydrogen peroxide-producing oxidases).

I.E.2. Fungi

Microbodies have been demonstrated and studied in a number of true fungi as well in ascomycetes (especially yeasts and yeast-like fungi such as *Neurospora*) and in slime molds (for a review see FREDERICK et al. 1975, GERHARDT 1978).

Yeasts are facultative anaerobes; catalase formation is repressed when they are grown on fermentable carbon sources (such as glucose) or during anaerobiosis (PARISH 1975 a, ROGERS and STEWART 1973, SZABO and AVERS 1969). Yeasts grown on methanol as the sole source of carbon possess relatively large peroxisomes and exhibit a high level of catalase activity (FUKUI et al. 1975 a, HAZEU et al. 1975, ROGGENKAMP et al. 1974, 1975). This can be seen in connection with the oxidation of methanol by the peroxisomal flavoenzyme alcohol oxidase (FUKUI et al. 1975 b, ROGGENKAMP et al. 1975):

$$CH_3OH + O_2 \rightarrow H_2C = O + H_2O_2.$$

The hydrogen peroxide formed is used for the coupled oxidation of a second molecule of methanol due to the peroxidatic action of catalase (ROGGENKAMP et al. 1974):

$$CH_3OH + H_2O_2 \rightarrow H_2C = O + 2 H_2O.$$

Increased catalase activity, which parallels the appearance of microbodies, is observed in *Candida* strains grown on *n*-alkanes (OSUMI et al. 1974, 1975 a, TANAKA and IIDA 1977, TERANISHI et al. 1974). These observations are con-

---

Fig. 22. Metabolic pathways in a generalized peroxisome (corresponding to the "ancestral peroxisome" of DE DUVE 1969): combination of reactions in glyoxysomes and leaf peroxisomes (glyoxylate cycle and glycolate metabolism). *AOAT* aspartate-oxoglutarate aminotransferase; *cat* catalase; *CS* citrate synthase; *GGLAT* glutamate-glyoxylate aminotransferase; *GO* glycolate oxidase; *HPR* hydroxypyruvate reductase; *ICDH* isocitrate dehydrogenase; *ICL* isocitrate lyase; *MDH* malate dehydrogenase; *MS* malate synthase; *SGLAT* serine-glyoxylate aminotransferase; *SHMT* serine hydroxymethyltransferase; *UO* urate oxidase.

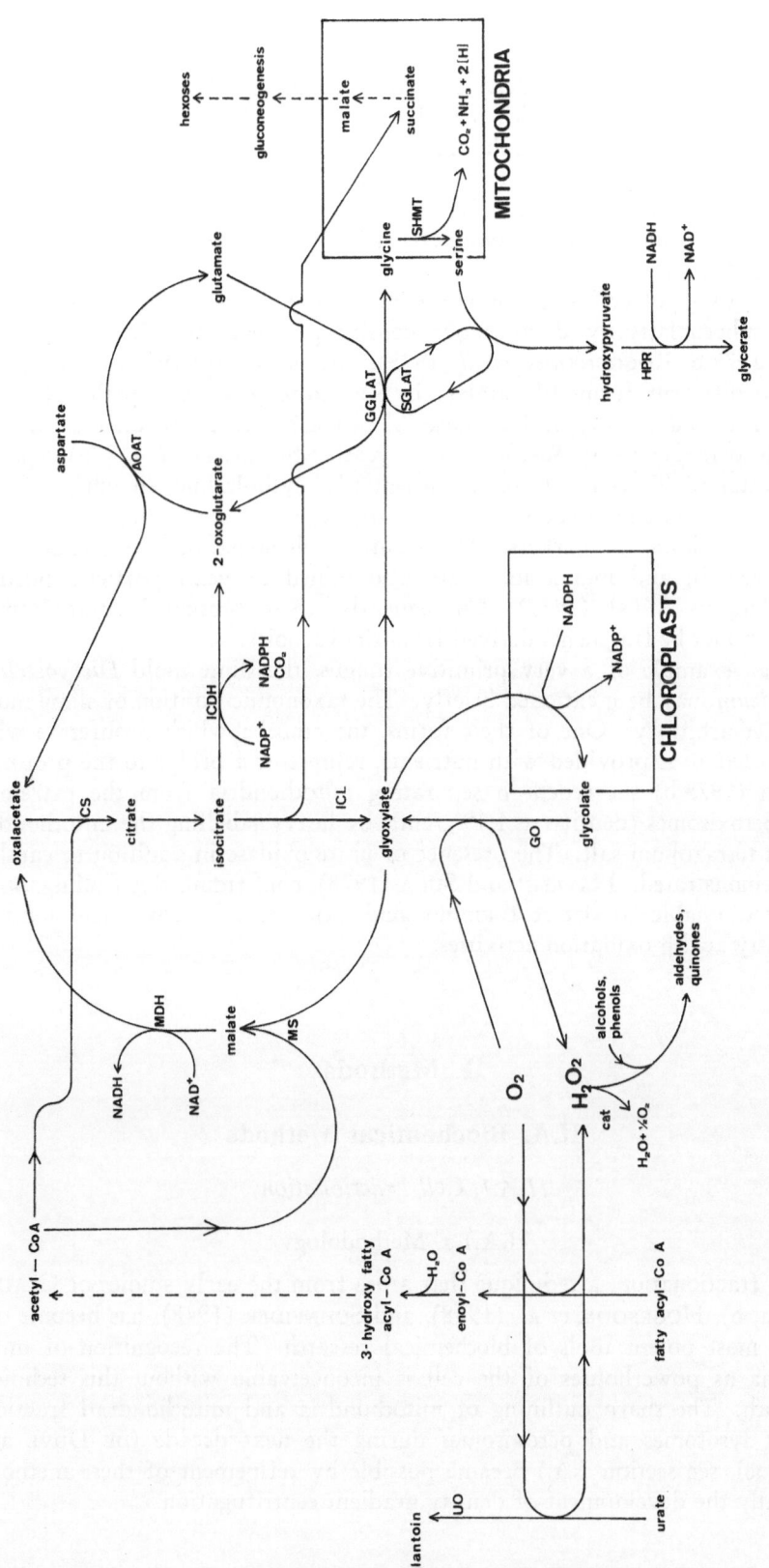

Fig. 22.

sistent with the known capacity of hydrocarbon-assimilating yeasts to oxidize alkanes to fatty acids, which may be degraded by peroxisomal β-oxidation and transformed into carbohydrate via the glyoxylate cycle. However, enzymes of these pathways have not yet been detected in *Candida* strains; in *Saccharomyces* part of the isocitrate lyase and malate synthase activity seems to be bound to peroxisomes (AVERS 1971; for contradictory results see PARISH 1975 a).

Peroxisomes of yeasts grown on alkanes and on methanol have been isolated from spheroplasts by discontinous density gradient centrifugation (OSUMI *et al.* 1975 b, ROGGENKAMP *et al.* 1975). Studies considering the occurrence of two different forms of catalase in *Saccharomyces*, namely the "typical" catalase T and the atypical catalase A, cast a severe doubt on the existence of true peroxisomes in *Saccharomyces* yeasts. SUSANI *et al.* (1976) reported that catalase T seems to be an essentially cytoplasmic enzyme, whereas catalase A is bound predominately to large vacuoles containing proteinase B and mannosidase as markers. These authors demonstrated that catalase A, proteinase B, and mannosidase are also joined in yeast particles purified according to AVERS (1971). Consequently, these presumed "peroxisomes" might in fact be fragments derived from the vacuoles.

As an example of a very primitive fungus, the slime mold *Dictyostelium discoideum* may be mentioned briefly. The taxonomic position of slime molds is rather arbitrary. One of their forms, the amoeba which proliferate when *Dictyostelium* is provided with nutrients, represents a bridge to the protozoa. PARISH (1975 b) succeeded in separating mitochondria from the extremely light peroxisomes (density = 1.19 g/cm^3) by heavy labelling the mitochondria with a tetrazolium salt. The presence of urate oxidase in addition to catalase was demonstrated. HAYASHI and SUGA (1978), confirming this finding, were, however, unable to detect D-amino acid oxidase, α-hydroxy acid oxidase, and fatty acid β-oxidation activities.

II. Methods

II.A. Biochemical Methods

II.A.1. Cell Fractionation

II.A.1.a. Methodology

Cell fractionation, a technique that arose from the early studies of CLAUDE (1946 a, b), HOGEBOOM *et al.* (1948), and SCHNEIDER (1948), has become one of the most potent tools of biochemical research. The recognition of mito-chondria as powerhouses of the cell is inconceivable without this technical approach. The sharp outlining of mitochondria and mitochondrial fractions against lysosomes and peroxisomes during the next decade (DE DUVE and his school, see section I.A.) became possible by refinement of these methods, especially the development of density gradient centrifugation.

Cell fractionation is the isolation of cell fractions from a tissue homogenate. Normally the fraction obtained contains more than one type of organelle and should not be confused with the organelles themselves: a catalase-rich preparation may be named "peroxisomal fraction" but the term "peroxisomes" should be avoided. The first step in fractionation, following the sampling, purifying, and cutting up of the tissue is always homogenization. This step is crucial and the method must be tried out and chosen empirically for each tissue under investigation because it depends on its rigidity, connective tissue content, vulnerability of the organelles, etc. For practically all preparations of liver peroxisomes and other catalase-containing particles homogenizers of the type designed by POTTER and ELVEHJEM (1936) fitted with Teflon or glass pestles (gap between pestle and tube 0.2–0.3 mm) have been used after appropriate prehomogenization of the tissue. Liver (LEIGHTON et al. 1968), kidney (BAUDHUIN et al. 1965 b, KITANO and MORIMOTO 1975), heart (HERZOG and FAHIMI 1974 a, 1975), brown fat tissue (PAVELKA et al. 1976), intestinal epithelium (CONNOCK 1973, CONNOCK and KIRCK 1973, CONNOCK and POVER 1970, CONNOCK et al. 1974), lung (GOLDENBERG et al. 1978 b), and various glandular tissues (BÖCK et al. 1975, GOLDENBERG et al. 1975 b) are examples for materials treated in this way. Generally the procedures chosen are mild and often only one up-and-downward movement of the tube is recommended. Blenders have not been applied to peroxisomal preparations because they are suspected of damaging organelles; for example, BERTHET and DE DUVE (1951) have shown that lysosomal enzymes are set free by blending.

The media for homogenization preferentially contain 0.25 M sucrose. They should include 0.1% ethanol in order to prevent formation of the inactive catalase compound II, which does not appear when compound I can react peroxidatically with ethanol (CHANCE 1950) (see section I.C.). Other additions sometimes used (e.g., see GOLDENBERG et al. 1978 a, b), such as EDTA (used for preserving glucose-6-phosphatase activity; BEAUFAY et al. 1954) or buffers are optional.

In all studies on peroxisomes the second step, the separation of cell fractions, is performed by centrifugation methods. Continuous free-flow electrophoresis according to HANNIG (1964), which makes use of the negative net charge of lysosomes and other cell organelles at physiological pH (SAWANT et al. 1964), has proved to be appropriate for the isolation of liver lysosomes. Unfortunately, peroxisomes cannot be separated from mitochondria by this method (for a more recent review see HANNIG and HEIDRICH 1974).

Exclusion chromatography (e.g., gel filtration on Sepharose gels; TANGEN et al. 1973) seems to be useful for the purification of microsomes that are excluded from the gel. Exclusion chromatography avoids the deterioration of particles brought about by high hydrostatic pressure during centrifugation (BRONFMAN and BEAUFAY 1973), but separation of the various cell organelles is not yet possible.

A comprehensive view of centrifugal methods is obtained by using the Svedberg equation (SVEDBERG and PEDERSEN 1940). The velocity v of a spherical particle (radius r, density ϱ_p) sedimenting in a medium of density ϱ_m and viscosity η at the radial distance x from the axis of the centrifuge (which runs with the angular velocity ω) becomes constant when the frictional force, according to Stokes' law, equals the centrifugal force:

$$6 r \pi \eta v = 4/3 r^3 \pi (\rho_p - \rho_m) \omega^2 x,$$

or

$$v = \frac{dx}{dt} = \frac{2 r^2}{9 \eta} (\rho_p - \rho_m) \omega^2 x \tag{1}$$

The first term of Equation (1) is the sedimentation coefficient s of the particle:

$$s = \frac{2 r^2}{9 \eta} (\rho_p - \rho_m) = \frac{dx/dt}{\omega^2 x} [sec] \tag{2a}$$

In the integrated form (DE DUVE and BERTHET 1953):

$$s = \frac{2 r^2}{9 \eta} (\rho_p - \rho_m) = \frac{3.5 (\log R_{max} - \log R_{min})}{\int_0^t (\text{rpm})^2 \, dt} \tag{2b}$$

The equation is useful for the determination of s by pelleting experiments. R_{max} and R_{min} are the distances from the axis of rotation to the surface of the pellet and to the fluid in the centrifuge tube; a particle with the coefficient s starts from R_{min} and is just pelleted after t minutes; rpm is revolutions per minute.

The centrifugal methods are classified according to two different sets of conditions that are practically important:

Differential Centrifugation

In this method the density of the sedimenting particles ϱ_p is always appreciably greater than the density of the medium ϱ_m. According to Equation (1), the velocity of the particles is proportional to the square of their radius; separation essentially ensues due to the size of the particles. In the most simple case it is achieved in a homogeneous medium by sedimentation of the larger particles, whereas the others more or less stay in the supernatant.

Differential pelleting (often called simply differential centrifugation). The integrated force (or the time) necessary for full pelleting can be assessed from Equation (2 b). At the beginning of the experiment various particles are homogeneously distributed over the medium. Complete separation is impossible: If there are two different types of particles such that the respective sedimentation coefficients and velocities are 2 : 1, we expect that in the moment when the larger particles are just fully pelleted, one-half of the smaller ones are still unsedimented, whereas the other half contaminates the pellet. Disregarding disturbances such as wall effects, we find that the ratio of the sedimentation coefficients of the smaller and larger particles (s_1/s_2) corresponds to the proportion of the smaller ones contaminating the pellet. As a rule, separation by differential pelleting is tolerably efficient when the sedimentation coefficients are different by at least one order of magnitude; the 10% of contaminating smaller particles can be removed by subsequent washings. It is evident from the *s* values in Table 8 and Fig. 23 that mitochondria, lysosomes, and peroxisomes cannot be resolved by differential pelleting.

Differential centrifugation in a density gradient (rate sedimentation). The contamination of pellets that normally occurs in differential pelleting method may be avoided by using this method. In this procedure the sample is loaded

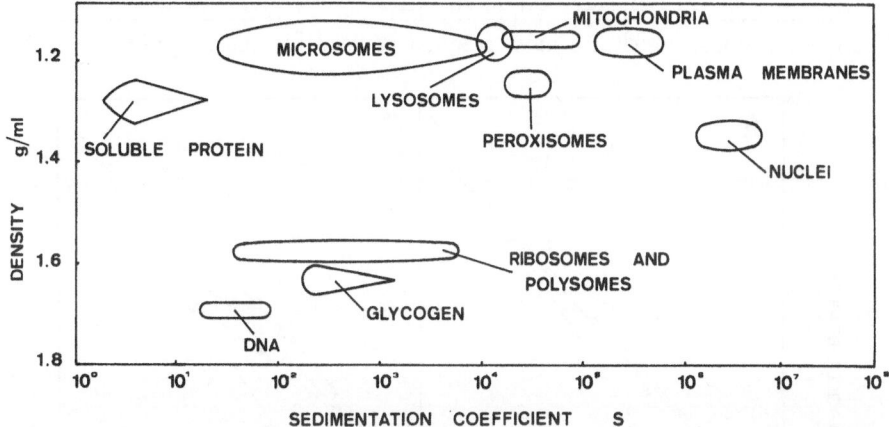

Fig. 23. Density versus sedimentation coefficient of liver cell particles. The diagram allows predictions on the separability of organelles by isopycnic or differential centrifugation. From ANDERSON *et al.* 1966.

Table 8. *Average sedimentation coefficients and equilibrium densities of rat liver mitochondria, lysosomes, and peroxisomes*

Organelle	Sedimentation coefficients (sec^{-1} × 10^{10}) = (Svedberg units × 10^{-3}) in 0.25 M sucrose [a]	Median equilibrium density [b] (g/cm^3)
Mitochondria	13.3 ± 1.0	1.19
Lysosomes	9.9 ± 0.3	1.20
Peroxisomes	9.2 ± 0.7	1.23

[a] Values (SLINDE and FLATMARK 1973) have been calculated from pelleting experiments in 0.25 M sucrose at 4 °C.

[b] Equilibrium densities (BEAUFAY *et al.* 1964) are valid for sucrose gradients and refer to cytochrome oxidase, catalase, and acid phosphatase as markers.

as a thin layer onto a stabilizing, shallow density gradient (mostly of rather low density in order that the particles never arrive at their isopycnic density in the gradient). The greater its sedimentation coefficient, the faster a particle migrates down, thus the separated fractions contain particles of equal size [*cf.*, Equation (1) and (2 a)]. In practice, however, sometimes the faster particles approach positions not too far from the isopycnic one and are slowed down remarkably.

Isopycnic Gradient Centrifugation

In a gradient whose heavy end is at least as dense as the particle with the highest density the centrifugation can be pursued until the different particles stop at their individual equilibria or buoyant densities, where ϱ_m becomes ϱ_p and $dx/dt = 0$ [Equation (1)]; naturally this final equilibrium position does not depend on the size of the particle.

It must be realized that the density of a particle is not constant but depends on the medium and changes while the particle migrates down the gradient. Sucrose solutions, which are most frequently used for the preparation of per-

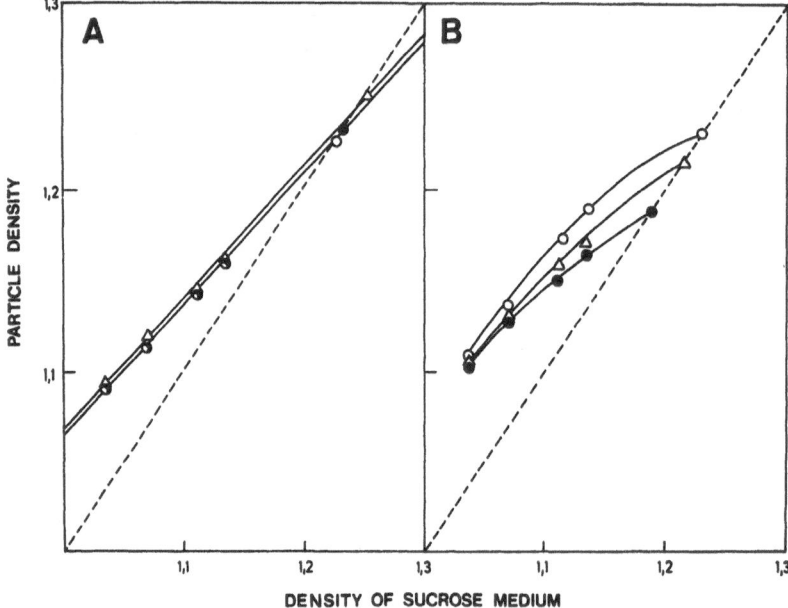

Fig. 24. Dependence of the density of rat liver particles on the density of the sucrose solution used as suspension medium. *A* Peroxisomes: ● catalase; ○ D-amino acid oxidase; △ urate oxidase. *B* Mitochondria and lysosomes: ● cytochrome oxidase; ○ acid deoxyribonuclease; △ acid phosphatase. From HINTON and DOBROTA (1976) according to BEAUFAY (1966).

oxisomes, exert great osmotic effects due to high osmolarity (as compared with macromolecular solutes such as glycogen). On the other hand, some of the intracellular membranes are permeable for sucrose; *e.g.*, the peroxisomal and the outer mitochondrial membranes are permeable, whereas the inner mitochondrial membrane cannot be passed by this molecule. These circumstances are most easily summarized by a formal model of cell organelles derived by DE DUVE *et al.* (1959) based on earlier observations on mitochondria (WERKHEISER and BARTLEY 1957). According to this model, membrane-bounded particles consist of a hydrated matrix containing the solid constituents of the particle, a space that is accessible only to water ("osmotic space"), and a space that is freely accessible to low molecular weight solutes such as the sucrose contained in the gradients ("sucrose space").

Peroxisomes possess virtually no osmotic space; due to the free influx of sucrose their density increases very markedly with the density of the medium, until in equilibrium the density of the matrix is obtained. Lysosomes and mitochondria, on the other hand, possess an osmotic space; in sucrose solutions the invasion of the solute into the particle's sucrose space is balanced in part by the osmotic water uptake, resulting in a somewhat lower isopycnic equilibrium density (Figs. 23 and 24).

Reasonable evaluation of the results of cell fractionation studies is essential for the critical examination of the method used. DE DUVE and his school, in particular, developed graphic representations (histograms). In this system the fractional activity of an enzyme (or the fractional amount of another component) in a subcellular fraction is defined as $Q/\Sigma Q$, where Q is the absolute amount found in the individual fraction and ΣQ is the sum of these amounts in all fractions; ΣQ is virtually equal to the total amount of the component at the beginning of the fractionation procedure. Sometimes instead of the fractional amount the percentage, $100 Q/\Sigma Q$ is used. The fractional amount (or percentage) is the yield of the component recovered in the individual fraction.

Generally, aligned along, the abscissa of the graph are values of Δx that are characteristic of the respective fraction (e.g., volume, protein content, etc.). On the ordinate the quotient of the fractional amount divided by Δx is plotted: $y = (Q/\Sigma Q)/\Delta x$. Now for each fraction a rectangle $y \cdot \Delta x$ is drawn. The surface area of the rectangle of width Δx and of height $y = Q/(\Sigma Q \cdot \Delta x)$ therefore represents the fractional amount $Q/\Sigma Q$ of the component in the respective fraction. Obviously, the sum of all these areas equals 1 (or 100%).

Fractionations by *differential pelleting* are usually presented by a plot of the relative specific activity (RSA) of an enzyme measured in the considered fraction *versus* its fractional (or percentage) protein (or nitrogen) content $P/\Sigma P$ (or $100 P/\Sigma P$). The relative specific activity is an appropriate measure of the enrichment or purification of an enzyme in the respective fraction; it is defined as the percentage of enzyme activity recovered in the fraction divided by the percentage of protein recovered in the fraction:

$$RSA = \frac{Q/\Sigma Q}{P/\Sigma P}$$

It can also be considered as the ratio of the specific activity in the fraction to the activity in the starting material:

$$RSA = \frac{Q/P}{\Sigma Q/\Sigma P}$$

As desired, the rectangles $\Delta x \cdot y$ equal the fractional activity:

$$(P/\Sigma P) \cdot \frac{Q/\Sigma Q}{P/\Sigma P} = Q/\Sigma Q$$

In the usual graphs ("Duvograms") these rectangles are aligned from left to right in the order of decreasing sedimentation constant (i.e., decreasing size). Examples of this type of plot are found in Figs. 1 and 56.

In *isopycnic density gradient experiments* the frequency $Q/(\Sigma\, Q \cdot \Delta\, \varrho)$ may be plotted on the ordinate, where $\Delta\, \varrho$ is the interval of density in the gradient that is kept by the fraction. Consequently, in this plot $\Delta\, \varrho$ values are aligned along the abscissa so that, again, the rectangles $\Delta\, x \cdot y$ represent the fractional amounts or activities $Q/\Sigma\, Q$ (as an example see Fig. 57).

In a graph generally suitable for *all density gradient experiments* the ordinate represents the relative concentration $C/\overline{C}$ of the component, where $\overline{C}$ is the concentration that the recovered component would have if it were homogeneously distributed in the recovered volume $\Sigma\, V$:

$$C/\overline{C} = (Q \cdot \Sigma\, V)/(V \cdot \Sigma\, Q)$$

where V is the volume of the individual fraction. In this plot the fractional (or percentage) volume $V/\Sigma\, V$ (or $100\, V/\Sigma\, V$) is taken for $\Delta\, x$; also in this case the area of one block

$$\Delta\, x \cdot y = (C/\overline{C}) \cdot (V/\Sigma\, V) = Q/\Sigma\, Q$$

gives the fractional activity. Figs. 2 and 52 are examples of this graphic presentation.

II.A.1.b. Liver Tissue

Due to overlapping sedimentation coefficients (see Section II.A.1.a.) the complete separation of large particles from liver is impossible by differential pelleting. Neither the "classical" scheme (Fig. 25) nor even the modified five-fraction scheme (APPELMANS et al. 1955, DE DUVE et al. 1955), which splits the crude mitochondrial fraction into a heavy (M) and a light (L) fraction (Fig. 26), is able to solve this problem (see Fig. 1). The L-fraction containing most of the peroxisomal activities carries inseparable cross contaminations of microsomal, lysosomal, and mitochondrial activities.

For further purification differential centrifugation must be followed by gradient centrifugation. Considering the similar size of "large" particles (Fig. 27), isopycnic gradient centrifugation seems to be the method of choice for separating mitochondria and peroxisomes, which in sucrose exhibit buoyant densities different enough to make then separable, whereas the density of lysosomes greatly overlaps that of peroxisomes (Fig. 23, Table 8). This difficulty may be overcome by an injection of Triton WR-1339 *in vivo*, which lowers the density of lysosomes appreciably (WATTIAUX et al. 1963).

As a standard procedure the method of LEIGHTON et al. (1968) has been widely accepted. It uses a special zonal rotor (the Beaufay automatic rotor) for gradient centrifugation. This special rotor differs from the usual zonal rotors in that it has an annular separation chamber near the outer wall; thus the separation path is extremely short (about 1 cm) and only low hydrostatic pressure develops. In spite of these appreciable advantages the Beaufay rotor is not commercially available.

Therefore, the isopycnic method described here is slightly modified for a larger zonal rotor of common shape with a cylindrical separation chamber divided into sector-shaped compartments by vanes; it possesses a larger separation path of 43 mm and a total capacity of 640 ml [B-14 aluminum zonal rotor manufactured by MSE, London, England; equivalent rotors are available from other sources].

Optimally, the procedure starts with a differential fraction rich in peroxisomes, namely, an L-fraction prepared from rat liver according to LEIGHTON et al. (1968), called a λ-fraction; more simply, large-particle (LP) frac-

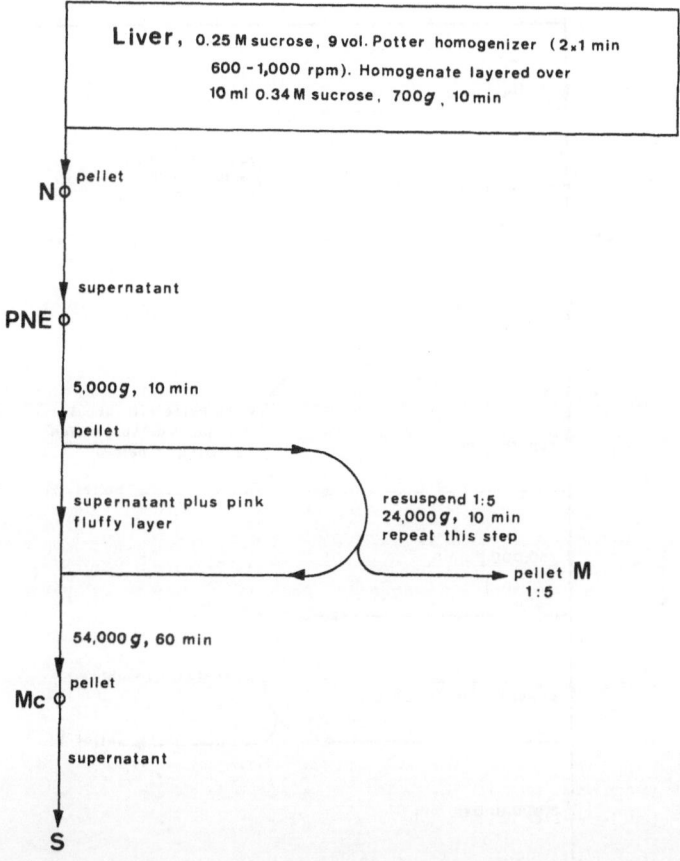

Fig. 25. "Classical" four-fraction scheme of cell fractionation for rat liver tissue (according to HOGEBOOM 1955). *PNE* postnuclear extract; *N* nuclear fraction; *M* mitochondrial fraction; *Mc* microsomal fraction; *S* supernatant.

tions also may be used. The respective fractionation procedures are depicted in Fig. 27.

A volume of λ- or LP-fractions corresponding to 20–30 g rat liver is layered onto 400 ml of a linear sucrose gradient (1.3–1.9 M sucrose and also containing 0.1% ethanol, 2% dextran-10, 1 mM EDTA, and 3 mM imidazole-HCl, pH 7.2; density 1.17–1.25 g/cm³). The gradient rests on a cushion of 50 ml 2.0 M sucrose; the sample may be covered with an overlay of 0.125 M sucrose or water.

After centrifugation for 4 hours at 25,000 rpm in an MSE SS 50 ultra-centrifuge, the gradient is displaced by 2.0 M sucrose. Then 20 fractions of 20 ml are collected from the rotor center. Most of the catalase activity (about

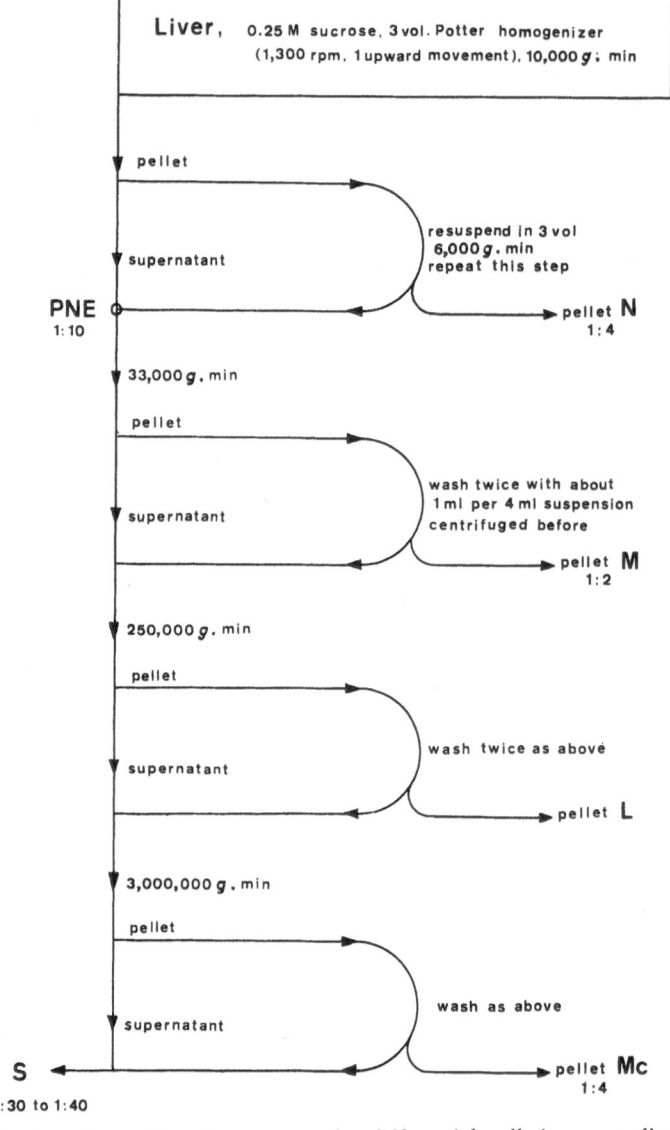

Fig. 26. Fractionation of liver homogenates by differential pelleting, according to the five-fraction scheme of APPELMANS *et al.* (1955) and DE DUVE *et al.* (1955). *PNE* postnuclear extract; *N* nuclear fraction; *M* heavy mitochondrial fraction; *L* light mitochondrial fraction; *Mc* microsomal fraction; *S* supernatant; *vol:* milliliters medium added per gram of original tissue. *Dilution:* a ratio of 1 : 5, for example, means that the material has been made up with sucrose medium to 5 ml per gram of original tissue. *Wash:* resuspension (generally with Potter homogenizer) followed by repelleting under the conditions used in the preceding step.

59% of the activity applied to the gradient) is found in fractions 10–15, corresponding to densities of 1.22–1.235 g/cm³. The peroxisomal content of these fractions is 20–30 times that of the original homogenate as judged by the relative specific activity of catalase. Cytochrome oxidase, glucose-6-phosphatase, and acid phosphatase are found at lower densities (about 1.18, 1.17, and 1.12 g/cm² respectively).

Obviously, the isopycnic gradient method can be performed on a smaller scale in a tube by using a swing-out rotor (BAUDHUIN et al. 1965 b, BEAUFAY

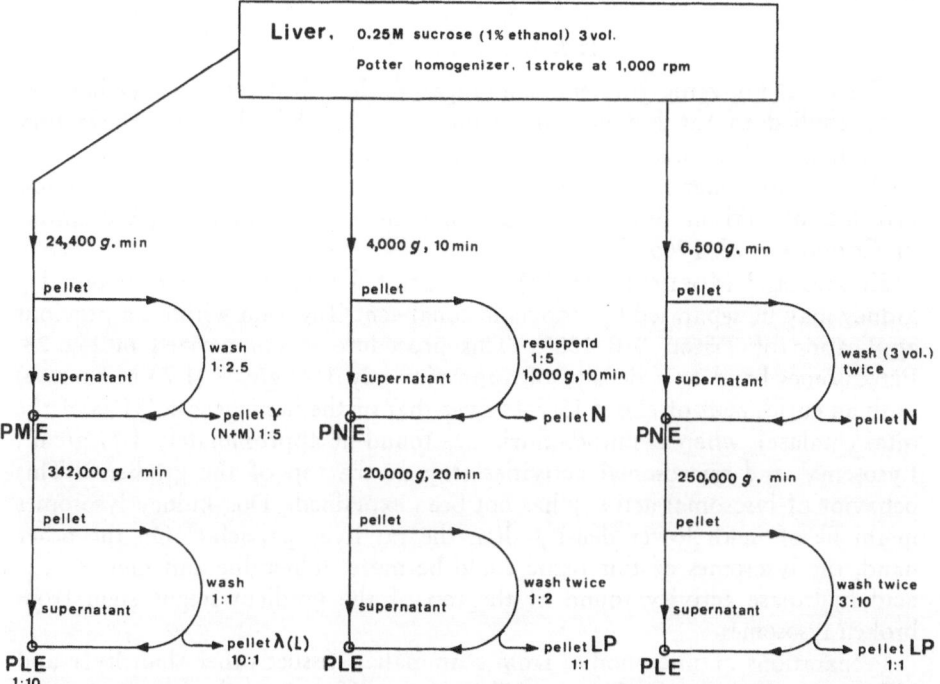

Fig. 27. Preparation of light mitochondrial fraction L (λ according to LEIGHTON et al. 1968) and of "large particles" (LP, corresponding essentially to the sum of the light and the heavy mitochondrial fractions; see BAUDHUIN 1974). PME, PLE postmitochondrial extracts (after "heavy" and "light" mitochondrial fractions, respectively). Other abbreviations as in Fig. 26.

et al. 1964). BAUDHUIN (1974) used sucrose gradients (plus ethanol and dextran-10, see above) ranging from 1.12 to 1.28 g/cm³ in a total volume of 5.0 ml. After centrifugation for 2 hours at 39,000 rpm in an SW 39 Spinco rotor, separation of peroxisomes, mitochondria, and lysosomes was demonstrable.

THOMPSON et al. (1974) introduced a rate zonal method (Spinco Ti-14 rotor, 400 ml) for isolating liver peroxisomes from mice pretreated with Triton WR-1339 and prednisolone; Triton WR-1339 depresses the density of lysosomes to a known extent, whereas the corticosteroid increases the sedimentability of the mitochondria. Under these conditions peroxisomes can be prepared by rate sedimentation, thereby avoiding the deteriorating effect of high hydrostatic pressures connected with isopycnic centrifugations. In fact, during a first zonal centrifugation of a "premicrosomal" liver preparation (20 minutes at 20,000 g), loaded onto a linear gradient 0.29–0.58 M sucrose, most of the mitochondria are removed due to their quicker sedimentation. In the second gradient, which is chosen to be somewhat denser (0.58–1.17 M sucrose), the less dense lysosomes are retarded appreciably and thus are separated to a great extent from the peroxisomes. The enrichment factor (relative specific activity referred to the homogenate) is found to be about 37 for urate oxidase and 27 for catalase.

II.A.1.c. Kidney Tissue

The usual isopycnic gradient centrifugation described for rat liver has also been applied to rat kidney (BAUDHUIN et al. 1965 b) but no satisfactory separation of lysosomal and peroxisomal activities has been achieved. This finding is not amazing because the density of rat kidney lysosomes is less affected by Triton WR-1339 than that of liver lysosomes (WATTIAUX-DE CONINCK et al. 1965).

KITANO and MORIMOTO (1975) demonstrated that peroxisomes from dog kidney may be separated by isopycnic zonal centrifugation without a previous application of Triton WR-1339. This procedure is summarized in Fig. 28. Peroxisomes band at a density of approximately 1.23 g/cm^3 (1.73 M sucrose) with an enrichment of about 15-fold over that of the homogenate (31% of the total catalase), whereas mitochondria are found at approximately 1.17 g/cm^3. Lysosomal and microsomal activities stay at the top of the gradient. This behavior of lysosomal activity has not been explained. Dog kidney lysosomes might be of much lower density than the rat liver particles. On the other hand, the lysosomes of this tissue could be more vulnerable and most of the acid hydrolase activity found at the top of the gradient might stem from broken lysosomes.

Preparations of microbodies from mammalian tissues other than liver and kidney are discussed in Chapter VI because different methods are used for their isolation. For example, the particles are generally of smaller size than liver and kidney peroxisomes, so that rate sedimentation is an important technique for their separation.

II.A.2. Assay of Peroxisomal Enzymes

In Table 6, all the enzymes have been enumerated that are known to be present in liver peroxisomes. Assay methods for peroxisomal enzymes are presented in this section as short recipes for easy reference. The theoretical consideration discussed here are far from complete; further enzymological data and theory are found throughout this volume.

II.A.2.a. Catalase

Hydrogen peroxide: hydrogen peroxide oxidoreductase, EC 1.11.1.6

$$H_2O_2 + H_2O_2 \rightarrow 2\,H_2O + O_2$$

Comment

In the liver of some mammalian species (mouse, rat, and rabbit) multiple forms of catalase are found, whereas in others (man, beef, sheep, pig, dog, cat, and guinea pig) only one form is found (HOLMES and MASTERS 1972). Mouse liver, the best studied example, exhibits 5 forms that may be interconverted by binding or splitting off sialic acid (JONES and MASTERS 1975); thus one gene seems to be responsible for all heteromorphs, as has been demonstrated for mouse kidney (HOFMAN and GRIESHABER 1976). The cytoplasmic forms with lower anodic mobility originate from the acidic form found in peroxisomes and microsomes by stepwise removal of sialic acid by neuraminidase. JONES and MASTERS (1974, 1975) speculate that newly synthesized catalase

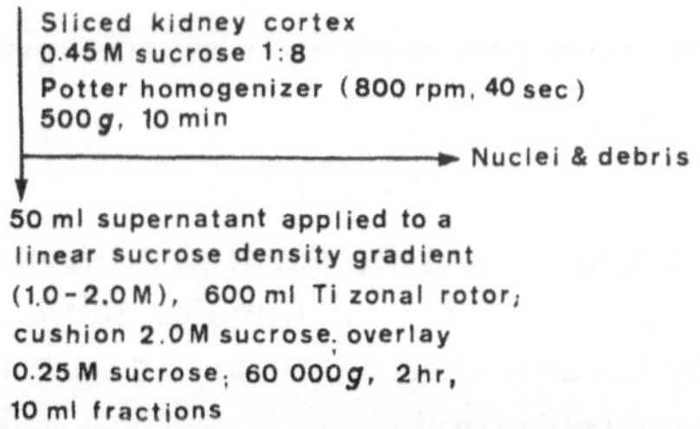

```
│  Sliced kidney cortex
│  0.45 M sucrose 1:8
│  Potter homogenizer (800 rpm, 40 sec)
│  500 g, 10 min
│                              ──────► Nuclei & debris
▼
50 ml supernatant applied to a
linear sucrose density gradient
(1.0 - 2.0 M), 600 ml Ti zonal rotor;
cushion 2.0 M sucrose, overlay
0.25 M sucrose; 60 000 g, 2 hr,
10 ml fractions
```

Fig. 28. Preparation of peroxisomes from dog kidney cortex by differential and subsequent zonal centrifugation (KITANO and MORIMOTO 1975). The peroxisomal peak (catalase and D-amino acid oxidase) is in fraction 50 (1.73 M sucrose).

is sialated in the Golgi apparatus and transferred into peroxisomes. By lysosomal desialation the peroxisomal enzyme is transformed into the more basic cytoplasmic heteromorphs. It follows from these considerations that part of the catalase activity recovered in the supernatant after cell fractionation corresponds to native cytoplasmic enzyme and is not due to peroxisomal leakage, as had been assumed by other investigators (*e.g.*, BEAUFAY et al. 1964, LEIGHTON et al. 1968). This view is confirmed by the observation that in different species catalase is differently distributed between cell sap and particles (HOLMES and MASTERS 1972) (Table 9). Similarly, after peroxisome proliferation induced by the hypolipidimic drug Su-13437, a fivefold increase in soluble catalase is observed, whereas the particle-bound activity is decreased (LEIGHTON et al. 1975).

Further proof of native cytoplasmic catalase came from histochemical studies by ROELS: catalase activity was demonstrated in the cytosol of sheep hepatocytes (ROELS 1976), as well as in rhesus monkey and guinea pig liver cells (ROELS et al. 1977). The presence of extraperoxisomal catalase in chloragogen cells of the earthworm was reported by FISCHER and HORVATH (1978). Recently, NOHL and HEGNER (1978) presented evidence for the occurrence of catalase in the matrix of rat heart mitochondria.

Animal catalase (molecular weight about 240,000) consists of 4 identical protomeres, each containing one ferriprotoporphyrine (TANFORD and LOVRIEN 1962, SUND et al. 1967). This hematin content can be used for spectrophotometric assay of the concentration of catalase at the wavelength of the Soret band [405 nm; extinction coefficient $3.40 \times 10^5/(M \cdot cm)$ for horse liver catalase, AGNER 1938; values for other tissues in CHANCE and MAEHLY 1955, GREENFIELD and PRICE 1956].

The assay of catalase is complicated by its unusual kinetics. The enzyme is inactivated by hydrogen peroxide concentrations below the saturation value by formation of compounds II and III (see Section I.c.), so that it is impossible to perform the assay at the optimal substrate concentration (the Michaelis constant for hydrogen peroxide is 1.1 M; OGURA 1955). Under experimental conditions the reaction rate never obeys a zero-order equation with respect to substrate concentration, so that in any case the overall reaction rate is given by (see CHANCE et al. 1952)

$$v = k_1 e [H_2O_2] \tag{3}$$

where e is the total enzyme concentration; e being given, v becomes of the first order with respect to $[H_2O_2]$

$$v = k [H_2O_2]. \tag{4}$$

The first-order rate constant k may be taken as a measure of catalase activity instead of international units (IU: micromoles substrate per minute under standard conditions) or katal (moles substrate per second), which both depend on hydrogen peroxide concentration:

$$k = \frac{1}{\Delta t} \ln \frac{[H_2O_2]_1}{[H_2O_2]_2} \tag{5}$$

where $\Delta t = t_2 - t_1$ is the period (usually 30 seconds during which the reaction is followed, and $[H_2O_2]_1$ and $[H_2O_2]_2$ are the hydrogen peroxide concentrations at t_1 and t_2 respectively.

The hydrogen peroxide concentration may be determined in different ways (see CHANCE and MAEHLY 1955):

1. Direct spectrophotometric assay of hydrogen peroxide at 230–250 nm (CHANCE and HERBERT 1950, BEERS and SIZER 1952).

2. Permanganate tiration of hyrogen peroxide (BONNICHSEN et al. 1947).

3. Reaction of hydrogen peroxide with Ti (IV) oxysulphate and photometric determination of the peroxititanium sulphate formed (at 410 nm; CHANTRENNE 1955, BAUDHUIN et al. 1964).

On the other hand, for the assay of catalase the formation of O_2 may be followed either polarographically or manometrically (BEERS and SIZER 1953, HALBACH 1977, JACOB 1964, KIRK 1963, RØRTH and JENSEN 1967).

In our hands the direct spectrophotometric measurement proved the most simple and accurate method; therefore, it is described in detail (cf., AEBI 1970). Instead of hydrogen peroxide the more stable sodium perborate may be used as substrate (THOMSON et al. 1978).

Assay

Reagents

1. Sodium potassium phosphate, 0.1 M, pH 7.0, containing 0.33% bovine serum albumin. Made up by adjusting the pH of 0.5 M KH_2PO_4 with concentrated NaOH and filling up with water after addition of albumin.

2. Hydrogen peroxide, approximately 30 mM, pH 7.0; 0.34 ml 30% H_2O_2 plus 50 ml reagent 1 filled up with water to 100 ml; must be freshly prepared daily. At 240 nm the optical density (OD) of a 1-cm layer of reagent 2 diluted with reagent 3 (1 : 3) should be about 0.5.

3. Reagent 1 diluted with water 1 : 2.

4. Triton X-100 1%, containing 0.33% bovine serum albumin and 2% NaCl; used as diluent for the sample (LEIGHTON et al. 1968). Dilution should

Table 9. *Comparison of liver catalase activity and its subcellular distribution in various mammalian species: "classical" four-fraction scheme (Fig. 25)*

Species	Specific activity (IU/mg protein)			Percent activity [a]	
	PNE	S	M	Mc	S
Man	160	190	1–3	1–2	93–97
Beef	390	410	6–11	1–3	88–93
Sheep	1,070	1,140	4–9	1–2	90–95
Pig	820	850	13–17	1–3	82–86
Dog	120	150	16–24	1–3	73–81
Cat	140	190	8–14	1–2	85–91
Rabbit	110	140	4–7	1–2	92–96
Guinea pig	940	1,310	3–6	1–3	94–96
Mouse	160	100	48–56	1–3	43–51
Rat	190	230	30–44	1–2	55–69

From: HOLMES and MASTERS 1972.

Abbreviations: PNE postnuclear extract; S supernatant; M mitochondrial fraction; Mc microsomal fraction; IU international units.

[a] Percent activities refer to PNE.

be done immediately before assay in order to avoid inactivation of catalase due to dissociation; the diluted sample is to give a ΔOD per minute of approximately 0.05 during the assay.

Performance

	Test (ml)	Blank (ml)	Conditions
Reagent 3	1.9	2.9	10-mm quartz cuvettes,
Diluted sample	0.1	0.1	240 nm, 25 °C, 30 seconds
Reagent 2 (start)	1.0	—	

The final concentration is phosphate 48.3 mM, hydrogen peroxide 10 mM, and bovine serum albumin 1.6 g/liter.

Calculation

Substitution in Equation (5) of the decadic for the natural logarithms and proportional optical densities for the hydrogen peroxide concentrations gives

$$k = \frac{2 \cdot 3}{\Delta t} \log \frac{(OD)_1}{(OD)_2} \ [\text{sec}^{-1}]. \tag{6}$$

The value for $(OD)_1$ (measured separately, omitting the sample and replacing it with diluent) is added to the decrease in optical density (Δ OD) determined over 30 seconds as described above to give $(OD)_2$. By substituting the respective OD values and $\Delta t = 0.5$ min in Equation (6), the activity of catalase in terms of the first-order rate constant k (expressed in min^{-1}) can be calculated.

Recording of ΔOD is done more easily on semilogarithmic paper; consequently the slope of the line drawn by the recorder is proportional to k (PRICE et al. 1962). From Equation (6), it follows that for $\Delta t = k = 1$, $(OD)_1/(OD)_2$ becomes 2.72; this means that 1 k unit of catalase corresponds to an activity cleaving 63.2% of a given concentration of hydrogen peroxide in 1 minute (k is given in min^{-1}; obviously an analogous definition holds true for k given in seconds).

For comparison, k values may be converted into international units. The reaction rate is

$$v = - \frac{d\,[H_2O_2]}{d\,t} = k\,[H_2O_2]$$

$[H_2O_2]$ is 10 mM in this assay (see Performance). $[H_2O_2]$ is given in mmol/liter and k in min^{-1}; therefore, v is in mmol/(liter · min) or μmol/(min · ml), i.e., v represents the usual international units (IU) per milliliter. Nevertheless, as v depends on the hydrogen peroxide concentration chosen for the assay, the international units are not an unequivocal measure of catalase activity; they may be calculated from the ΔOD measured during the assay using an extinction coefficient for hydrogen peroxide at 240 nm of 43.6/(M · cm).

II.A.2.b. Urate Oxidase (Uricase)

Urate: O_2 oxidoreductase, EC 1.7.3.3

uric acid + 2 H_2O + $O_2 \rightarrow$ allantoin + CO_2 + H_2O_2

This is the overall reaction; for intermediates in nonborate buffers see PRAETORIUS (1948) and MAHLER et al. (1956).

Comment

In animal tissue the enzyme seems to be bound mostly to the peroxisomal structure, although in mouse liver it has also been detected in endoplasmic reticulum and Golgi apparatus (YOKOTA and NAGATA 1974). Most vertebrates are able to dispose of uric acid by means of urate oxidase, which is abundant in the liver; important exceptions are lizards, snakes, birds, and primates, which are devoid of urate oxidase and therefore are uricothelic animals (FLORKIN and DU-CHÂTEAU-BOSSON 1943; for review see HRUBAN and RECHCIGL 1969, LASKOWSKI 1951). The enzyme has been found less frequently in kidney: it is present in dog and beef kidney but seems to be missing in the kidneys of most mammalian species, e.g., rat kidney (BAUDHUIN et al. 1965 b, HRUBAN and RECHCIGL 1969).

Regularly insoluble urate oxidase is associated with the crystalline core (for references see TSUKUDA et al. 1971), which has been supposed to be nothing but crystalline urate oxidase (HRUBAN and SWIFT 1964). Using SDS-polyacrylamide gel electrophoresis, HAYASHI et al. (1976) found only one protein band (molecular weight 35,000) in purified cores. They suggested that this band probably corresponds to the subunit of urate oxidase. On the other hand, the specific activity of purified urate oxidase is higher by one order of magnitude than that of purified cores (BAUDHUIN et al. 1965 a); thus the above assumption cannot be supported.

AFZELIUS (1965) proposed that there is a positive correlation between the occurrence of crystalline cores in the microbodies of certain tissues and the presence of urate oxidase. Although in some cases this is true (e.g., no cores are present in human liver peroxisomes or in microperoxisomes of extrahepatic tissues), in carp liver microbodies cores are barely demonstrable although urate oxidase is abundant (KRAMAR et al. 1974). Most probably the enzyme is loosely attached to the peroxisome membrane (GOLDENBERG 1977 a, b).

In sedimentation rate experiments the pig liver enzyme exhibits a molecular weight of about 100,000 (MAHLER et al. 1955); PITTS et al. (1974), using sedimentation equilibrium, found a molecular weight of 125,000 and were able to demonstrate four subunits (approximately 32,000 daltons each). The activity of urate oxidase may be determined by measuring the urate-dependent oxygen uptake either manometrically (KEILIN and HARTREE 1936 a) or polarographically with a Clark-type oxygen electrode (KRAMAR et al. 1974). For its simplicity, we prefer to use the photometric method (BROWN et al. 1966, KALCKAR et al. 1947, SCHNEIDER and HOGEBOOM 1952) as described by BERGMEYER et al. (1970), which is based on the decrease in OD at 293 nm due to the disappearance of urate.

Assay

Reagents
1. Potassium borate-boric acid, pH 9.5, containing 0.1% Triton X-100.
2. Uric acid, 12 mM, in reagent 1.

Performance

	Mixture (ml)	Final concentration (mM)	Conditions
1. Borate	2.8	18.7	10-mm quartz cuvettes,
Sample	0.1	—	293 nm, 37 °C,
2. Uric acid	0.1	0.4	start: uric acid

The blank used is reagent 1 instead of reagent 2.

Calculation

As the molar extinction coefficient for uric acid is $\varepsilon_{293} = 12.6 \times 10^3 \cdot M^{-1} \cdot cm^{-1}$, the decrease in OD/min is multiplied by 2.38 to give the activity in $\mu mol/(min \cdot ml)$ (for the 0.1-ml sample used in the test).

II.A.2.c. D-Amino Acid Oxidase

D-Amino acid: O_2 oxidoreductase, EC 1.4.3.3

$$D\text{-amino acid} + O_2 + H_2O \rightarrow 2\text{-oxoacid} + NH_3 + H_2O_2$$

Comment

This enzyme is mainly found in kidney tissue; it is also found in smaller concentrations in vertebrate liver (for review see KREBS 1951, HRUBAN and RECHCIGL 1969). It has FAD as a coenzyme (WARBURG and CHRISTIAN 1938) and turns over only the unphysiological D-α-amino acids. Its activity with different D-α-monoaminomonocarboxylic acids shows great variations; D-proline gives the highest velocity, whereas acidic and basic amino acids react sluggishly or not at all; glycine is turned over rather slowly (KREBS 1951, DIXON and KLEPPE 1965). The optimum pH is about 8.8 (KREBS 1935).

The molecular weight of the dimeric enzyme is approximately 90,000; the 45,000-dalton subunit contains one FAD. Further association to a tetramere is possible (ANTONINI *et al.* 1966).

The assay to D-amino acid oxidase can be performed by manometric measurement of the oxygen uptake with D-alanine as substrate (BURTON 1955), preferably in the presence of catalase, so that the gross reaction may be written as follows:

$$\text{D-amino acid} + \frac{1}{2} O_2 \xrightarrow[\text{catalase}]{\text{oxidase}} \text{2-oxoacid} + NH_3.$$

Otherwise, especially in purified preparation containing no or little catalase, a higher oxygen consumption should be observed. Alternatively, the oxygen consumption may be measured polarographically in a Clark-type oxygen electrode (KRAMAR *et al.* 1974).

In a very sensitive test the 2-oxoacid formed is reacted with 2,4-dinitrophenylhydrazine. The resulting 2,4-dinitrophenylhydrazone can be extracted with organic solvents and measured photometrically at 450 nm (TONHAZY 1950; modified according to BAUDHUIN *et al.* 1964).

Assay

Reagents
    1. Sodium pyrophosphate, 50 mM, pH 8.5.
    2. D-alanine, 0.5 M, pH 8.5.
    3. FAD, 27 µM.
    4. Trichloroacetic acid (TCA), 6.2 M.
    5. 2,4-Dinitrophenylhydrazine, 1 g/liter (in 2 M HCl).
    6. Toluene, saturated with water.
    7. Ethanolic potassium hydroxide, 25 g/liter (in 95% ethanol).

Performance

	Mixture (ml)	Final concentration (mM)	Conditions
1. Sodium pyrophosphate	0.7	33.3	37 °C, 30–240 minutes, start: sample, stop:
2. D-Alanine	0.1	47.6	1 ml TCA and centri-
3. FAD	0.05	0.027	fugation
4. Sample	0.2		

The blank used is D-alanine added *after* TCA.

After low-spin centrifugation, the TCA supernatant is decanted carefully from the sediment and mixed with 2 ml reagent 5. After 5 minutes at 37 °C, 2 ml reagent 6 is added. The biphasic mixture is vigorously shaken in order to extract the 2,4-dinitrophenylhydrazone. Then 1 ml of the upper phase is mixed with 5 ml reagent 7. After 5 minutes, 1 ml water is added and the OD measured at 450 nm.

Calculation

Preparation of a calibrating curve with a series of pyruvate solutions is advisable. In our laboratory OD 0.1 corresponds to 0.11 µmol pyruvate in the original mixture. Assuming the sample volume is 0.2 ml, the increase in OD/min may be converted into units (U):

$$\text{activity (U/ml)} = (\text{OD/minute}) \cdot (1.1/0.2).$$

II.A.2.d. L-α-Hydroxy Acid Oxidase

Glycolate: O_2 oxidoreductase, EC 1.1.3.1

L-α-Hydroxy acid: O_2 oxidoreductase, EC 1.1.3.15

$$\text{glycolate} + O_2 \rightarrow \text{glyoxylate} + H_2O_2$$

$$\text{L-α-hydroxy acid} + O_2 \rightarrow \text{2-oxoacid} + H_2O_2$$

Comment

Multiple forms of this enzyme are found in mammalian tissues. The liver enzyme is specific for short-chain aliphatic α-hydroxyacids; it is called glycolate oxidase, or hydroxy acid oxidase A (KUN et al. 1954, NAKANO et al. 1968 a, SCHUMAN and MASSEY 1971). Form A has also been found in pig kidney (ROBINSON et al. 1962). In contrast, the enzyme found in rat kidney turns over only long-chain α-hydroxy acids and, remarkably, L-amino acids; it is called hydroxy acid oxidase B or L-amino acid oxidase (BLANCHARD et al. 1944, NAKANO and DANOWSKI 1966, SALVATORE et al. 1966).

Both enzymes are considered to be tetrameres with a molecular weight at infinite dilution of about 150,000, consisting of four protomeres of approximately 40,000 (PHILIPS et al. 1976). Further association seems to be possible: NAKANO et al. (1968 a, b) reported a molecular weight of about 300,000 for both enzymes. Although FMN acts as a prosthetic group in both the rat kidney and liver enzymes (NAKANO and DANOWSKI 1966, USHIMA and NAKANO 1969), addition of FMN has been reported to cause no increase in the liver enzyme activity (NAKANO et al. 1968 a). DULEY and HOLMES (1974) were able to demonstrate that in mice the liver and the kidney enzymes trace back to two different genetic loci.

The assay described for D-amino acid oxidase may be adapted. For example, in the 2,4-dinitrophenylhydrazine test 50 mM sodium pyrophosphate, pH 8.0, and 1 M sodium glycolate (or another α-hydroxy acid) are used instead of reagents 1 and 2. Flavins can be omitted.

II.A.2.e. Allantoinase

Allantoin amidohydrolase, EC 3.5.2.5

$$\text{Allantoin} + H_2O \rightarrow \text{allantoate}$$

Comment

Allantoinase and allantoicase are found in the liver of those nonuricothelic vertebrates that are able to degrade allantoin, the first product of urate catabolism (fish and amphibia; LASKOWSKI 1951), and hence excrete urea and glyoxylic acid as the terminus of purine disposition.

In vertebrates allantoicase generally seems to be localized in soluble cytoplasm, whereas allontoinase in frog liver is a peroxisomal enzyme (SCOTT et al. 1969). This location of allantoinase, however, might be an exception; in other cases studied it turned out to be soluble, as in avian and fish liver (GOLDENBERG 1977 a, b, SCOTT et al. 1969).

The assay is based on the nonenzymatic hydrolysis of allantoic acid, the product of allantoinase, to urea and glyoxylac acid by heating at low pH; glyoxylic acid is reacted with 2,4-dinitrophenylhydrazine (BROWN 1964 a, BROWN et al. 1966).

Assay

Reagents
1. Allantoin, 15 mM, in 50 mM Tris, pH 8.0 (to be freshly prepared daily).
2. TCA, 6.2 M.
3. 2,4-Dinitrophenylhydrazine, 1 g/liter, in 6 M HCl.

Performance

The 0.1-ml sample is incubated for 30 minutes at 37 °C with 0.9 ml reagent 1 (final concentration 4.5 mM allantoin, 15 mM Tris). For a blank add the sample after incubation.

The enzymatic reaction is stopped with 0.1 ml TCA (reagent 2). After clearing by low-speed centrifugation, the mixture is heated to 60 °C for 15 minutes in order to hydrolyze allantoin. In the cooled solution glyoxylic acid is determined by reaction with 2 ml 2,4-dinitrophenylhydrazine (reagent 3). All subsequent steps for determination of the 2,4-dinitrophenylhydrazone are performed as described for D-amino acid oxidase.

II.A.2.f. Carnitine Acetyltransferase (Short-Chain Carnitine Acyltransferase)

Acetyl-CoA: carnitine O-acetyltransferase, EC 2.3.1.7, and other carnitine acyltransferases

$$\text{acetyl-CoA} + \text{L-carnitine} \rightarrow \text{CoA} + \text{O-acetylcarnitine}$$

Comment

Carnitine acyltransferases for short-, medium-, and long-chain acyl groups (carnitine acetyltransferase, carnitine octanoyltransferase, and carnitine palmitoyltransferase) are known to occur in mitochondria (NORUM and BREMER 1967, SOLBERG 1971), where they take part in the shuttle system for activated acyl groups across the membrane (BRESSLER and KATZ 1965). An additional role has been proposed for carnitine acetyltransferase as a buffer against rapid changes in acetyl-CoA levels (PEARSON and TUBBS 1967, SNOS-WELL and KOUNDAKJIAN 1972).

In liver endoplasmic reticulum and peroxisomes short- and middle-chain, but no long-chain, activities are present (MARKWELL and BIEBER 1976, MARKWELL et al. 1973). The peroxisomal activities correspond to two individual enzymes localized in the peroxisomal matrix. Microsomal activities, although presumably tracing back to the same enzymes as the peroxisomal ones, are bound to the membrane (MARKWELL et al. 1976). Interestingly, acyl groups not belonging to the fatty acid series (malonyl and acetoacetyl groups) are also actively transferred by the short-chain enzyme (Table 10).

Table 10. *Substrate specificity of peroxisomal carnitine acyltransferase system and purified carnitine acetyltransferase*

Acyl-CoA	Relative initial velocity	
	Rat liver peroxisomes	Purified peroxisomal carnitine acetyltransferase
Acetyl	100	100
Propionyl	140	110
Butyryl	212	82
Hexanoyl	220	14
Octanoyl	100	3
Decanoyl	60	0
Lauryl	32	0
Myristoyl	10	0
Palmitoyl	0	0
Malonyl	84	55
Succinyl	0	0
Acetoacetyl	100	74
β-Hydroxy-β-methylglutaryl	10	0

From: MARKWELL et al. 1976.

The function of peroxisomal carnitine acyltransferases is not clear; it might be connected to peroxisomal fatty acid β-oxidation. The finding that both the carnitine acetyltransferase and the β-oxidation systems are induced by CPIB seems highly significant in this respect (LAZAROW and DE DUVE 1977, MOODY and REDDY 1974).

The assays of carnitine acyltransferases are based on the measurement of CoA set free during the acyl transfer from acyl-CoA to added carnitine; the free CoA is determined with the sulfhydryl reagent 5,5'-dithiobis (2-nitrobenzoic acid) (DNB or Ellman's reagent). Sulfhydryl compounds liberate a colored anion from this reagent; the anion is measured spectrophotometrically at 412 nm (FRITZ et al. 1963, MARKWELL et al. 1973).

A special assay for carnitine acetyltransferase not described here uses the formation of NADH from malate; the oxaloacetate produced in the malate dehydrogenase reaction in the presence of citrate synthase accepts acetyl groups from acetyl-CoA, which is formed from acetylcarnitine and CoA during the transferase reaction (FRITZ et al. 1963).

Assay

Reagents

1. DNB, 0.5 mM in 116 mM Tris-chloride, 2.5 mM EDTA, 0.2% Triton X-100, pH 8.0.
2. Acetyl-CoA, lithium salt, 2 mM (pH must not excede 7).
2 a, 2 b. Octanyol-CoA and pamitoyl-CoA, 2 mM.
3. D,L-carnitine, 1 M.

Performance

	Mixture (ml)	Final concentra-tion (mM)	Conditions
1. DNB-Tris-EDTA	0.85	0.425-99-2.13	10-mm cuvettes,
Sample	0.1 (or less)		(1 ml), 412 nm, 25 °C
2. Ac(et)yl-CoA	0.05	0.1	wait after 2 until OD
3. Carnitine	0.005	5.0	is constant, start:
			carnitine

As a blank the sample is run without carnitine.

Calculation

As the molar extinction coefficient of the reaction product at 412 nm is $\varepsilon = 1.36 \times 10^4/(M \cdot cm)$, the Δ OD/min measured is multiplied by 0.735 to give the activity in μmol/(min $\cdot$ ml) (for the 0.1-ml sample used in the test).

II.A.2.g. Glyoxylate Aminotransferase

L-Amino acid: glyoxylate aminotransferase, EC 2.6.1.x

L-amino acid + glyoxylate → 2-oxoacid + glycine

Comment

This activity has been detected in the soluble matrix of rat liver and kidney peroxisomes. It differs from the glyoxylate aminotransferases of leaf peroxisomes in that it uses L-leucine and L-phenylalanine as preferred amino donors (Table 11). The specificity with glyoxylate as amino acceptor is rather high; the activities with pyruvate and oxalacetate are less than 10% of that found with glyoxylate (HSIEH and TOLBERT 1976, McGROARTY and TOLBERT 1973, TOLBERT 1973). During the several steps of purification the activities with leucine, phenylalanine, and histidine as donors were not separable, whereas the activity with L-alanine (Table 11) seemed to be due to mitochondrial contamination.

The assay of glyoxylate aminotransferase with leucine or phenylalanine as amino group donors may be done by reacting α-ketoisocaproate or phenyl-pyruvate, the newly generated 2-oxoacids with 2,4-dinitrophenylhydrazine (*cf.*, α-hydroxy acid oxidase and D-amino acid oxidase assays). The specificity of the test rests on the selective solubility of the 2,4-dinitrophenylhydrazones formed in cyclohexane (HSIEH and TOLBERT 1976, TAYLOR and JENKINS 1966).

For amino acids other than leucine and phenylalanine the formation of glycine from ^{14}C-glyoxylate may be used (KISAKE and TOLBERT 1969, REHFELD and TOLBERT 1972).

Assay

Reagents

1. Sodium pyrophosphate, 50 mM, pH 8.4, containing 0.05% Triton X-100.
2. L-Leucine, 0.5 M.
2 a. L-Phenylalanine 0.5 M.
3. Pyridoxal phosphate, 0.25 mM.
4. Sodium glyoxylate, 50 mM.
5. TCA, 6.2 M.
6. 2,4-Dinitrophenylhydrazine, 3 g/liter in 2 M HCl.
7. Cyclohexane.
8. Sodium carbonate, 100 g/liter.
9. Sodium hydroxide, 1 M.

Performance

	Mixture (ml)	Final concentration (mM)	Conditions
1. Sodium pyrophosphate	1.0	40	40 °C, 60 minutes,
2. L-Leucine, or			start: sodium-glyoxy-
2 a. L-Phenylalanine	0.05	20	late, stop: 0.2 ml TCA
3. Pyridoxal phosphate	0.05	0.01	and centrifugation
Sample	0.1		
4. Sodium glyoxylate	0.05	2	

For a blank the sample is boiled before addition.

After low-speed centrifugation the supernatant is decanted from the sediment and mixed with 0.5 ml reagent 6. After 10 minutes at room temperature, 2.5 ml cyclohexane (reagent 7) is added. The biphasic mixture is vigorously shaken in order to extract the 2,4-dinitrophenylhydrazone. After low-speed centrifugation, 2 ml of the upper phase is shaken with 1.5 ml reagent 8 and centrifuged. To 1 ml of the reddish brown lower phase, which now contains the 2,4-dinitrophenylhydrazone, 1 ml reagent 9 is added. After 10 minutes the OD is measured at 440 nm.

Calculation

According to HSIEH and TOLBERT (1976) an optical density of 0.1 corresponds to 0.0469 μmol α-ketoisocaproate or 0.0549 μmol phenylpyruvate in the original mixture.

II.A.2.h. Xanthine Dehydrogenase

Xanthine: NAD$^+$ oxidoreductase, EC 1.2.1.37

xanthine $+$ NAD$^+$ $+$ H$_2$O $\rightarrow$ urate $+$ NADH

xanthine oxidase, xanthine: O$_2$ oxidoreductase, EC 1.2.3.2

xanthine $+$ O$_2$ $+$ H$_2$O $\rightarrow$ urate $+$ H$_2$O$_2$

Comment

Xanthine oxidase found in rat liver supernatant is in fact an NAD$^+$-dependent dehydrogenase (type D), which may be converted into an oxidase (type O) by storage at $-20\,^\circ$C, incubation at $37\,^\circ$C, or treatment with solvents (STIRPE and DELLA CORTE 1969). Whereas in all vertebrates species checked thus far (including frogs and carp; see Section I.D.2.) this enzyme was found to be located in the soluble fraction of the liver, in chicken liver and kidney it appears to be connected with peroxisomes (SCOTT *et al.* 1969). Since dehydrogenase activity (type D) is inhibited by NADH, which competes with NAD$^+$ (DELLA CORTE and STIRPE 1970), it is advisable to reoxidize NADH (*e.g.*, by pyruvate plus LDH) in the dehydrogenase assay. In this case, instead of NADH, the urate formed is measured photometrically (at 293 nm); this may also be done in assaying the oxidase (type O) activity (ROWE and WYNGAARDEN 1966).

Assay (xanthine dehydrogenase test)

Reagents
1. Tris-HCl, 0.1 M, pH 8.3, saturated with air.
2. NAD$^+$, 15.5 mM.
3. LDH (from pig heart), 0.25 mg/ml.
4. Sodium pyruvate, 50 mM.
5. Xanthine, 1.8 mM.

Performance

	Mixture (ml)	Final concentration (mM)	Conditions
1. Tris-HCl	2.6	86.4	10-mm quartz cuvettes,
2. NAD$^+$	0.1	0.51	293 nm, 37 $^\circ$C, start:
3. LDH	0.01	0.83 µg/ml	xanthine
Sample	0.1		
4. Sodium pyruvate	0.1	1.66	
5. Xanthine	0.1	0.06	

The blank is run without xanthine.

Calculation

As the molar extinction coefficient of the reaction product is $\varepsilon_{293} = 12.6 \times 10^3/(\text{M}\cdot\text{cm})$, the increase in OD/minute due to the formation of uric acid is multiplied by 2.38 to give the activity in µmol/(min·ml) (for the 0.1-ml sample used in the test). For calculation of the net dehydrogenase activity the oxidase activity must be substracted. The assay of xanthine oxidase is performed in the same way except that NAD$^+$, LDH, and pyruvate are omitted and the volume of Tris-HCl (solution 1) is 2.8 ml.

II.A.2.i. NADH-Cytochrome c Reductase (Antimycin Insensitive)

NADH- ferricytochrome b$_5$ oxidoreductase, EC 1.6.2.2?

NADH + 2 ferricytochrome c $\rightarrow$ NAD$^+$ + 2 ferrocytochrome c

Comment

This peroxisomal activity is antimycin insensitive, as is the microsomal one. It has been detected in rat liver and dog kidney peroxisomes. It is bound to the peroxisomal membrane (DONALDSON *et al.* 1972). Triton X-100 and digitonin inactivate the enzyme, which has a broad pH optimum of 6–9. As the peroxisomal membrane also exhibits NADH-cytochrome b_5 reductase

Table 11. *Substrate specificity of glyoxylate aminotransferase reaction in a peroxisomal fraction from rat liver*

Amino group donor	Relative activity
L-Leucine	100 *
D-Leucine	0
L-Phenylalanine	98
L-Alanine	90
L-Histidine	75
L-Asparagine	82
L-Glutamine	54
L-Serine	28
L-Proline	7
L-Tryptophan	4
L-Glutamate	0
L-Threonine	6
L-Cysteine	4
L-Tyrosine	0
L-Aspartate	15
L-Lysine	7
L-Arginine	3
Glycine	Trace
L-Isoleucine	2
L-Valine	2

From: HSIEH and TOLBERT 1976.

* The rate with leucine as amino group donor is 181 n mol/(min · mg) peroxisomal protein.

activity, both the b_5 and the c activity might be due to the same enzyme. The presence of antimycin-insensitive NADH-cytochrome c reductase in microsomal as well as in peroxisomal membranes supports speculations that the peroxisomal membranes might be derived from the endoplasmic reticulum. The assay of NADH-cytochrome c reductase is based on the increase in optical density at 550 nm due to the reduction of cytochrome c by NADH.

Assay

Reagents
1. Potassium-phosphate, 50 mM, pH 7.5.
2. Potassium cyanide, 0.1 M.
3. Oxidized cytochrome c, 862 μM.
4. NADH, 4 mM.

Performance

	Mixture (ml)	Final concen-	Conditions
1. Potassium phosphate	0.8	39.8	10-mm cuvettes,
2. Potassium cyanide	0.005	0.5	550 nm, 25 °C,
3. Cytochrome c	0.05	0.043	start: NADH
Sample	0.05 or less		
4. NADH	0.1	0.4	

For a blank the sample is run without NADH.

In order to inhibit the mitochondrial NADH-cytochrome c reductase, 5 µl 0.2% ethanolic solution of antimycin A may be added to the assay.

Calculation

At 550 nm the molar extinction coefficient for cytochrome c (reduced minus oxidized) is $2.11 \times 10^4/(M \cdot cm)$ (VAN GELDER and SLATER 1962). The Δ OD/min measured is multiplied by 0.935 to give the activity in µmol/(min $\cdot$ ml) (for the 50-µl sample used in the test).

II.A.2.j. Glycerol-3-Phosphate Dehydrogenase (NAD)

sn-Glycerol-3-phosphate: NAD^+ 2-oxidoreductase, EC 1.1.1.8

sn-glycerol-3-phosphate + NAD^+ ⇄ dihydroxyacetone phosphate + NADH

Comment

GEE et al. (1974) detected the NAD^+-dependent ("soluble") glycerol-3-phosphate dehydrogenase in liver peroxisomes of some vertebrate species; considerable specific activities were found. In rat liver the peroxisomal activity is even higher than the cytoplasmic activity (326 and 147 nmol/(min $\cdot$ mg protein), respectively. On the other hand, in carp liver the enzyme has been shown to be exclusively cytoplasmic (GOLDENBERG et al. 1978 b), and it has not yet been detected in catalase-positive particles from extrahepatic tissues; PAVELKA et al. (1976) proved its absence in rat brown fat peroxisomes.

The assay of glycerol phosphate dehydrogenase is usually done by NADH-dependent reduction of DHAP (GEE et al. 1974). Large amounts of sucrose in gradient fractions inhibit the enzyme and are removed by dialysis against 100 mM Tris (pH 7.5), 40 mM citrate, 16 mM magnesium chloride, and 1 mM NAD^+; correction factors may be established.

Assay

Reagents

1. Glycylglycine 300 mM, pH 7.5, containing 0.025% Triton X-100.
2. Potassium cyanide, 100 mM.
3. NADH, 5 mM.
4. DHAP, 10 mM (prepared by acid hydrolysis from dihydroxyacetone phosphate-dimethylketal, which is commercially available).

Performance

	Mixture (ml)	Final concentration (mM)	Conditions
1. Glycylglycine	2.6	260	1-mm cuvettes, 340 nm,
2. Potassium cyanide	0.1	3.33	25 °C, start: DHAP
3. NADH	0.1	0.167	
Sample	0.1		
4. DHAP	0.1	0.333	

Calculation

The calculation is made from the decrease in OD. For the factor used see under Isocitrate Dehydrogenase: Calculation.

II.A.2.k. Isocitrate Dehydrogenase (NADP)

Threo-D_5-isocitrate: NAD^+ oxidoreductase (decarboxylating), EC 1.1.1.42

threo-D_5-isocitrate $+ NADP^+ \rightarrow$ 2-oxoglutarate $+ CO_2 + NADPH$

Comment

Contrary to the NAD-dependent isocitrate dehydrogenase, which is found only in the mitochondria, this enzyme is multilocular; the main fraction is cytoplasmic and minor activities occur in mitochondria and peroxisomes (LEIGHTON *et al.* 1968, 1969). The assay described below is the usual optical test (KORNBERG 1955).

Assay

Reagents

1. Potassium phosphate, 0.1 M, pH 7.6, containing 0.1% Triton X-100.
2. Magnesium chloride, 1 M.
3. $NADP^+$, 20 mM.
4. D,L-Isocitrate trisodium salt, 0.1 M.

Performance

	Mixture (ml)	Final concentration (mM)	Conditions
1. Potassium phosphate	2.8	—	10-mm cuvettes,
2. MgCl$_2$	0.025	8.3	340 nm, 25 °C, start:
3. NADP$^+$	0.1	0.67	isocitrate
Sample	0.1		
4. Isocitrate	0.05	1.67	

For a blank the sample is run without isocitrate.

Calculation

As the molar extinction coefficient for NAD(P)H at 340 nm is $\varepsilon = 6.22 \times 10^3/(M \cdot cm)$, the Δ OD/min measured is multiplied by 4.82 to give the activity in μmol/(min · ml) (for the 0.1-ml sample used in the test).

II.A.2.l. Palmitoyl-CoA Oxidizing System (Fatty Acid β-Oxidation)

$$\text{Palmitoyl-CoA} + 7 \text{ CoA} + 7 \text{ FAD} + 7 \text{ NAD}^+$$
$$\rightarrow 8 \text{ acetyl-CoA} + 7 \text{ FADH}_2 + 7 \text{ NADH} + 7 \text{ H}^+$$

or (for peroxisomes) assuming formation of hydrogen peroxide in the fatty acyl-CoA dehydrogenase reaction (see below):

$$\text{palmitoyl-CoA} + 7 \text{ CoA} + 7 \text{ O}_2 + 7 \text{ NAD}^+$$
$$\rightarrow 8 \text{ acetyl-CoA} + 7 \text{ H}_2\text{O}_2 + 7 \text{ NADH} + 7 \text{ H}^+.$$

Recycling of half of the oxygen by catalytic decomposition of hydrogen peroxide gives

$$\text{palmitoyl-CoA} + 7 \text{ CoA} + 7/2 \text{ O}_2 + 7 \text{ NAD}^+$$
$$\rightarrow 8 \text{ acetyl-CoA} + 7 \text{ H}_2\text{O} + 7 \text{ NADH} + 7 \text{ H}^+.$$

Comment

It has been well established that the inner compartment (inner membrane-matrix compartment) of animal mitochondria houses the enzymes of β-oxidation (BEATTIE 1968, BRDICZKA *et al.* 1968), presumably loosely attached to the inner side of the inner membrane (LANDRISCINA *et al.* 1970, WIT-PEETERS *et al.* 1971). COOPER and BEEVERS (1969) also detected β-oxidation in glyoxysomes from castor bean endosperm and suggested the involvement of molecular oxygen and formation of hydrogen peroxide during the acyl-CoA dehydrogenase step (*cf.*, the gross equations above):

According to BIEGLMAYER *et al.* (1973), the β-oxydation enzymes are bound to the glyoxysomal membrane.

Remarkably enough, LAZAROW and DE DUVE (1976) proved the presence of a comparable β-oxidation system in animal (rat liver) peroxisomes which can be induced by application of the hypolipidemic drug clofibrate. The NAD$^+$ reduction depends on oxygen; hydrogen peroxide is formed and cyanide does not inhibit the system. LAZAROW (1978) demonstrated three single enzymes of the β-oxidation pathway, namely, 3-hydroxyacyl-CoA dehydrogenase, enoyl-CoA hydratase, and thiolase. With an excess of CoA he found a ratio of $1:5:5$ between palmitoyl-CoA oxidation, acetyl-CoA accumulation, and NAD$^+$ reduction. In contrast to mitochondria, peroxisomes are not able to turn over short-chain fatty acyl-CoA.

The results of LAZAROW and DE DUVE (1976) in rat liver seem to point to an exclusively peroxisomal location of the β-oxidation system. KRAHLING *et al.* (1978) explained the failure of these experiments to detect β-oxidation in mitochondria: Mitochondria, having been exposed to high sucrose concentrations in the gradient, do not exhibit palmitoyl-CoA-dependent respiration when incubated in isotonic buffers; only in dilute media an appreciable respiration is observed. Under these conditions the ratio of mitochondrial to peroxisomal palmitoyl-CoA oxidation is found to be approximately 3; with an appropriate correction for peroxisomal damage (assuming that all supernatant catalase activity comes from broken peroxisomes), this ratio becomes 1.2. Obviously, a fair comparison of these quantities of palmitoyl-CoA-dependent oxygen consumption is only possible if the acetyl-CoA formed accumulates in the mitochondria instead of being oxidized further by the tricarboxylic acid cycle and respiratory chain; therefore, malate must be excluded. Furthermore, the figures for mitochondrial respiration are to be divided by 2. This makes an allowance for the fact that in mitochondria two oxygen atoms are consumed per acetyl-CoA molecule, whereas in the peroxisomes only one atom of oxygen is necessary, because NADH generated during the β-hydroxyacyl-CoA dehydrogenase step presumably is not oxidized in the peroxisomes.

In any event, considering the relative number of mitochondria and peroxisomes in the original tissue, the peroxisomal β-oxidation capacity seems to be surprisingly high.

The peroxisomal β-oxidation is assayed simply by determining the palmitoyl-CoA-dependent accumulation of NADH. This test is described here as modified by LAZAROW and DE DUVE (1976), but with the acetylpyridine analogue of NAD (APAD) as hydrogen acceptor (COOPER and BEEVERS 1969). CoA must be added in order to reach a ratio of NAD^+ reduction to palmitoyl-CoA consumption exceeding unity [about 3; according to LAZAROW and DE DUVE (1976) the theoretical ratio of 7 is not obtained].

An assay that uses (under anaerobic conditions) 2,3,5-triphenyltetrazolium salt as common hydrogen acceptor for both dehydrogenases has been designed (ALLMANN et al. 1966, MII and GREEN 1954). As this test also involves enzymes that are not present in peroxisomes, it should be modified.

Respiration methods using the Clark electrode have been mentioned above (see KRAHLING et al. 1978).

Of course, the acetyl-CoA formed may also be determined by the citrate synthase reaction; oxalacetate, the acetate acceptor, is delivered by the malate dehydrogenase reaction (DECKER 1970, LAZAROW 1978). Single enzymes of the β-oxidation pathway may be determined most simply by a spectrophotometric test using the light absorption of the intermediate CoA derivates (LAZAROW 1978, LYNEN and OCHOA 1953).

Assay

Reagents

1. Potassium phosphate, 0.1 M, pH 7.4, containing bovine serum albumin (0.2 g/liter) and Triton X-100 (0.1 g/liter).
2. Dithiothreitol, 0.6 M; CoA, 5 mM.
3. APAD, 6 mM.
4. Palmitoyl-CoA, 1 mM.

Performance

	Mixture (ml)	Final concen- tration (mM)	Conditions
1. Potassium phosphate	0.9	90	10-mm cuvettes (1 ml),
2. Dithiothreitol-CoA	0.02	12–0.1	365 nm, Start (after
3. APAD	0.05	0.3	1–2 minutes)
Sample	0.02 or less		palmitoyl-CoA
4. Palmitoyl-CoA	0.01	0.01	

For a blank the test is run without palmitoyl-CoA.

Calculation

The molar extinction coefficient for reduced APAD is $\varepsilon_{365} = 9.0 \times 10^3/$ (M $\cdot$ cm); therefore, for the 0.02-ml sample the increase in OD/minute is multiplied by 5.56.

II.A.3. Preparation of Sub-Peroxisomal Fractions

II.A.3.a. Crystalline Cores

Nucleoids of liver peroxisomes are of higher density than the particles themselves. In sucrose gradients densities are at least 1.257 g/cm³ (LEIGHTON et al. 1968, HAYASHI et al. 1971). Nevertheless, the usual procedures used to purify nucleoids do not involve density gradients.

Thanks to the remarkable resistance of cores against detergents, they survive as the only subcellular structure when a rough cell fraction is treated with Triton X-100 or sodium deoxycholate; the washed cores may be pelleted simply by high-speed centrifugation. The procedures are simple and only two methods will be briefly described.

1. HAYASHI et al. (1976 b) start with light mitochondrial fraction from rat liver (for preparation see Section II.A.1.b.), suspend it in 0.5% sodium deoxycholate-0.25 M sucrose, and leave it at 0 °C for 1 hour. The pelleted cores (30 minutes at 20,000 g) are suspended in an appropriate volume of the deoxycholate-sucrose mixture and stirred for 30 minutes. After this procedure is repeated four times, the pellet is washed 2 or 3 times with 0.25 M sucrose and finally suspended in 0.25 M sucrose. For Wistar rat liver the purified sample of urate oxidase is 446 times as pure as the homogenate: homogenate, 0.017 ∓ 0.0006 units/mg protein; purified cores, 7.59 ∓ 1.12 units/mg protein.

2. TSUKADA et al. (1971) prepare a fraction rich in peroxisomes (comparable to the light mitochondrial fraction) from the liver of male guinea pigs, cats, and rabbits. The liver tissue is homogenized with 9 volumes of 0.25 M sucrose [10% polyvinylpyrrolidone (PVP), 1 mM EDTA, pH 7.6]; and the fraction, pelleted at 7,000–20,000 g for 15 minutes, is stirred with 10 volumes of 0.25 M sucrose-0.15% Triton X-100 (pH 7.4). The cores are pelleted (15 minutes at 50,000 g) and the treatment with Triton X-100 is repeated 4 times. After one wash with 0.25 M sucrose the cores are suspended in 0.4% sucrose. The

purification of urate oxidase reported by Tsukada *et al.* (1971) is approximately 100–200-fold:

Activity of urate oxidase (units/mg protein)

	Guinea pig	Cat	Rabbit
Homogenate	0.0082	0.0118	0.0069
Purified cores	1.71	1.00	0.72

II.A.3.b. Membranes

It is crucial to break up the purified particles without loss of membrane proteins. Detergents such as Triton X-100, which cause marked discontinuities in the peroxisomal membrane and leakage of catalase (Baudhuin 1969 a, Yokota and Fahimi 1978), are not used because the gradual stripping of the membrane is accompanied by the onset of solubilization of integral proteins

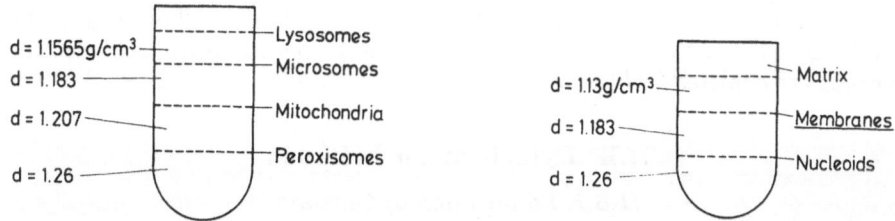

Fig. 29. Preparation of membranes from rat liver peroxisomes by osmotic shock with pyrophosphate and collection of the membranes on a discontinuous saccharose gradient. 1. Treatment of (male) rats with Triton WR-1339 (85 mg/100 g body weight) 85 hours before sacrifice. 2. Preparation of a "λ"-fraction (24,000 g · min < λ < 350,000 g · min) from a liver homogenate by differential centrifugation according to Leighton *et al.* (1968). 3. Subfractionation of λ on a discontinuous sucrose density gradient. 4. Dilution of peroxisomal fractions with 2 vol. 10 mM pyrophosphate, pH 9.0 and incubation at room temperature for 4 hours. 5. Isolation of membranes on a second density gradient.

(*e.g.*, Triton X-100, sodium deoxycholate; Helenius and Simons 1975). Freezing and thawing was found to be inefficient and to inactivate the oxidases appreciably (Leighton *et al.* 1969).

Liver peroxisomes possess no osmotic space (see Section II.A.1.a.) and therefore are hardly damaged when transferred to water; the internal low molecular weight solute (*e.g.*, sucrose) is simply replaced by water. Glyoxysomes and leaf peroxisomes, by contrast, are sensitive to osmotic shock (dilution with water; Brown *et al.* 1974, Gerhardt and Beevers 1970).

Leighton *et al.* (1969) observed that the peroxisomal membrane is broken up and separated from the particles when a suspension of peroxisomes in sucrose solution is diluted with alkaline sodium pyrophosphate (0.01 M, pH 9.0). Donaldson *et al.* (1972) adopted this method to obtain membrane preparations as follows:

Rat liver peroxisomes are purified as usual (Leighton *et al.* 1968) (see Section II.A.1.b.). An aliquot (*e.g.*, 1 ml) of the peroxisomal fraction is diluted with the same volume of 0.01 M sodium pyrophosphate, pH 9.0 (final pH about 8.5). After standing on ice overnight, the suspension is layered on top of a dicontinuous sucrose gradient prepared from equal volume (5 ml each when the Spinco SW-39 rotor is used) of 56, 49, 47, 45, 42, 38, and 30% (w/w) sucrose. After 60 minutes at 39,000 rpm, 5-ml fractions are collected from the bottom of the tube. Catalase, along with most of the protein, becomes solubilized and is found on top of the gradient, whereas urate oxidase, the marker of the cores, is recovered at a density of 1.23 g/cm³; this is less than the buoyant density of the cores found by other authors (see Section II.A.3.b.). NADH-cytochrome c reductase, which indicates the position of the membranes, is distributed in between, with a peak at about 1.17 g/cm³. It is seen from the enzyme distribution diagram presented by LEIGHTON et al. (1969) that NADH-cytochrome c reductase is smeared over the gradient. Furthermore, the peak fraction is not free of urate oxidase activity [about 0.2 μmol/ (min · ml) is found here, which is about 60% of the maximal activity found in the gradient]; on the other hand, catalase seems to be perfectly removed.

A quick and simple procedure that yields membrane preparations of comparably low catalase content has been designed in our laboratory. The essentials of this method of obtaining membrane preparations for SDS electrophoresis are depicted in Fig. 29.

II.B. Cytochemical Methods

II.B.1. Localization of Catalase

II.B.1.a. DAB Methods

Peroxisomes are identified by incubation of tissue sections in alkaline media containing DAB and hydrogen peroxide, thus demonstrating the peroxidatic activities of catalase. The reaction is blocked by 3-amino-1,2,4-triazole. Alkaline hydrolysis of catalase brakes down the enzyme to peroxidatically active subunits that—unlike intact catalase—oxidize DAB or benzidine (see Section II.B.1.b.). In parallel with this mechanism, peroxisomes of acatalasemic mice contain in vivo a partially "degraded" form of catalase that readily oxidizes DAB (GOLDFISCHER and ESSNER 1970).

All the various DAB incubation media are derived from the GRAHAM and KARNOVSKY method (1966) for demonstrating horseradish peroxidase. Since it has been observed that peroxidatic activity of catalase can also be demonstrated in this medium (FAHIMI 1968, 1969), modifications of the original medium have been introduced, including increased concentrations of DAB and hydrogen peroxide, and alkaline pH.

A widely used incubation medium is the pH 9.0 medium of NOVIKOFF and GOLDFISCHER (1969), which is prepared as follows:

Combine

10 ml 50 mM 2-amino-2-methyl-1,3-propanediol buffer, pH 9.4.

0.2 ml 1% hydrogen peroxide, freshly diluted.

20 mg DAB.

Filter, if necessary, and adjust the pH to 9.0.

Tissue slices (frozen sections or chopper slices) are incubated for 2 hours at 37 °C; 5% sucrose may be added to the individual medium. Postfixation is performed with osmium tetroxide as usual, as well as dehydration and embedding.

Control incubations include the following:

1. Incubation without hydrogen peroxide.

2. Incubation in hydrogen peroxide-free medium with 2 mM sodium pyruvate and 0.05% catalase added.

3. Incubation with 10, 20, or 100 mM aminotriazole in the complete medium.

4. Incubation with 1 or 10 mM potassium cyanide in the complete medium.

5. Incubation with 4.6 mM 2,6-dichlorophenolindophenol in the complete medium.

Table 12. *Composition and function of various DAB test media for peroxidatic activities*

Medium	Composition	Function
A	Sörensøn phosphate buffer, pH 7.3	Incubation time and temperature control
B	0.05 M propanediol buffer, 0.03% free base DAB, 0.12% H_2O_2, adjusted to pH 9.0	Primary DAB reagent for "peroxidatic catalase"; high H_2O_2 supposed to inhibit reaction in mitochondria
C	DAB reagent B plus 0.06 M 3-amino-1,2,4-triazole (AT), adjusted to pH 9.0	Inhibition of primary DAB stain reaction in peroxisomes (by AT) and in mitochondria (by high H_2O_2)
D	DAB reagent B plus 5×10^{-4} M KCN, adjusted to pH 9.0	Mitochondrial DAB inhibition; peroxisomes less affected
E	0.03% DAB, 0.02 M Hepes buffer, pH 7.5, 0.06% H_2O_2, adjusted to pH 7.5	Maximized mitochondrial DAB stain reaction; less peroxisome reaction than in medium B at high pH
F	DAB reagent E plus 0.06 M triazole, adjusted to pH 7.5	DAB stain reaction strongly inhibited in peroxisomes; maximal in mitochondria

From: HANNA et al. 1976.

Addition of pyruvate and catalase to the incubation media significantly diminishes the reaction but does not completely prevent it. Faint staining of peroxisomes is also observed with an incubation medium devoid of hydrogen peroxide. Most probably, these reactions are due to hydrogen peroxide that is endogeneously generated either before or during the incubation (FAHIMI 1969). High concentrations of hydrogen peroxide (0.02%) reduce or prevent mito-chondrial staining as well as staining of endogeneous peroxidatic activities.

Aminotriazole at a concentration of 100 mM blocks DAB staining of per-oxisomes; lower concentrations correspondingly decrease the staining reaction without complete inhibition. The reaction of peroxisomes is completely in-hibited by 4.6 mM 2,6-dichlorophenolindophenol. Potassium cyanide may only reduce peroxisomal staining, whereas it completely suppresses the re-activity of mitochondria.

In their study on peroxisomes of the firefly lantern, HANNA et al. (1976) used the incubation scheme shown in Table 12 (mainly based on the work of LOCKE and McMAHON 1971), which may be useful in practice.

For light microscopic studies tissue fixation with osmium tetroxide after

DAB-incubation might be unsuitable, because many substances other than oxidized DAB reduce the fixans. In cases in which the dye contrast of oxidized DAB is too low, sections can be briefly rinsed with Tris-buffer and treated with 0.5% copper (II) nitrate in 0.5 M Tris-buffer, pH 7.6; this treatment intensifies the staining of oxidized DAB (HANKER and ROMANOVICZ 1977).

Tissue fixation with aldehyde solutions is necessary for the demonstration of catalase by means of DAB media (GOLDFISCHER and ESSNER 1969). No staining is observed in unfixed tissue (SIES et al. 1972, ROELS and WISSE 1973). This aspect of the DAB reaction mechanism has been investigated in detail by HERZOG and FAHIMI (1974 b). Catalatic activity of catalase was found to be markedly inhibited by glutaraldehyde (1% glutaraldehyde inhibits 70% of activity, 6% glutaraldehyde about 90% of activity). On the other hand, peroxidatic activities of catalase increased with glutaraldehyde treatment, reached a maximum with 6% glutaraldehyde, and declined with a further increase in glutaraldehyde concentrations (DAB as electron donor). In contrast, when ethanol is used as a substrate, glutaraldehyde causes marked inhibition of catalase peroxidatic activity. Similar results were obtained with pyrogallol as a substrate. Correspondingly, unfixed mitochondrial fractions containing peroxisomes, when incubated with DAB and hydrogen peroxide, show positively reacting mitochondria but no stained peroxisomes; the reverse occurred when glutaraldehyde-fixed fractions were incubated.

The importance of these findings for understanding the demonstration of catalase with cytochemical DAB procedures is evident. HERZOG and FAHIMI (1974 b) conclude that inactivation of catalatic activity by glutaraldehyde is the essential mechanism that allows sufficient concentrations of hydrogen peroxide to be available for peroxidatic activities. This effect is further enhanced by the use of alkaline DAB media after glutaraldehyde fixation. HERZOG and FAHIMI (1973) have shown that oxidation of DAB by glutaraldehyde-denatured catalase is optimal at pH 10.5.

The peroxidatic activity of catalase can be distinguished from **peroxidatic** activities of endogeneous peroxidases (lacto-, sialo-, or myeloperoxidase), peroxidatic activities of other hemoproteins (hemoglobin, myoglobin, or cytochrome c), and exogeneous horseradish peroxidase by means of various fixation procedures and/or inhibitors. HERZOG and MILLER (1972) observed that 10^{-2} M aminotriazole inhibits the staining of exogeneous catalase and endogeneous peroxidase, but leaves unaffected the reaction of hemoglobin as well as the staining of injected horseradish peroxidase. In a histochemical method recently introduced p-phenylenediamine and pyrocatechol are used to demonstrate peroxidatic activities. This method readily demonstrates injected horseradish peroxidase but does not stain endogeneous catalase and endogeneous peroxidase (HANKER et al. 1977 c). Methanol markedly inhibits granulocyte peroxidase activity, but in vitro leaves horseradish peroxidase unaffected (STREEFKERK and VAN DER PLOEG 1974). On the other hand, methanol inhibits injected horseradish peroxidase, as well as granulocyte peroxidase and peroxidatic activities of hemoglobin (STRAUS 1971); horseradish peroxidase can also be inhibited by phenylhydrazine (STRAUS 1972).

As already mentioned, aldehyde fixation—especially prolonged aldehyde fixation—increases DAB staining of peroxisomes. ROELS et al. (1973) took advantage of this phenomenon. They found that incubation of unfixed tissue at pH 7.3 and 23 °C with 0.003% hydrogen peroxide allows peroxidase to react, but leaves catalase unreactive. Aldehyde fixation, increased pH (to 9.0 or 9.7), increased concentrations of hydrogen peroxide (0.03% to 0.07%), and raised temperature (37 °C) cause DAB reactivity of catalase. Peroxidases are sensitive to aldehyde fixation. Salivary peroxidase is sensitive to glutaraldehyde, and less sensitive to formaldehyde: after glutaraldehyde fixation and incubation with 0.07% hydrogen peroxide at 37 °C in pH 9.7 media, only eosinophilic granules in granulocytes strongly react; neutrophilic granules (myeloperoxidase) and exogeneous peroxidase (horseradish peroxidase) show decreased reactivity; and all the other peroxidases are inhibited (ROELS et al. 1973). Catalase reacts under these conditions. Corresponding results have been reported by ROELS and WISSE (1973), as well as by ROELS et al. (1975).

Extraperoxisomal catalase has been identified histochemically in sheep hepatocytes by ROELS (1976). Sheep were chosen because there is about nine times as much catalase in the cytosol of this species as in rat liver cells (HOLMES and MASTERS 1970). A reaction product has been observed between cisternae of the rough endoplasmic reticulum. Although ribosomes are obscured by the reaction products, the lumina of the cisternae remain free, as well as the perinuclear cisterna. The entire cytosol is marked with reaction product, which also fills microvilli at the surfaces of bile canaliculi and sinusoids. This mode of distribution of extraperoxisomal catalase is somewhat different from that which has been observed with immunocytochemical methods (YOKOTA and NAGATA 1974 b). As a technical comment, ROELS (1976) noted that DAB reactivity of catalase seems to be more sensitive after formaldehyde fixation than after glutaraldehyde fixation; lead staining improves the electron contrast of the reaction product (BÖCK 1973) (Fig. 60).

In a second paper on this topic, ROELS et al. (1977) reported on positive cytoplasmic staining of rhesus monkey and guinea pig hepatocytes; they found no cytoplasmic catalase staining in liver cells of rats. Cytoplasmic catalase has been demonstrated histochemically in chloragogen cells of the earthworm (FISCHER and HORVÁTH 1978).

In DAB histochemistry considerable difficulties (besides positive staining of the entire cytosol) arise in interpreting and preventing diffusion artifacts. When diffusion artifacts are considered the question arises as to whether catalase, oxidized DAB, or both have been diffused. FAHIMI (1973) showed that ribosomal staining near positively reacting peroxisomes is caused by diffusion of catalase rather than by diffusion of oxidized DAB. In tracer studies ribosomal staining has also been observed near strongly reactive sites of intravenously injected horseradish peroxidase (BÖCK 1972 a); it is believed to show diffusion of peroxidatic activities. In more detailed studies FAHIMI (1974) showed that buffer storage of aldehyde-fixed tissue causes microbody membranes to become morphologically discontinuous (buffer storage longer than 18 hours, postfixation with either osmium tetroxide or permanganate).

Furthermore, DAB cytochemistry revealed catalase activity to diffuse out of leaky particles, whereas peroxisomes with continuous membranes were never seen to be surrounded with reaction product. It is important to realize that diffusion artifacts and, consequently, ribosomal staining have been reported predominantly in studies on peroxisomes at timed intervals after an experiment (*e.g.*, after stimulation with clofibrate, LEGG and WOOD 1970, during recovery from partial hepatectomy, GOLDENBERG *et al.* 1975 a). Diffusion artifacts may simply be explained by the assumption that tissue samples were pooled after glutaraldehyde fixation to be reacted together at the end of the experiment; storage of tissues in buffer or other aqueous media, however is now known to cause diffusion of aldehyde-fixed catalase (FAHIMI 1974).

It has been shown that oxidized DAB specifically stains heme enzymes, particularly peroxisomal catalase (HIRAI 1968, 1971); similarly, this staining reaction is inhibited by specific enzyme inhibitors, as well as by boiling of the tissues. Oxidized DAB has also been used to stain nucleic acids (ROELS and GOLDFISCHER 1971).

Part of a recent review on histochemical DAB methods (LITWIN 1979) deals with cytochemical techniques for demonstrating catalase.

II.B.1.b. Benzidine

Peroxidatic activities of catalase have been studied by GOLDFISCHER and ESSNER (1969) using benzidine as substrate. Liver and kidney tissue blocks are fixed in 3% glutaraldehyde for 3 hours, the aldehyde solution being buffered to pH 7.4 with 0.1 M cacodylate. Frozen sections 10 μm thick are incubated at 37 °C for 50–90 minutes; 30-μm thick sections are used for electron microscopy. The incubation medium is prepared as follows:

Combine

20 mg benzidine · HCl dissolved in 10 ml distilled water.
0.2 ml 3% hydrogen peroxide, freshly diluted. Carefully add sodium hydroxide to reach pH 9.0.

Oxidized benzidine is osmiophilic. For electron microscopy, sections are postfixed in 1% osmium tetroxide in phosphate buffer, pH 7.4. Addition of aminotriazole to the medium, omission of hydrogen peroxide, embedding of the tissue in paraffin, or heating of the tissue for 5 minutes to 90 °C inhibits the brownish staining of microbodies. Under the electron microscope the contrast of microbody matrix material is enhanced; the reaction product is similar to that obtained with DAB. DAB media, however, are superior for the visualization of microbodies because incubation times are shorter and diffusion artifacts are less frequently observed. GOLDFISCHER and ESSNER (1969) claim that the staining reaction is based on the effect of alkaline pH either in the fixation medium or in the incubation medium, which causes hydrolysis of the catalase molecule to peroxidatically active subunits. Microbodies can be stained at neutral pH values when the preceding aldehyde fixation has been performed at strongly alkaline pH, or when the tissue has been fixed at neutral pH but treated with alkaline buffer before being incubated at neutral pH. HERZOG and FAHIMI (1974 b), however, have shown that inactivation of catalatic activity by fixation with glutaraldehyde is the crucial step for the cytochemical demonstration of peroxidatic activities of catalase (see Section II.B.1.a.).

II.B.1.c. Antibody Techniques

Catalase-specific fluorescence has been localized in fine granules and in the cytoplasm of liver cells by means of fluorescent antibody techniques (MORIKAWA and HARADA 1969, NISHIMURA et al. 1964, YASUDA and SUZUKI 1963, YOKOTA 1972). More detailed information could be obtained by applying immunohistochemical techniques to ultrathin frozen sections using ferritin-conjugated antibodies as electron-dense markers (YOKOTA and NAGATA 1974 b). In mouse hepatocytes catalase antigenicity was observed in microbodies, Golgi vesicles, endoplasmic reticulum, and cytoplasm. In microbodies the regions of crystalline cores were significantly less marked than the remaining matrix. In the endoplasmic reticulum ribosomes were more strongly marked than the cisternae. Controls included preincubation of the sections with unlabeled anticatalase antibody (which prevented any staining with ferritin-labeled anticatalase antibody) and staining with ferritin-labeled normal rabbit γ-globulin (which did not evince any staining).

Obviously, catalase antigenicity demonstrated in the cytoplasm and in the endoplasmic reticulum corresponds to diffuse cytoplasmic fluorescence observed in light microscopic studies. It seems probable that enzymatically inactive forms of catalase are also demonstrated by immunocytochemical techniques, thus explaining the different results obtained with various DAB methods.

II.B.2. Localization of Xanthine Oxidase

II.B.2.a. Tetrazolium Salts

In the presence of oxygen, xanthine oxidase oxidizes hypoxanthine and xanthine, forming uric acid and hydrogen peroxide. Electrons can also be transferred to suitable tetrazolium salts, such as nitro blue tetrazolium (NBT) or tetranitro blue tetrazolium (TNBT). The original method of BOURNE (1953) has been modified by SACKLER (1966):

Dissolve

10 mg TNBT.
3.4 mg hypoxanthine.
In 10 ml 0.2 M phosphate buffer pH 7.4.

Tissue is fixed in glutaraldehyde (2.5%) in cacodylate buffer (0.1 M, pH 7.4, containing 1% calcium chloride) at 4 °C for 1 hour. Frozen sections are incubated in the above medium for 10 minutes at 37 °C. Incubation time was extended to 60 minutes by WOHLRAB (1974). No particulate localization has been observed. The enzyme is inhibited in the presence of 16 µM ammeline

(FRIDOVICH 1965). Histochemical findings correlate well with biochemical observations that indicate that only insignificant amounts of xanthine oxidase are associated with organelle fractions (GOLDENBERG 1977 a, b).

II.B.2.b. Autoradiography

Xanthine oxidase, a molybdenoflavoprotein, was labeled with ^{99}Mo during its *post partum* synthesis period (WERNER *et al.* 1971). Radioactivity was localized in the subcortical region of the kidney and in hepatocytes. Von Kupffer cells were found to be significantly more strongly labeled than hepatocytes. No particulate localization could be observed.

II.B.2.c. Cerium (IV) Precipitation

As outlined by VEENHUIS and VENDELAAR BONGA (1977), hydrogen peroxide generated by peroxisomal oxidases can be used to precipitate cerium peroxihydrate. This method has not yet been applied to demonstrate xanthine oxidase activity. The appropriate incubation medium is given in Section II.B.6.c.

II.B.3. Localization of Urate Oxidase

II.B.3.a. Coupled Peroxidase

When urate oxidase oxidizes uric acid (by molecular oxygen) to allantoin, the hydrogen peroxide that (in addition to carbon dioxide) is generated can be used to supply a coupled peroxidase reaction. The method was introduced by GRAHAM and KARNOVSKY (1965) and has been used successfully by HAMOR and GARSIDE (1973) to demonstrate uricase-containing organelles in the *zona radiata* and yolk of the Atlantic salmon. Cryostat sections postfixed in ice-cold acetone are incubated at 37 °C for 30–60 minutes. The medium is prepared as follows:

Combine
3 mg 3-amino-9-ethylcarbazole dissolved in 1 ml dimethylformamide.
10 ml 0.05 M Tris-HCl buffer, pH 8.5.
Shake, wait 3 minutes, then filter. Add
3 mg sodium urate,
5 mg EDTA,
3 mg horseradish peroxidase.

After the incubation, sections are postfixed in formalin and mounted. A red precipitate develops near hepatic microbodies. The reaction is inhibited by preincubation of the sections for 15 minutes in 2 mM 2,6,8-trichloropurine dissolved in Tris-HCl buffer, pH 8.0. 5,6,7,8-Tetrahydro-1-naphthylamine in the incubation medium enhances the development of color by the oxidation product of 3-amino-9-ethylcarbazole. It is necessary to exclude the possibility of nonenzymatic oxidation of the chromogen (incubation without peroxidase), which has been observed in rabbit liver and kidney (GRAHAM and KARNOVSKY 1965).

II.B.3.b. Antibody Techniques

Mouse liver urate oxidase has been localized at the fine structural level by means of ferritin-labeled anti-urate oxidase antibodies in ultrathin frozen sections (YOKOTA and NAGATA 1974 a). Urate oxidase antigenicity has been demonstrated in microbodies, cisternae of the endoplasmic reticulum (especially in terminal flattened areas), and Golgi vesicles. Crystalline cores of microbodies are stained more intensively than matrix material. Pretreatment of the sections with unlabeled anti-urate oxidase antibodies prevents any staining with labeled specific antibodies. These findings confirm earlier light microscopic studies (YOKOTA 1973), which had shown a granular distribution of urate oxidase antigenicity and a faint cytoplasmic fluorescence (in mouse hepatocytes).

II.B.3.c. Cerium (IV) Precipitation

VEENHUIS and WENDELAAR BONGA (1977) suggested the possibility of localizing urate oxidase activity by cerium (IV) precipitation. Their methodology is given in Section II.B.4.c. Uric acid or urate have been used as saturated solution in the medium described there.

II.B.4. Localization of L-α-Hydroxy Acid Oxidase

II.B.4.a. Tetrazolium Salts

The first histochemical attempts to localize L-α-hydroxy acid oxidase were made by ALLEN and BEARD (1965). They used a tetrazolium method that also has been found to be suitable for the demonstration of the enzyme in acrylamide gels. The reaction mixture consists of the following (in final concentrations):

0.05 M Sörensen's phosphate buffer pH 7.5.
0.5 M sodium-L-lactate or 0.1 M any other D,L-α-hydroxy acid.
78 µM phenazine methosulfate.
2.0 mg/ml NBT.

L-α-Hydroxy acid oxidase activity has been demonstrated in microbodies of the distal convoluted tubules in rat kidney. D,L-α-Hydroxyvaleric acid, D,L-α-hydroxybutyric acid, and L-lactic acid are oxidized. The reaction is inhibited when the corresponding α-keto acid is present in equimolar concentrations in the incubation medium.

The medium has been improved (SHNITKA and TALIBI 1971) by using TNBT instead of NBT, by increasing the pH to 7.8, and by reducing the concentrations of substrates. This medium consists of the following:

10.0 ml Sörensen's phosphate buffer, pH 7.8.
15 mg TNBT.
37 mg D,L-α-hydroxybutyric acid.
10 mg phenazine methosulfate.

Mounted cryostat sections are incubated for 20–30 minutes at 37 °C; afterward the incubation sections are washed, postfixed in aldehyde solutions, and mounted in water-miscible resin.

II.B.4.b. Formation of Hatchett's Brown

The reaction scheme has been given by Shnitka and Talibi (1971). α-Keto acids developed by the enzyme reduce potassium ferricyanide, which precipitates in the presence of cupric ions, forming Hatchett's brown. The original incubation medium has been used to demonstrate L-α-hydroxy acid oxidase in rat kidney microbodies. The incubation medium, which works well at the light and electron microscopic level, consists of the following:

6.50 ml 0.1 M Sörensen's phosphate buffer, pH 7.2.
0.50 ml 0.04 M sodium potassium tartrate.
1.0 ml 0.03 M copper sulfate.
1.0 ml double distilled water.
1.0 ml 0.005 M potassium ferricyanide.
37 mg D,L-α-hydroxybutyric acid, sodium salt.
5 mg phenazine methosulfate.
1.0 ml dimethyl sulfoxide.

Mounted cryostat sections are incubated for 40 minutes at 37 °C. Peroxisomes in kidney tubules appear pale brown. If the contrast of this staining is not sufficient for direct light microscopic studies, it may be enhanced either by conversion of copper (II) ferrocyanide to copper rubeanate (Lindquist 1969, Uzman 1956), or by treatment with DAB solutions and postosmication (Hanker et al. 1972 a, b).

For electron microscopy the tissue is fixed at 4 °C for 10 minutes in Karnovsky's (1965) glutaraldehyde-paraformaldehyde mixture, reduced to slices by means of a tissue chopper, and incubated in the above medium for 30–45 minutes at 37 °C. Thereafter, tissue slices are washed in cacodylate buffer (0.1 M, pH 7.4), postfixed in osmium tetroxide, and processed as usual.

Burke and Trelease (1975) used a slightly modified medium (Shnitka, personal communication) to identify glycolate oxidase activity in microbodies of cucumber cotyledons and in fractions obtained from this tissue. Cotyledon segments were fixed and preincubated as described for malate synthase (Trelease et al. 1974) (see Section II.B.6.). Suspensions of organelles were fixed by addition of glutaraldehyde, up to a final concentration of 3%, for 10 minutes. The suspensions were then precipitated in a 3 mM solution of potassium ferricyanide before being placed into the incubation medium prepared as follows:

Combine
8.5 ml 25 mM potassium phosphate, pH 7.2.
0.5 ml 60 mM copper sulfate in 40 mM sodium potassium tartrate, pH 7.2.
1.0 ml 15 mM potassium ferricyanide.
To 10.0 ml solution add
5 mg phenazine methosulfate.
2 mg flavin mononucleotide.
12.5 mg sodium glycolate.
Incubation is performed for 30 minutes in the dark at room temperature. Postfixation with 2% osmium tetroxide in 50 mM sodium cacodylate, pH 7.1.

BURKE and TRELEASE (1975) observed in microbodies of cucumber cotyledons either homogeneous staining or only staining of matrix regions near the limiting membrane. The same results are obtained when isolated organelles are used. Faint needle-shaped deposits are evenly distributed in the background.

A similar medium was used by GRUBER and FREDERICK (1977), who studied the distribution of glycolate oxidase in the filamentous green alga *Klebsormidium flaccidum*. Algae were fixed in 1% purified glutaraldehyde, buffered to pH 7.2 with 0.05 M potassium phosphate for 20 minutes in the dark at room temperature. Comparable staining reactions were observed in microbodies

Table 13. *Oxidation of L-α-hydroxy acids by Hydra*

	H. pseudoligactis		H. littoralis	
	nmol keto acid/ (mg protein · hr) [a]	Cytochemical reaction product	nmol keto acid/ (mg protein · hr) [b]	Cytochemical reaction product
L-Lactic acid	71.9	+++ [c]	144	++
D,L-α-Hydroxyvaleric acid	73.9	++	77.2	++
L-α-Hydroxyisocaproic acid	62.9	++ to +++	85.0	++
D,L-α-Hydroxybutyric acid	58.0	++ to +++	49.5	+ to ++
D,L-α-Hydroxycaproic acid	18.1	++	37.9	+
Glycolic acid	13.8	+ to ++	16.7	0
D,L-α-Hydroxyisovaleric acid	7.78	+	15.1	0
No substrate	10.8	0	23.6	0

From: HAND 1976.

[a] Average of 3 determinations.

[b] Average of 2 determinations.

[c] +++ reaction product has diffused into cytoplasm, ++ reaction product fills most of matrix, + spotty reaction product in matrix, 0 no reaction product or slight increase in membrane density.

of *Klebsormidium* after incubation with either glycolate or L-lactate. The staining reaction, however, was less intensive with L-lactate. When glycolate was used as a substrate the reaction product was often seen as a dense profile surrounding the microbodies. These halos are assumed to represent glycolate oxidase activity because they were absent when the substrate was omitted from the medium or when D-lactate was used as a substrate.

It has been suggested that these artifacts occur when the substrate is oxidized too rapidly and hence reduced ferricyanide is not immediately precipitated by copper (II) ions in the microbody (HAND 1975, 1976). Obviously, diffusion artifacts that are caused in this manner can be limited by lowering the turnover rate of the enzyme under investigation (prolongation of the aldehyde fixation and/or lowering of the incubation temperature). GRUBER and FREDERICK (1977) found that phenazine methosulfate is essential for microbody staining in *Klebsormidium*. It seems that phenazine methosulfate facilitates the electron transfer from glycolate to ferricyanide by glycolate oxidase.

After incubation the specimens are rinsed in cacodylate/sucrose; copper ferrocyanide deposits may be amplified by incubation in 0.05% DAB in 0.05 M acetate buffer, pH 5.6, for 30 minutes at room temperature (HANKER et al. 1972 a, b). The tissue is then processed as usual.

In his study on L-α-hydroxy acid oxidase in microperoxisomes of two species of *Hydra*, HAND (1976) compared the chemical activity determinations and results of cytochemical staining using 7 different L-α-hydroxy acids as substrates. The results obtained with these two methods were found to be in good agreement; they are presented in Table 13 to illustrate the reliability of the method.

It is important to realize that considerable differences exist among the oxidation rates of various α-hydroxy acids in different species and even in different organs of the same species. Therefore, in cytochemical studies of L-α-hydroxy acid oxidase a wide spectrum of substrates should be used.

II.B.4.c. Cerium (IV) Precipitation

This method has been described by VEENHUIS and WENDELAAR BONGA (1977). The incubation medium is given in Section II.B.6.c. L-Lactate and glycolate are used in concentrations of 50 mM as substrates. Unfortunately, the method has been applied to tissue devoid of L-α-hydroxy acid oxidase activity.

II.B.5. Localization of D-Amino Acid Oxidase

II.B.5.a. Tetrazolium Salts

In the presence of phenazine methosulfate, electrons from the D-amino acid oxidase system can be transferred to tetrazolium salts (FARBER et al. 1958). This method, however, is useful only at the light microscopic level (BEARD and NOVIKOFF 1969). The following incubation medium has been used by SZMIGIELSKI (1972) to demonstrate D-amino acid oxidase activity in granulocytes (leukocytes were separated from erythrocytes by sedimentation):

0.1–0.3 M substrate.
0.05 M phosphate buffer, pH 8.3.
1 mM phenazine methosulfate.
5 mM FAD.
2 mg/ml TNBT.
6% dimethylsulfoxide.

II.B.5.b. Coupled Peroxidase

Alternatively, the coupled peroxidase method (GRAHAM and KARNOVSKY 1965) can be used to demonstrate hydrogen peroxide developed by D-amino acid oxidase as described for urate oxidase (see Section II.B.3.a.).

II.B.5.c. Cerium (IV) Precipitation

The reaction scheme given by BRIGGS et al. (1975 a, b) to demonstrate hydrogen peroxide (Ce^{3+} is oxidized and cerium peroxihydrate, $Ce(OH)_4$, precipitates) has been modified for the cytochemical localization of D-amino acid oxidase by VEENHUIS and WENDELAAR BONGA (1977). Unfixed tissue or tissue fixed for 10 minutes at 0 °C in either 3⁰/₀ glutaraldehyde or 3⁰/₀ formaldehyde, buffered with 0.1 M cacodylate to pH 7.2, is used. The specimens are preincubated in 0.1 M Tris-maleate buffer, pH 7.5, containing 50 mM aminotriazole and 5 mM cerium (III) chloride, for 30 minutes at 25 °C. The tissue is then transferred into the definite incubation medium, which consists of the following (final concentrations):

100 mM Tris-maleate buffer pH 7.5.
50 mM aminotriazole.
5 mM cerium (III) chloride.
50 mM substrate (D-alanine).

Incubation is performed for 2 hours at 25 °C under continuous aeration. The tissue is then washed in 0.1 M cacodylate buffer, pH 6.0, postfixed with osmium tetroxide, and processed as usual.

VEENHUIS and WENDELAAR BONGA (1977) successfully used this medium to identify D-amino acid oxidase activity in microbodies in kidney cells of sticklebacks (*Gasterosteus aculeatus trachurus*), and ARNOLD et al. (1977) succeeded in demonstrating this enzyme in rat kidney peroxisomes and in microperoxisomes of rat cerebellum. Obviously this method can also be used with appropriate substrates, to demonstrate other hydrogen peroxide-generating enzymes, such as urate oxidase or L-α-hydroxy acid oxidase.

II.B.6. Localization of Malate Synthase

Malate synthase has been located in glyoxysomes of cucumber and sunflower seedlings by TRELEASE et al. (1974). The method is based on the formation of Hattchet's brown in regions where CoA-SH reduces ferricyanide to ferrocyanide. CoA-SH is formed where malate synthase catalyzes the following reaction:

$$\text{acetyl-CoA} + \text{glyoxylate} \rightarrow \text{L-malate} + \text{CoA-SH}.$$

Tissue fixation is performed with a mixture of 4⁰/₀ formaldehyde and 1⁰/₀ glutaraldehyde, buffered to pH 7.1 with 0.05 M sodium cacodylate, at 4 °C for 5–50 minutes. The tissues are washed in three changes of 20 mM potassium phosphate, pH 6.9, preincubated with 3.0 mM potassium ferricyanide (in 20 mM potassium phosphate, pH 6.9) at room temperature for 30 minutes, and rinsed in the same buffer. The incubation medium consists of the following:

0.3 ml 65 mM potassium phosphate, pH 6.9.
0.2 ml solution containing 50 mM copper sulfate and 500 mM sodium potassium tartrate, pH 6.9.
0.25 ml double-distilled water.
0.03 ml 50 mM potassium ferricyanide.
0.1 ml 50 mM magnesium chloride.
0.02 ml 150 mM sodium glyoxylate.
0.1 ml 10 mM acetyl-CoA.

Dimethyl sulfoxide can be added, 0.05 ml, after ferricyanide and before magnesium chloride, in which case the amount of water is reduced to 0.2 ml. Incubation is performed at 37 °C for 40 minutes in corked glass tubes.

After incubation tissues are rinsed in 50 mM sodium cacodylate, postfixed in buffered osmium tetroxide, and processed as usual. Controls are performed by omitting either glyoxylate or acetyl-CoA. The incubation medium is derived from that given by HIGGINS and BARRNETT (1970) for the localization of carnitine acetyltransferase (see Section II.B.7.).

II.B.7. Localization of Carnitine Acetyltransferase

A technique for the cytochemical localization of carnitine acetyltransferase was introduced by HIGGINS and BARRNETT (1970), who studied rat heart mitochondria. The method is based on reduction of ferricyanide by the thiol group of CoA-SH; ferrocyanide is then precipitated by uranium ions. Either unfixed or fixed tissues are used. Fixation is performed with a mixture of 1% glutaraldehyde and 4% formaldehyde in 0.05 M cacodylate-nitric acid buffer, pH 7.4, with 4.5% (w/v) dextrose added. After fixation the tissues are washed in the same buffer for 15–18 hours. Thereafter, tissues are preincubated for 15 minutes in a solution of 15 mg potassium ferricyanide-2.5 ml maleate buffer to remove endogeneous reducing activities. The incubation medium consists of the following:

5 mg potassium ferricyanide.
2.5 mg uranyl acetate.
2 mg acetyl-CoA.
4 mg carnitine.
2.5 ml maleate buffer, 0.05 M, pH 7.0, with 4.5% dextrose.

Incubation is performed at room temperature for 30–60 minutes. After incubation tissues are rinsed in maleate buffer, postfixed in osmium tetroxide, and embedded as usual. Preincubation of the tissues in 4×10^{-4} M mercuric chloride for 5 minutes prevents the development of any reaction product.

HIGGINS and BARRNETT (1970) also reported on the possibility of precipitating CoA-SH directly with cadmium ions. In the case of carnitine acetyltransferase this method was found to be less reliable than the uranyl ferricyanide technique just described. Incubation for the cadmium method is performed in a medium consisting of the following:

2.5 ml 0.05 M cacodylate-nitric acid buffer, pH 7.0, with 4.5% (w/v) dextrose.
10 mg cadmium nitrate.
2 mg acetyl-CoA.
4 mg carnitine.

Incubation is performed at room temperature for 30–60 minutes. After incubation the tissues are not washed but are immediately fixed with 3% glutaraldehyde in cacodylate-nitric acid buffer containing cadmium nitrate,

4 mg/ml. Postfixation is performed in 1% osmium tetroxide in 50% ethanol-50% cacodylate buffer, and dehydration is done with absolute ethanol only to prevent cadmium precipitates from dissolving (MIZUTANI and BARRNETT 1965).

II.B.8. Identification of Peroxisomal Systems

It remains a matter of dispute whether the identification of peroxidatic activity of catalase in certain cell organelles allows one to classify these organelles as peroxisomes. Biochemical data give only over all information, and nothing is known concerning the concomitant presence of catalase and hydrogen peroxide-producing oxidases in single organelles. A cytochemical method that demonstrates the peroxisomal system (DE DUVE and BAUDHUIN 1966) is therefore highly desirable. Peroxisomes offer the following possibility:

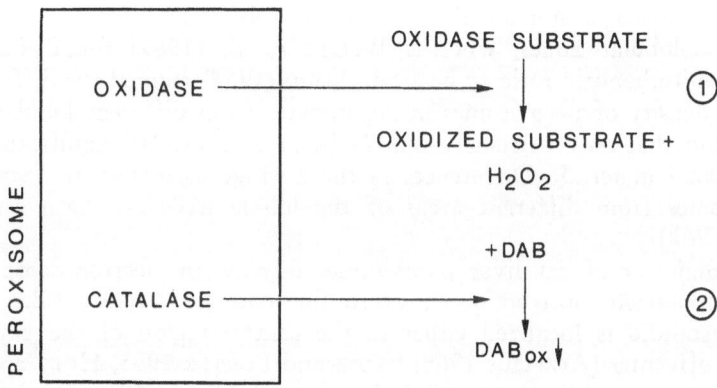

The pH optima of peroxisomal oxidases, such as D-amino acid oxidase, α-hydroxy acid oxidase, or urate oxidase, are in the range of pH 8–9; this is a sufficient pH for the peroxidatic activity of catalase to develop. In fact, positive staining of cell organelles with DAB media has been reported with hydrogen peroxide substituted for by an oxidase substrate (granulocyte granules, BRIGGS et al. 1975 a; microbodies, VAN DIJKEN et al. 1975, VEENHUIS et al. 1976).

Obviously, the final reaction product indicates the localization of the peroxidatic activity and not the site where hydrogen peroxide is generated. There is no proof that the oxidase that generates hydrogen peroxide is also localized in the same organelle; however, combined studies with direct identification of hydrogen peroxide-generating sites [by means of cerium (IV) precipitation methods] will furnish interesting results.

III. Mammalian Peroxisomes

III.A. Rat Liver Peroxisomes

III.A.1. Morphology

The first detailed morphological description of rat liver peroxisomes was given by ROUILLER and BERNHARD (1956), who reported on ovoid or round granules with diameters ranging from 0.1 to 0.5 μm. These organelles include a dense, finely granular matrix, are bounded by a single well-defined membrane, and exhibit an opaque, homogeneous core. During the following years these initial observations were confirmed and extended. Rat liver peroxisomes have become the most important objects of morphological and biochemical peroxisome studies. The original description is essentially still valid (BRUNI and PORTER 1965, GÄNSLER and ROUILLER 1956, HRUBAN and RECHCIGL 1969, MASTERS and HOLMES 1977, NOVIKOFF and SHIN 1964, ROUILLER and BERNHARD 1956, SHNITKA 1966, TSUKADA et al. 1966).

Three-dimensional reconstruction of serial sections revealed a mean diameter of 410 nm, and a mean volume of 0.04 μm³ (WEDEL and BERGER 1975). LOUD (1968) reported a numerical ratio of microbodies : mitochondria of 1 : 2 in all sublobular zones, whereas WEIBEL et al. (1969) found the microbodies : mitochondria ratio to be 1 : 4. LOUD (1968) studied the differences in volume density of peroxisomes in hepatocytes from different lobular regions and found that only extremely centrolobular cells display significantly more peroxisomal material. Differences in the average diameters and volumes of peroxisomes from different areas of the lobule have not been established (LOUD 1968).

The majority of rat liver peroxisomes display an electron-dense area of highly organized substructure—a crystalline core (see Section I.B.2., Fig. 4). The crystalloid is localized either in the central region of the particle, or slightly offcenter (AFZELIUS 1965, BRUNI and PORTER 1965, HOLT and HICKS 1961, HRUBAN and RECHCIGL 1969, HRUBAN and SWIFT 1964, SHNITKA 1966, TSUKADA et al. 1966). The presence of crystalloids in peroxisomes is closely related to the occurrence of urate oxidase (AFZELIUS 1965, HAYASHI et al. 1971, HRUBAN and RECHCIGL 1969, HRUBAN and SWIFT 1964, HUANG and BEEVERS 1973, LATA et al. 1977, LEIGHTON et al. 1969, MASTERS and HOLMES 1977, SHNITKA 1966, TSUKADA et al. 1966). Structures resembling nucleoids of rat hepatic peroxisomes have been found in urate oxidase preparations (HRUBAN and SWIFT 1964, LATA et al. 1977). Highly purified urate oxidase, however, is noncrystalline (MAHLER et al. 1955). Recent investigations revealed that, in addition to urate oxidase, D-amino acid oxidase and L-α-hydroxy acid oxidase are also associated with the nucleoid (HAYASHI et al. 1971, 1973, 1976 a).

Isolated nucleoids consist of parallel bundles of highly dense tubules with outer and inner diameters of approximately 15 and 5 nm, respectively (TSUKADA et al. 1966). Ten of these tubules are arranged around a 15 × 20 nm longitudinal space (a so-called 1 : 10 pattern; see Section I.B.2.). The mean size of nucleoids has been estimated as 250 × 270 nm. In longitudinally sectioned nucleoids the parallel arrangement of tubules is evident (Fig. 4); in cross sections the nucleoid appears as a honeycomb-like structure (TSUKADA *et al.* 1966) (see Section I.B.2.). Recently LATA *et al.* (1977) have shown that changes in rat liver urate oxidase activity occur in response to prolonged star-

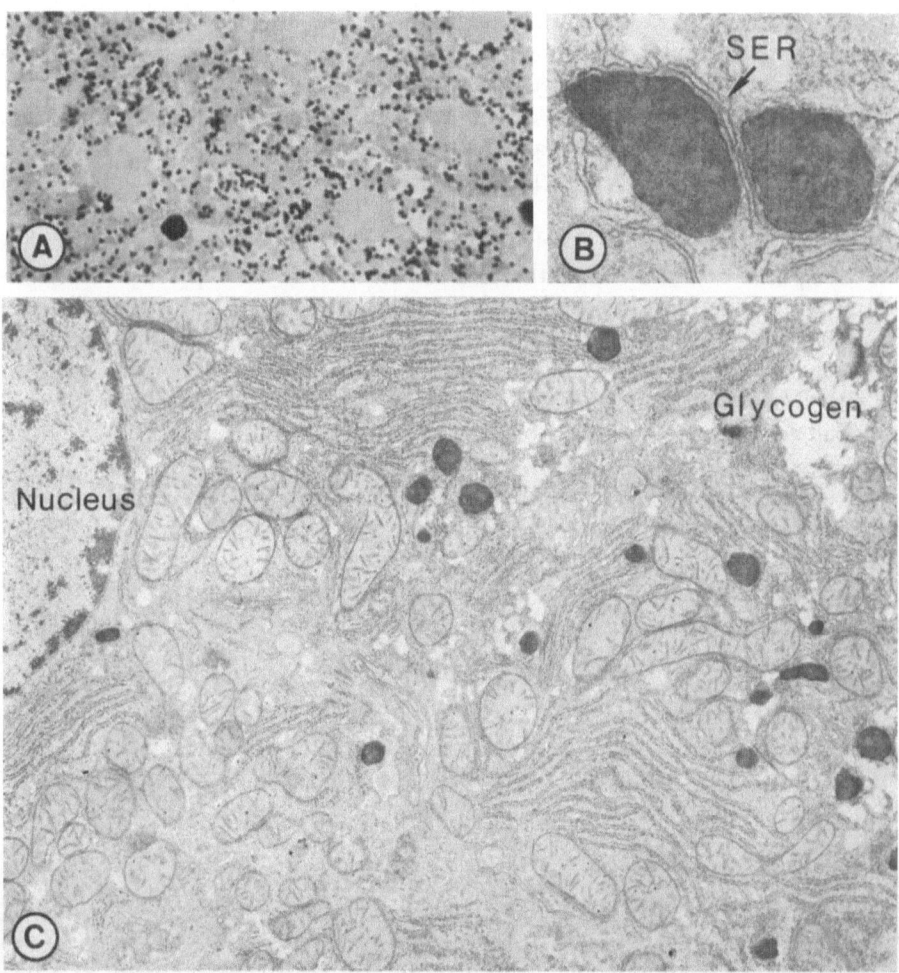

Fig. 30 A–C. Cytochemical identification of peroxidatic activity of catalase in rat hepatocytes (pH 9.0, DAB medium). *A* Peroxisomes appear as small particles that are evenly distributed in the cytoplasm of hepatocytes. Semithin section. ×500. *B* Electron micrograph of two peroxisomes shows the reaction product to be restricted to the organelles. Peroxisomes are bounded by a single membrane. Proximity to profiles of the smooth endoplasmic reticulum (*SER*) is evident. ×40,000. *C* Low-power electron micrograph shows positively stained peroxisomes evenly distributed throughout the cytoplasm. No reaction is seen in mitochondria, in the Golgi apparatus, or in the endoplasmic reticulum. ×10,000.

vation. These changes are paralleled by morphological alterations in peroxisomes. Nucleoids are diminished during the early days of starvation and reappear on the fifth day of the experiment.

In addition to peroxisomes with nucleoids, smaller particles devoid of nucleoids are also present in rat liver hepatocytes. These particles have been classified as microperoxisomes and their role as progenitors of larger nucleoid-containing peroxisomes has been considered (NOVIKOFF et al. 1973 a, SAITO et al. 1973). However, morphometric analysis of the response of rat hepatocytes to nafenopin revealed that peroxisomes and microperoxisomes behave differently; the results of STÄUBLI et al. (1977) do not support the assumption that microperoxisomes are progenitors of peroxisomes.

Peroxisomes are randomly distributed throughout the entire cytoplasm. They are observed as single particles or are gathered into groups (AFZELIUS 1965, FAHIMI 1968, 1969, NOVIKOFF et al. 1973 a, SHNITKA 1966). LEGG and WOOD (1970 a) found that a favored location for peroxisomes is in proximity to the membranes of the rough endoplasmic reticulum. This proximity has been emphasized by many authors and continuity of the peroxisome membrane with smooth endoplasmic reticulum has been described repeatedly (see Section I.B.3., Figs. 8 and 9). Various types of continuities exist: some with and some without matrix material within the connecting stalk (NOVIKOFF et al. 1973 a, REDDY and SVOBODA 1971 a). In contrast to these observations, membranous continuities were not observed by other authors (LEGG and WOOD 1970 a).

In addition to direct membranous continuities, intermembranous thread-like junctions have been described at membranes of peroxisomes and endoplasmic reticulum oriented in parallel (KARTENBECK and FRANKE 1974); these junctions have been interpreted as indicative of the structural similarity and functional relationship of these membranes.

III.A.2. Cytochemistry

In 1968 FAHIMI, HIRAI, and NOVIKOFF and GOLDFISCHER independently published a staining method for catalase that allows the identification of microbodies (peroxisomes) at the light and electron microscopic levels (Fig. 30). The histochemical method is based on the oxidation of DAB, which had been introduced by GRAHAM and KARNOVSKY (1966) as a substrate for peroxidases. The reaction is ascribed to peroxidatic activities of catalase; it is inhibited by the catalase inhibitor aminotriazole (MARGOLIASH and NOVO-GRODSKY 1958). The original incubation medium has been modified by an increase in the pH and the concentration of hydrogen peroxide (FAHIMI 1969, NOVIKOFF and GOLDFISCHER 1969). The fine structural localization of hydroxy acid oxidase activity in rat liver peroxisomes has been described by HAND (1975).

It has been suggested that the oxidation of benzidine compounds in peroxisomes is due to the splitting of catalase into peroxidatically active subunits by strongly alkaline pH (GOLDFISCHER and ESSNER 1969); however, incubation of unfixed peroxisomes at pH 8.5 does not reveal any peroxidatic activity. It has been shown by HERZOG and FAHIMI (1974 b) that glutaraldehyde fixation accelerates the oxidation of hydrogen donors and inhibits catalatic activity. Thus, pretreatment with glutaraldehyde seems to be the most important step for the peroxidatic reaction of microbodies.

No differences have been seen between the reaction intensities of peroxisomes within a single hepatocyte or within hepatocytes in different regions of the hepatic lobule (FAHIMI 1969, HIRAI 1969, NOVIKOFF and GOLDFISCHER 1969, SAITO et al. 1973). The distribution pattern of DAB-reactive peroxisomes closely resembles that observed after staining for urate oxidase activity (GRAHAM and KARNOVSKY 1965) or after immunofluorescence staining to demonstrate catalase (MORIKAWA and HARADA 1969, NISHIMURA et al. 1964). The electron-opaque reaction product is restricted to round or oval particles (microbodies); it completely fills the interior of the organelles and consequently often obscures the limiting membrane (FAHIMI 1969, SAITO et al. 1973) (Figs. 4 and 30). Finger-like protrusions filled with reaction product are interpreted as connections with the endoplasmic reticulum.

Storage of chopped liver sections in buffer (after glutaraldehyde fixation and before postfixation with osmium tetroxide) 18 hours or longer makes the limiting membrane of some of the peroxisomes become discontinuous. These leaky organelles are then found side by side with normal looking peroxisomes within the same cell (FAHIMI 1974). The phenomenon is paralleled by the cytochemical finding that peroxisomal catalase diffuses out of certain microbodies after prolonged storage of liver tissue in buffer. It might indicate differences in the stability of the limiting membranes and suggests a certain heterogeneity of the peroxisome population in rat hepatocytes (YOKOTA and FAHIMI 1978). In this context it is noteworthy that staining of ribosomes that lie adjacent to peroxisomes has been interpreted as evidence that catalase might be synthesized by these ribosomes and might be transferred into peroxisomes through the cytosol (LEGG and WOOD 1970 a). However, newly synthesized catalase is not enzymatically active outside of peroxisomes (LAZAROW and DE DUVE 1973 a, b), and critical examination of the histochemical method revealed that the ribosomal staining is due to the diffusion of peroxisomal catalase (FAHIMI 1973, 1974).

The cytochemical localization of catalase in sinusoidal cells (von Kupffer cells, endothelial cells) and in fat-storing cells shows stained particles (microperoxisomes) in all of these cells (FAHIMI et al. 1976).

III.B. Peroxisomes in Rat Kidney Epithelium

The presence of numerous microbodies (peroxisomes) in the proximal tubules of rat kidney has been established by electron microscopic and cytochemical studies (BARRETT and HEIDGER 1975, BEARD and NOVIKOFF 1969, ERICSSON and TRUMP 1966, LANGER 1968, MAUNSBACH 1966, REDDY et al. 1976 a). In earlier morphological studies nucleoids had generally been noted in these microbodies (ERICSSON and TRUMP 1966, LANGER 1968, MAUNSBACH 1966, TRUMP and ERICSSON 1965). TISHER et al. (1968, 1969) emphasized that nucleoids of renal microbodies are artifacts due to immersion fixation. This question has been analysed in detail by BARRETT and HEIDGER (1975), who confirmed the interpretation of TISHER et al.

After perfusion fixation, two types of microbodies are identified, Mb-I and Mb-II. Mb-I are characterized by circular and/or tubular profiles within the matrix; in Mb-II, marginal plates or crystalloid inclusions are present in addition to circular and/or tubular profiles. Mb-I are usually 0.8–1.2 μm in diameter and rarely reach diameters up to 2.0–3.0 μm. Mb-II are 1.1–3.0 μm in diameter. BARRETT and HEIDGER (1975) reported that all microbodies tend to be basally located within the cell, mainly in a perinuclear position. Granules 5–6 nm in diameter (LANGER 1968) are assembled to form circular and tubular profiles close to the microbody membrane. Tubular profiles often cause deformation of the organelle's spherical shape.

It seems that there is no precise distinction between marginal plates and crystalline inclusions. BARRETT and HEIDGER (1975) refer to inclusions that are up to 0.2 μm wide and 3.0 μm long as "crystalline inclusions", while inclusions less than 0.1 μm wide and shorter than 1.7 μm are termed "marginal plates". Both types of inclusions are tightly attached to the limiting membrane of the microbody, thus causing a certain deformation of the organelle's shape. Both marginal plates and crystalline inclusions usually display a periodicity of 30 nm; less frequently, periodicities of 20 or 10 nm are observed. Tubular inclusions, as well as marginal plates, stain positively with the cytochemical catalase method (BARRETT and HEIDGER 1975, KALMBACH and FAHIMI 1978) (Fig. 31). Structures that resemble tubular inclusions may be seen free in the cytoplasm without connection to a microbody (LANGER 1968) (see Section III.F.).

Catalase and L-α-hydroxy acid oxidase activity have been demonstrated (the latter at the light microscopic level) in peroxisomes of the rat nephron (BEARD and NOVIKOFF 1969). Peroxisomes are larger and more frequent in the P_3 segment (pars descendens) than in the P_1 or P_2 segments (pars convoluta). Microperoxisomes are present in epithelial cells of Henle's loop (CHANG et al. 1971, NOVIKOFF et al. 1973 b). Fine structural localization of L-α-hydroxy acid oxidase has been achieved in rat kidney peroxisomes by SHNITKA and TALIBI (1971), who introduced a copper ferricyanide method instead of the tetrazolium salt methods that had previously been used (Section II.B.4.). Copper (II) ferrocyanide reaction product is evenly distributed in the peroxisome matrix and the peroxisome population of an individual cell is uniformly reactive. No L-α-hydroxy acid oxidase activity can be demonstrated in other cellular compartments.

The angular shape and tubular inclusions of large microbodies in rat proximal kidney tubules (P_3 segment) enable their identification in freeze-etch preparations (KALMBACH and FAHIMI 1978) (Fig. 14 B). Tubular inclusions with diameters between 100 and 125 nm cause corresponding impressions upon the limiting membranes. Few membrane particles are seen on the E-face of the membrane, whereas many more particles are seen on the P-face. In regions where tubular inclusions underlie the E-face of the microbody membrane, a crystalline pattern with a periodicity of 9.2 ± 0.8 nm perpendicular to the axis of these inclusions is noted (KALMBACH and FAHIMI 1978). Catalase staining of tubular inclusions is much more intensive than the catalase reaction of homogeneous matrix material (BARRETT and HEIDGER 1975, KALMBACH and FAHIMI 1978).

The effect of aminotriazole on renal peroxisomal enzymes and plasma renin activity has been studied by MORIMOTO *et al.* (1978). Administration of aminotriazole (1 g/kg, i.p., daily) to rats for 3 or 5 days causes catalase activities to decrease to about 40% of the control level, but increases D-amino acid oxidase activity to 190% of the control value on the 5th day. Plasma renin activities are significantly decreased within 3 days of drug administration, whereas renal renin activities remain constant.

III.C. Liver and Kidney Tissues of Other Species

A comprehensive description of microbodies in all classes of animals was given by HRUBAN and RECHCIGL in 1969. This section covers only more recent papers, or former results that either are of historical interest or are related to the material in other chapters of this book.

III.C.1. Liver

Urate oxidase is present in hepatic peroxisomes of all mammals except man and the apes, which are not uricolytic. Histochemical methods for detecting urate oxidase activity thus can be used to identify peroxisomes in uricolytic animals and have been used to define the boundary between uricolytic and nonuricolytic primates (NAKAJIMA and BOURNE 1970). The enzyme has been demonstrated in prosimians and in platyrrhine monkeys (tree shrew, slow loris, potto, galago, and marmoset; owl monkey, squirrel monkey, capuchine monkey, and spider monkey). Old World monkeys (Java and rhesus monkeys) and apes (orangutan and chimpanzee) are devoid of urate oxidase. The presence of urate oxidase is often indicated by the presence of crystalline cores within the peroxisomes (AFZELIUS 1965). Correspondingly, human hepatic microbodies are devoid of cores. Marginal densities, which recently have been demonstrated in hepatic microbodies of human liver tissue (STERNLIEB and QUINTANA 1977), are not equivalents of crystalline cores: marginal densities have been shown to stain when the cytochemical catalase reaction is applied, which is not the case with crystalline cores (BARRETT and HEIDGER 1975) (*cf.*, Figs. 4 B and 31 B). It is, however, necessary to mention that microbodies with cores have been observed in human hepatocytes during early stages of development (6th week of gestation; DE LA IGLESIA *et al.* 1966) and under pathological conditions (DE LA IGLESIA 1969). In accord with the results mentioned above, DE LA IGLESIA *et al.* (1966) found dense cores in hepatic microbodies of the squirrel monkey.

Marginal plates have been observed in cow liver microbodies (SHNITKA 1966) and are a common feature of hepatic microbodies in pigs (in addition to crystalloids, SINGH *et al.* 1978). It is surprising that SINGH *et al.* (1978) found no space between the marginal plate and the limiting membrane. The thickness of marginal plates in pig hepatic microbodies significantly increases in animals fed rapeseed oil (SINGH *et al.* 1977).

Mouse hepatic microbodies display several peculiarities that distinguish these organelles as favorable samples for morphological and biochemical studies. The percentage of small and coreless particles (*i.e.*, microperoxisomes) is relatively high. Continuities between these small peroxisomes and smooth endoplasmic reticulum are frequently seen even in single sections. Mouse hepatic peroxisomes are often arranged in clusters (DAEMS 1966), which facilitates studies on a possible spatial relationship among the microbodies themselves, and between microbodies and other cellular compartments. Peroxidatic activities of catalase and the possible involvement of the endoplasmic reticulum in the origin of peroxisomes have been studied in hepatocytes of fetal and early postnatal mice (ESSNER 1967, 1969). Starting from day 18 of gestation, the increasing number of peroxisomes seems to reflect increasing activities of peroxisomal enzymes (catalase, urate oxidase, and L-α-hydroxy acid oxidase have been studied; HOLMES 1971, LEE 1973, MILES and HOLMES 1975).

Acatalasemic mice have been used as a model to acquire a better understanding of the mechanism of the cytochemical catalase reaction (GOLDFISCHER and ESSNER 1970) (see Section II.B.1.).

III.C.2. Kidney

Microbodies were described for the first time in the mouse kidney (RHODIN 1954) as "spheric or oval bodies between the mitochondria". The organelles are about 0.1 × 0.3 μm in size; a mean number of about 10 is seen in each cell. The matrix material is finely granular; its electron density is similar to that of the mitochondrial matrix. The organelles are devoid of crystalline inclusions. Continuities can be seen between microbodies and between microbodies and the endoplasmic reticulum (Fig. 12). Cytochemical catalase staining is demonstrated in Fig. 14 A. Mouse kidney peroxisomes must be regarded as microperoxisomes (Fig. 31 A). Their development has been studied during ontogenesis (GOECKERMANN and VIGIL 1975, PIPAN and PSENIČNIK 1975) (see Section III.E.).

Microbodies are present in the entire proximal tubule in human kidney. They are seen in greatest number in the *pars convoluta* (TISHER et al. 1966). The organelles are located close to the endoplasmic reticulum, and membrane continuities are occasionally observed between microbodies and endoplasmic reticulum (Fig. 7 C). Marginal plates have been described as a "linear density continuous with the outer limiting membrane of the organelle" (TISHER et al. 1966). Additional densities, which may correspond to nucleoids, have been repeatedly reported. However, the results of BARRETT and HEIDGER (1975)

Fig. 31 A, B. Peroxisomes in kidney epithelium. *A* Mouse kidney, distal part of the proximal convolute. Positively reacting, coreless microperoxisomes are evenly distributed in the cytoplasm of epithelial cells. No spatial relationship is seen between these organelles and Golgi regions (G), mitochondria, or endoplasmic reticulum. *Arrows:* larger organelles that are not completely filled with reaction product; under these conditions granularity of the reaction product becomes visible. *BM* basement membrane. Fixed with formaldehyde, incubated in alkaline DAB medium for peroxidatic activities of catalase, and postfixed with osmium tetroxide. Brief staining with lead citrate. ×12,000. *B* Rat kidney, proximal part of the proximal convolute. Brief incubation for peroxidatic activity of catalase reveals stronger staining of marginal plates. Central regions of peroxisomes remain less reactive. These areas may correspond to "nucleoids" which artificially develop in these organelles during immersion fixation; most probably, however, less substrate has reached the central parts of the organelles. This type of peroxisome frequently displays irregular shape. Unstained section. ×35,000.

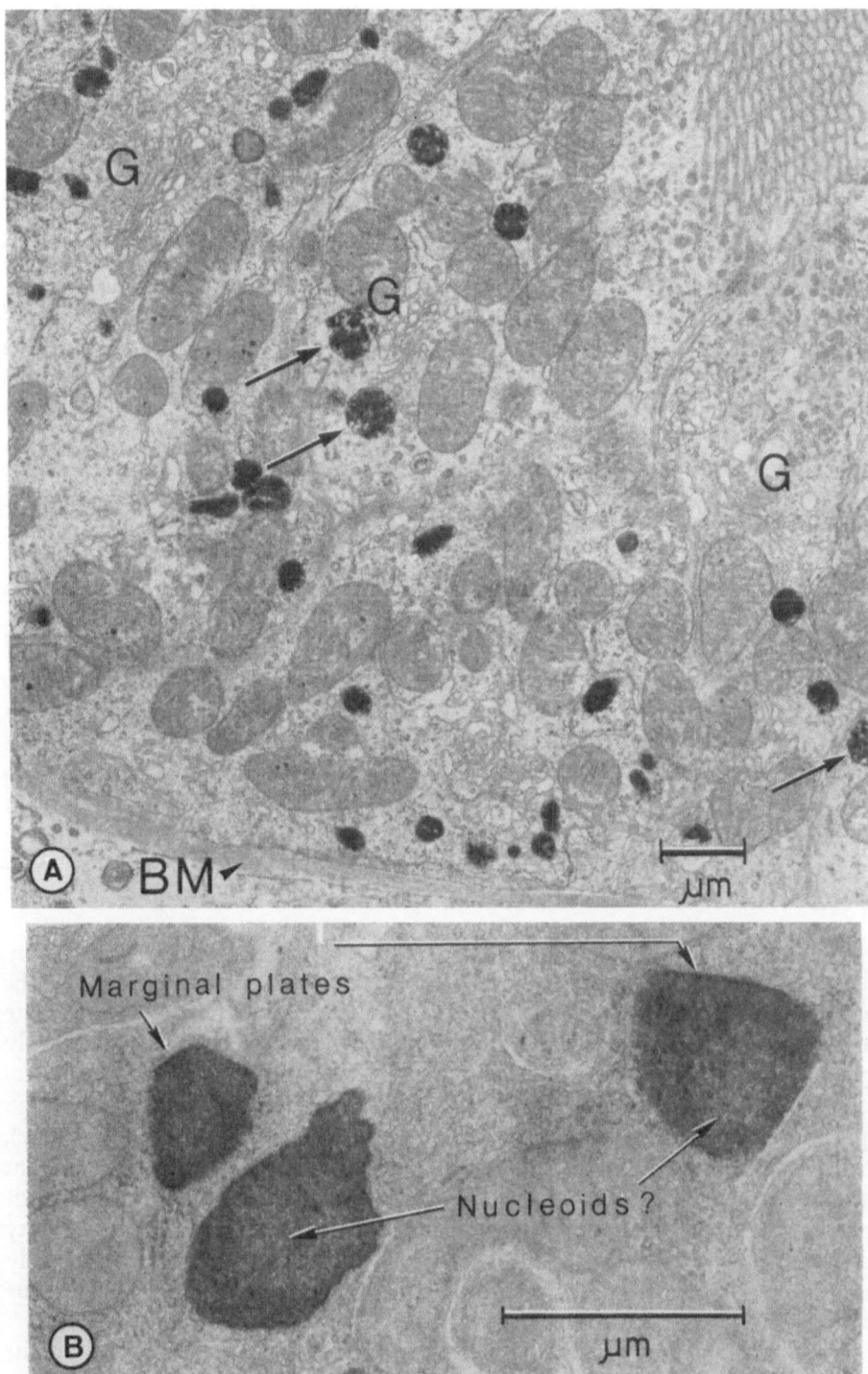

Fig. 31.

suggest that these densities result from fixation by immersion. Nucleoids are not seen in well-preserved material (Fig. 7 B, C). The matrix material is finely granular.

Marginal plates have been noted in renal microbodies of the cat (REHG et al. 1975, TIEDEMANN 1972), horse (HRUBAN and RECHCIGL 1967), sheep (TIEDEMANN 1972), dog (MAUNSBACH and WIRSEN 1966), rabbit (the first description of a marginal plate, BREEMAN and MONTGOMERY 1960) and rhesus monkey (TISHER et al. 1968, 1969).

Large lipid droplets consisting of triglycerides, cholesterol, and cholesterol esters are frequently found in the cat proximal convoluted tubule. Peroxisomes with marginal plates are numerous in the proximity of these lipid droplets. REHG et al. (1975) and MONTALI et al. (1974) have drawn attention to this phenomenon, which might be of significance in connection with the identification of carnitine acetyltransferase as a microbody enzyme (MARKWELL et al. 1973).

Renal microbodies have been studied in the developing chicken and mouse (ESSNER 1970, GOECKERMANN and VIGIL 1975). In mouse kidney catalase activity increases during the first three weeks of postnatal life, as does the number of microbodies. In the chicken, catalase activity already reaches the adult level 1 day after hatching and microbodies are numerous (see Section III.E.).

Most recently, glycogen particles have been identified in peroxisomes of riboflavin-deficient mouse kidney (KOBAYASHI and YAMAMOTO 1978). These glycogen particles can be digested with amylase, and it has been shown by means of DAB cytochemistry that areas that had been occupied by glycogen are free of catalase. These results substantiate early observations that microbodies might contain glycogen particles (hepatic microbodies in the Syrian hamster; HRUBAN and RECHCIGL 1964).

III.D. Biochemical Features of Mammalian Liver Peroxisomes

Unlike mitochondria and lysosomes, the characterization of peroxisomes by enzymes is restricted to a few markers (see Section I.D.1.). Even among the group of enzymes involved in hydrogen peroxide metabolism (catalase, urate oxidase, D-amino acid oxidase, and L-α-hydroxy acid oxidase), the exclusively peroxisomal location of one—catalase—is not fully accepted (see Section II.A.2.a.). Catalase and urate oxidase represent a surprisingly high proportion of total peroxisomal protein (about 16 and 10% respectively), whereas D-amino acid oxidase and L-α-hydroxy acid oxidase together amount to only about 5% (LEIGHTON et al. 1969). Considering the activities of marker enzymes in isolated peroxisomes (Table 14), the vast catalase activity is conspicuous. The biological meaning of this high activity is not fully understood as yet. Due to low hydrogen peroxide concentrations in the tissue the in situ turnover rate of catalase is rather low (see Section I.C.).

An important peroxisomal capacity, the β-oxidation of fatty acids, exhibits activity similar to that of other peroxisomal markers: according to LAZAROW and DE DUVE (1976), the specific activity for rat liver peroxisomes is

100 nmol/(min · mg protein) (expressed in terms of NAD$^+$ reduced; compare the lower activity reported by Shindo and Hashimoto 1978). The relative specifity for long-chain fatty acids (Lazarow 1978, Thinès-Sempoux and Bormann 1978) suggests that peroxisomal β-oxidation plays a preferen-

Table 14. *Specific activities of peroxisomal marker enzymes in a peroxisomal fraction prepared from rat liver by density equilibration*

Enzyme	Specific activity nmol/(min · mg protein)
Catalase	11.4 × 10^6
Glycolate oxidase	32
Urate oxidase	133

From: Huang and Beevers 1973.

Table 15. *Phospholipid composition of rat liver peroxisomes, mitochondria, and microsomes purified by density equilibration in a zonal rotor*

Phospholipid	Peroxisomes	Mitochondria	Microsomes
Total phospholipid (mg/mg protein)	0.086	0.20	0.32
	Percent of total phospholipid		
Phosphatidylcholine	55.1	44.5	49.8
Phosphatidyl ethanolamine	16.0	28.1	18.8
Phosphatidyl inositol and sphingomyelin	19.7	7.1	19.7
Phosphatidylserine	7.4	1.9	8.5
Cardiolipin	1.6	18.4	3.1

From: Donaldson et al. 1972.

tial role in chain shortening of longer fatty acids (see also Christiansen 1978).

Most liver peroxisomal enzymes have been found in the matrix (see Table 6). Enzymes of β-oxidation also seem to be located in this compartment; other than in mitochondrial β-oxidation, even acyl-CoA dehydrogenase seems to be a matrical enzyme (M. Hüttinger, H. Goldenberg, R. Kramar, unpublished experiments, 1979).

NADH-cytochrome c reductase, which is the only enzyme of rat liver peroxisomes known to be bound to the membrane, is of interest as an enzyme common to peroxisomal and microsomal membranes; Donaldson et al. (1972) consider this as evidence for the derivation of the peroxisomal membrane from the endoplasmic reticulum. Another enzyme activity found in the endoplasmic reticulum as well as in peroxisomes is acyl-CoA synthetase (SHINDO and HASHIMOTO 1978). On the other hand, the microsomal marker glucose-6-phosphatase is not present in peroxisomes (DONALDSON et al. 1972). A strict selection of proteins seems to take place during the formation of the peroxisomal membrane (see Section III.I.), so that the enzymic relationship between peroxisomal membranes and endoplasmic reticulum membranes might be less close than formerly expected. Lipid analysis reveals closer similarities between peroxisomes and microsomes (Table 15).

DNA was never detected in animal peroxisomes and it is assumed that the biogenesis of these organelles contrary to that of mitochondria, depends exclusively on nuclear DNA (see Section III.J.). Nevertheless, this assumption must not be viewed as being definitely correct. Reports on the occurrence of DNA in microbodies of yeasts make it advisable to be cautious until unequivocal evidence of the absence of DNA in animal peroxisomes is available. For example, in a recent study OSUMI et al. (1978) were able to extract DNA from microbodies purified from the protoplasts of the hydrocarbon-utilizing yeast *Candida tropicalis;* this DNA has a buoyant density different from that of nuclear and mitochondrial DNA. In a previous electron microscopic study, OSUMI (1976) found a contour length of about 15 μm for this presumably microbody-linked DNA.

In general, proteins from liver peroxisomes have not been examined extensively, whereas the polypeptide pattern of the membrane of glyoxysomes (BIEGLMAYER and RUIS 1974, BOWDEN and LORD 1976, BROWN et al. 1974), as well as that of leaf peroxisomes (LUDWIG and KINDL 1976), has been studied by SDS-polyacrylamide gel electrophoresis. BIEGLMAYER and RUIS detected at least 11 bands; the most intensive one corresponds to a molecular weight of 66,000. The corresponding proteins are essentially insoluble in 1 M potassium chloride or 0.5% sodium cholate and therefore must be considered intrinsic.

Fig. 32/1–6. SDS-polyacrylamide gel electrophoresis of rat liver peroxisomal and microsomal membranes (for preparation see Section II.A.3.b.). From HÜTTINGER et al. 1979. Membrane preparations are solubilized for 5 minutes at 100 °C with urea-SDS-mercaptoethanol (final concentration: 1 mg protein/ml, 1% SDS, 4 M urea, 1% mercaptoethanol, 0.5 M sucrose, 20 mM sodium phosphate, pH 7.0). 10–50 μl plus tracking dye (bromphenol blue) for electrophoresis. Polyacrylamide gels: 2.4 g acrylamide, 0.04 g N,N'-methylene-bis-acrylamide, 0.3 ml 10% SDS, 15 ml 0.2 M sodium phosphate, pH 7.0, 0.04 ml tetramethylethylenediamine, 2 ml 0.5% ammonium persulfate. Electrophoresis buffer: 0.1 M sodium phosphate, pH 7.0, 0.1% SDS. Electrophoresis: approximately 140 V, until bromphenol blue reaches lower end of gel. Staining: 1% amido black in 7% acetic acid for 12 hours; destaining with 7% acetic acid. Quantitation: scanning with "Chromoscan". Molecular weight: calibration of gels with appropriate standard proteins. Membrane extraction: treatment of membrane pellets with 1% sodium deoxycholate in 50 mM Tricine, pH 7.5 (0.5 ml/mg protein) for 2 hours; resedimentation (60 minutes at 300,000 g); solubilization of the pellet; dialysis and lyophilization of the supernatant. *1* Peroxisomal membranes, control. *2* Peroxisomal membranes of rat fed for 3 weeks a diet containing 0.75% clofibrate. *3* Peroxisomal membranes, extracted with deoxycholate. *4* Peroxisomal membranes, lyophilized deoxycholate extract. *5* Smooth endoplasmic reticulum membranes, control. *6* Smooth endoplasmic reticulum membranes after clofibrate feeding. Start: on the right. Ordinate is relative light absorption. Approximate molecular weights are given at the top.

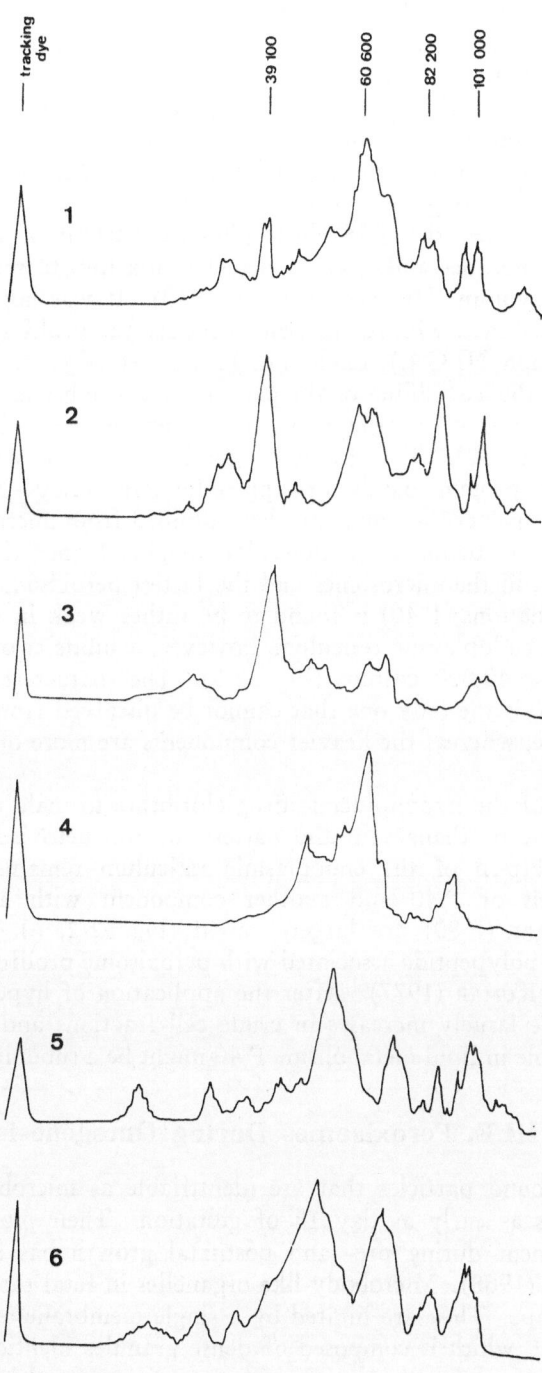

Fig. 32.

Deoxycholate or combinations of potassium chloride and cholate solubilize practically all proteins with the exception of the main polypeptide (molecular weight 66,000), which is anchored in the membrane with particular strength.

LUDWIG and KINDL (1976) found a comparable main component, called SP-63 (molecular weight 63,000) in leaf peroxisomes. In contrast, BROWN et al. (1974) reported that in pumpkin or castor bean glyoxysomes the 28,000-dalton component is unextractable by 0.15–1.0 M potassium chloride; it is also present in leaf peroxisomes.

We have examined the SDS electrophoretic pattern of solubilized liver peroxisomal membranes and tried to detect similarities to membranes of the endoplasmic reticulum (HÜTTINGER et al. 1979). It was hoped that the application of clofibrate, which is known to induce the proliferation of peroxisomes (see Section III.G.1.), might aid in the detection of typical proteins connected with the assembling of the peroxisomal membrane. It was assumed that proteins bound very strongly to the membrane were not extracted by deoxycholate (Fig. 32). The graph of solubilized peroxisomal membranes shows 9 main protein bands of approximately 40,000–100,000 daltons (Fig. 32/1). This pattern is similar to that obtained from microsomes, with the exception that the bands with molecular weights higher than 100,000 are more prominent in the microsomes and the fastest peroxisomal main fraction (about 40,000 daltons; P 40) is found to be rather weak in the endoplasmic reticulum. The endoplasmic reticulum, however, exhibits two strong bands in the range below 40,000 daltons (Fig. 32/5). The characteristic peroxisomal component P 40 is the only one that cannot be dissolved from the membrane by deoxycholate, whereas the heavier components are more or less extractable (Fig. 32/3, 4).

Application of the hypolipidemic drug clofibrate to male rats for 2 weeks produces significant changes in the pattern of the peroxisomal membrane, whereas the pattern of the endoplasmic reticulum remains relatively unchanged. Levels of P 40 and another component with about twice the molecular weight (P 80) are largely raised (Fig. 32/2, 6). P 80 might be identical to the polypeptide associated with peroxisome proliferation described by REDDY and KUMAR (1977). After the application of hypolipidemic drugs this polypeptide largely increases in crude cell fractions and probably plays an important role in lipid metabolism. P 40 might be a subunit of P 80.

III.E. Peroxisomes During Ontogenesis

Membrane-bound particles that are identifiable as microbodies appear in rat hepatocytes as early as day 15 of gestation. Their morphogenesis and enzyme equipment during pre- and postnatal growth has been studied by TSUKADA et al. (1968). Microbody-like organelles in fetal rat hepatocytes are of irregular shape. They are limited by a single membrane and contain little matrix material, which is composed of dense granules identical to those seen in microbodies of adult animals. Granular matrix material is enriched during further steps of development. The limiting membrane often shows a continuity into the rough endoplasmic reticulum; i.e., the developing microbody appears

Table 16. *Number of microbodies, incidence of agranular tubular projections of microbody membranes, and size of microbody nucleoids in rat hepatocytes*

Age of rats	Total no. of microbodies	No. of microbodies containing nucleoids	Incidence of agranular projections (%)	Size of nucleoids (nm)
Fetuses (days of gestation)				
15 (12 [a])	2.29 ± 0.224 [b]	0.21 ± 0.014		
18 (10)	4.56 ± 0.412	1.06 ± 0.109		
21 (10)	4.08 ± 0.390	2.21 ± 0.276	4.32 (16/371) [c]	
Neonates (days after birth)				
1 (7)	6.41 ± 0.553	2.63 ± 0.238	7.45 (42/564)	234.8 ± 1.00 [d]
5 (7)	8.22 ± 0.606	3.18 ± 0.334	6.07 (36/597)	226.6 ± 8.08
10 (5)	8.51 ± 0.520	4.10 ± 0.396	3.30 (14/425)	210.4 ± 6.16
20 (5)	8.63 ± 0.470	4.75 ± 0.217	1.50 (6/405)	255.7 ± 8.84
40 (5)	8.76 ± 0.803	5.62 ± 0.216	0 (0/309)	332.2 ± 9.42
60 (5)	7.11 ± 0.624	3.75 ± 0.286	0.14 (1/387)	274.0 ± 2.40

From: Tsukada *et al.* 1968 a.

[a] Figure in parentheses is the number of determinations.

[b] Mean ± SE in 100 µm² cytoplasmic area of hepatocytes. Determinations were made on cytoplasmic areas of approximately 2,000 µm².

[c] Figures in parentheses: number of microbodies bearing agranular projections/number of total microbodies.

[d] Mean ± SE. Approximately 200 nucleoids were examined in each determination.

Table 17. *Activities of urate oxidase, catalase, and D-amino acid oxidase in rat liver*

Age of rats	Urate oxidase [a]	Catalase [b]	D-Amino acid oxidase [c]
Fetuses (days of gestation)			
18	0.53 ± 0.086 [d]	7.53 ± 0.840	0
21	1.21 ± 0.180	18.60 ± 0.620	0.30 ± 0.021
Neonates (days after birth)			
1	6.80 ± 0.320	33.7 ± 1.60	0.81 ± 0.034
5	4.95 ± 0.230	35.1 ± 0.90	1.28 ± 0.153
10	6.00 ± 0.126	37.8 ± 1.52	1.87 ± 0.070
20	15.80 ± 0.555	47.1 ± 0.85	5.64 ± 0.460
40	19.15 ± 0.450	92.8 ± 3.66	6.12 ± 0.668
60	16.75 ± 0.310	109.8 ± 1.40	5.65 ± 0.260

From: Tsukada *et al.* 1968 a.

[a] Uric acid oxidized, µg/(5 min · mg dry tissue).

[b] k/(min · 100 mg dry tissue).

[c] Pyruvic acid produced, µg/(20 min · mg dry tissue).

[d] Mean ± SE. Each value was obtained from seven determinations.

as a balloon-like protrusion of the endoplasmic reticulum. The surface of the outgrowing organelle only rarely is provided with ribosomes.

Continuities between microbodies and endoplasmic reticulum are also remarkable during the perinatal period. In 1-day-old rats hook- or ring-like smooth-surfaced tubular segments project from the surfaces of microbodies (ESSNER 1970) (Fig. 14). These projections were seen most frequently during days 1–5 *post partum*. They seem to decrease in number with age. The number of microbodies that contain nucleoids increases continually up to the 40th day after birth. These results are summarized in Table 16. Morphological observations are in good agreement with biochemical findings during this phase of development (TSUKADA *et al.* 1968 a); these data are presented in Table 17.

The concept that microbodies originate as protrusions of the rough endoplasmic reticulum was derived from these findings. These protrusions lose ribosomes that had been attached to the membrane, become enriched in matrix material, acquire a nucleoid during the latest gestation period, and continue to grow during the first days of postnatal life.

Morphometric analysis has ascertained that the volume density of microbodies increases at birth and remains relatively constant during the first postnatal week, although reduplication of liver cells occurs during the first 3 days after birth and two further divisions are completed within the first week (ROHR *et al.* 1971) (Table 18). This means that the percentage of young microbodies that grow from the endoplasmic reticulum becomes relatively high during this period. If it is assumed that microbodies become enriched in catalase before urate oxidase, and that D-amino acid oxidase increases later on, the relative decrease in oxidases and increase in catalase during this period (Table 17) can be explained.

An earlier hypothesis that the Golgi apparatus is involved in the formation of peroxisomes (DVOŘÁK *et al.* 1967 a, b, WOOD 1969) now seems to be unlikely.

The origin of hepatic microbodies in mouse fetuses was studied by ESSNER (1967, 1969). Dilatations of terminal regions of the rough endoplasmic reticulum occur. Amorphous material accumulates in such dilatations and microbodies arise in this manner. Two types of microbodies—both reactive with the cytochemical catalase stain—can be distinguished: a population of smaller particles without a nucleoid (which is present by day 16 of gestation), and a population of larger particles with nucleoids (which are observed in older fetuses along with the smaller organelles). The smaller organelles are often pleomorphic and sometimes are difficult to distinguish from profiles of the endoplasmic reticulum (ESSNER 1969).

During early postnatal life the amount of peroxisomal catalase increases six- to eightfold, whereas cytoplasmic catalase remains relatively constant (HOLMES 1971). This should be reflected by an increase in the volume density of peroxisomes; unfortunately, morphometric data are not available. Two classes of microbodies—one bearing nucleoids, the other devoid of nucleoids— are also observed in adult mouse hepatocytes (DAEMS 1966, SHNITKA 1966) (Figs. 3, 5, 8 C, D, and 9).

Hepatic and renal peroxisomes have also been studied in the developing chicken (ESSNER 1970). Microbodies with nucleoids are more frequent in hepatocytes of the chicken both before and after hatching than in adult animals. The number of peroxisomes in proximal kidney tubules increases continually after hatching to reach its maximum in adult animals. ESSNER (1970) found no topological relationship between peroxisomes and glycogen areas. He did not refer to a possible spatial relationship between peroxisomes and endoplasmic reticulum in developing chicken liver and kidney tissue.

Peroxisomes in the mouse kidney have been studied during prenatal and postnatal development (GOECKERMANN and VIGIL 1975). Catalase activity and the number of peroxisomes rapidly increase during the first 3 weeks of

Table 18. *Microbodies in perinatal rat liver: Morphometric parameters*

Volume density/ml tissue (mean ± SE)	SE in percent of mean	Specific value per 100 g body wt. (cm³)	Value per hepatocyte (mononuclear) (μm³)	Time
0.0061 ± 0.0011	17.18	0.03	42	Immediately before delivery
0.0081 ± 0.0012	14.78	0.06	36	1 day post partum
0.0120 ± 0.0018	15.08	0.05	40	3 days post partum
0.0102 ± 0.0016	15.3	0.04	37	8 days post partum

From: ROHR et al. 1971.

postnatal life; this contrasts with the unchanging numbers of mitochondria and lysosomes during this period. It has therefore been suggested that the development of peroxisomes is related to the onset of specific function (GOECKERMANN and VIGIL 1975, PIPAN and PŠENIČNIK 1975). Similar conclusions were drawn by REDDY et al. (1976 a), who studied the regenerating epithelial outline of rat proximal kidney tubules.

In the developing mouse kidney wet weight and protein linearly increase from day 17 of gestation to day 20 *post partum.* In contrast, catalase activity dramatically increases during the first 3 weeks of postnatal life to reach a plateau; D-amino acid oxidase behaves like catalase, but with a delay of about 14 days (GOECKERMANN and VIGIL 1975). These observations indicate that the coreless peroxisomes (microperoxisomes) of mouse proximal kidney tubules are true peroxisomal systems in adult animals and include gradually decreasing ratios of catalase activity/D-amino acid oxidase activity during the first 2 weeks of postnatal life. Morphologically, the proximity of and continuities between smooth endoplasmic reticulum and peroxisomes have been observed during this phase of postnatal development. Clusters of peroxisomes are present during the time these organelles increase in number (GOECKERMANN and VIGIL 1975). The morphometric and biochemical data are shown in Table 19.

The differentiation of microperoxisomes has been studied in adrenocortical cells of fetal guinea pigs (BLACK and BOGART 1973). Membrane-bound particles (0.1–0.45 μm in diameter) filled with fairly homogeneous, somewhat coarsely particulate content appear as early as day 27 of gestation (guinea pig gestation period is 65–70 days). Most of these particles are round or oval; a few are elongated and look similar to dilatated regions of the endoplasmic reticulum. The limiting membranes of spherical particles occasionally are seen to be continuous with the endoplasmic reticulum. Ribosomes are occasionally present on the membranes of these granules, but more frequently ribosomes

Table 19. *Comparison of marker enzyme specific activities and number of peroxisomes and mitochondria in proximal tubule cells during pre- and postnatal mouse kidney development*

Time (days)	Peroxisomes				Mitochondria			
	Catalase specific activity ± SE μmol H_2O_2/(min · mg protein)	Significance (p)	No./100 μm²	Significance (p)	Cytochrome oxidase specific activity ± SE μmol cytochrome c oxidized/(min · mg protein)	Significance (p)	No./100 μm²	Significance (p)
17 (prenatal)	66.13 ± 3.03				83.58 ± 2.67			
18	68.34 ± 2.87	NS [a]			93.62 ± 2.45	0.02		
19	70.81 ± 4.31	NS			126.42 ± 4.89	0.001		
20 (birth)	86.61 ± 4.81	0.05	10.00 ± 0.33		192.64 ± 6.28	0.001	60.98 ± 2.35	
7 (postnatal)	182.25 ± 8.00	0.001	13.37 ± 0.84	0.001	168.29 ± 3.68	0.01	55.44 ± 3.37	NS
14	297.11 ± 6.40	0.001	27.07 ± 1.45	0.001	231.61 ± 9.47	0.001	64.49 ± 3.03	NS
21	593.63 ± 14.80	0.001	34.16 ± 2.22	0.01	190.60 ± 7.00	0.01	60.63 ± 4.22	NS
28	633.48 ± 14.78	NS	40.50 ± 3.72	NS	310.31 ± 3.70	0.001	72.23 ± 5.07	NS
84 (adult)	610.44 ± 9.66	NS	35.29 ± 1.90	NS	337.77 ± 12.86	NS	71.52 ± 2.84	NS

From: GOECKERMANN and VIGIL 1975.

The degree of significance between the value determined at one age and that at the preceding age is shown for catalase and cytochrome oxidase activities as well as the number of respective organelles. A correlation is seen between catalase specific activity and the number of peroxisomes in the proximal tubule between birth and 21 days.

[a] NS: not significant.

are seen on the endoplasmic reticulum membranes at various distances from the continuities between endoplasmic reticulum and microperoxisomes. The matrix material of the granules stains positively with the cytochemical DAB method for catalase. Microperoxisome membranes are provided with ribosomes in young fetuses (day 26–30 of gestation) more often than in older fetuses. Continuities between microperoxisomes and endoplasmic reticulum also seem to be more frequent in younger fetuses. All together, the findings of BLACK and BOGART (1973) are in good agreement with those of TSUKADA *et al.* (1968), who studied the genesis of liver peroxisomes.

Comparable observations were made by PIPAN and PŠENIČNIK (1975) in mouse kidney tubules and mouse enterocytes. These authors also observed connections between microperoxisomes and rough endoplasmic reticulum in embryonic cells. The organelles can first be identified on gestation day 14. As the rough endoplasmic reticulum becomes less prominent in older fetuses, microperoxisomes display connections to the smooth endoplasmic reticulum.

There are few microperoxisomes in mouse differentiating granular pneumocytes until birth. The number increases about threefold immediately before birth, and this increase continues during the first postnatal week (SCHNEEBERGER 1972 b); the ratio of peroxisomes to mitochondria also increases during

Table 20. *Correlation of ratio of mitochondria to peroxisomes in granular pneumocytes with stage of pre- and postnatal development (mouse)*

Fetal and neonatal age (days)	Crown–rump length range (cm)	Fetal wt. range (g)	Ratio of mitochondria to peroxisomes
Fetuses			
15	1.1–1.3	0.225–0.265	ND [a]
16	1.3–1.5	0.330–0.445	2.2
17	1.6–1.8	0.510–0.715	4.2
18	2.1–2.4	1.045–1.295	5.0
19	2.2–2.5	1.155–1.445	1.7
Neonates			
1	2.3–2.6	1.290–1.540	1.3
3	2.5–2.8	1.510–1.760	1.7
5	3.0–3.5	2.050–2.250	1.1
7	3.6–3.9	4.020–4.750	0.7
16		6.050–6.830	0.7

From: SCHNEEBERGER 1972 b.

[a] ND: not determined.

this time (Table 20). The increasing number of peroxisomes prior to birth indicates that elevated oxygen tension *per se* does not stimulate their formation, and it takes about 1 week for the high ratio of peroxisomes to mitochondria to be reached, long after the animals have begun to breath air. No connections between developing microperoxisomes and endoplasmic reticulum were noted during this study (SCHNEEBERGER 1972 b).

CALVERT and MENARD (1978) have determined the number of microbodies and the catalase activity in the epithelial lining of the small intestine in mouse fetuses, as well as newborn and young mice. A few microperoxisomes are first seen at 15 days of gestation. At 17 days the number of microperoxisomes increases significantly, reaching a maximum at 18 days. The number then remains constant during the first 4 weeks of life. The number of microperoxisomes is greatest in duodenal enterocytes, and it decreases toward the ileum. Biochemical measurements of catalase activity seem to parallel these results; except that catalase activity constantly increases during the perinatal period, whereas the number of microperoxisomes remains constant. The authors suggest that catalase accumulates in microperoxisomes during this time, up to the definite level. Membrane continuities between endoplasmic reticulum and microperoxisomes are frequently seen. CALVERT and MENARD (1978) assume that microperoxisomes of enterocytes are involved in lipid metabolism.

MOORIDAN and CUTLER (1978) studied the frequency distribution of microperoxisomes in the developing submandibular gland of the rat. Fetuses at day 15 of gestation and older, newborn animals and animals up to 12 weeks old were used. About 9 microperoxisomes are present in early secretory and striated duct cells. In secretory cells the number of microperoxisomes decreases (only 3.5 are found in mature acinar cells), whereas in striated duct cells the number of microperoxisomes continually increases (up to 40 per cell). The number of microperoxisomes in the convoluted granular tubule cells varies during the course of differentiation. The authors conclude that the frequency of microperoxisomes changes as a function of the particular path of differentiation.

In conclusion, the number of peroxisomes and/or microperoxisomes generally seems to increase during the late gestation period and during early postnatal life. It has been suggested that this numerical increase is related to the onset of specific cellular functions. In contrast, microperoxisomes in the rat central nervous system are relatively numerous during the first 2 postnatal weeks and decrease in number thereafter (ARNOLD and HOLTZMAN 1978). Along with the observation of numerous microperoxisomes in oligodendrocytes, these results indicate that microperoxisomes are of metabolic significance for myelination. In view of this possibility, the observations are not contradictory but agree with the view mentioned above.

PŠENIČNIK and PIPAN (1977) reported that treatment with nafenopin causes the proliferation of microperoxisomes in mouse enterocytes, analogous to the proliferative effect of nafenopin on hepatic peroxisomes. They observed a significant increase in microperoxisome volume density, surface density, and numerical density. Unfortunately, they did not report any morphological details, such as the presence or absence of connections between microperoxisomes and endoplasmic reticulum. Peroxisomal profiles are not convincingly demonstrated in the figures presented by PŠENIČNIK and PIPAN (1977) and their observation requires further investigation.

More recently CALVERT et al. (1979) reported on proliferation of microperoxisomes and increased catalase activity in the epithelial lining of fetal small intestine after subcutaneous injection of clofibrate into pregnant mice. The effect is dose-dependent up to a certain limit, and is only seen when the drug is injected not before day 18 of gestation (*i.e.,* 24 hours before delivery).

III.F. Φ Bodies

A unique, new type of cell organelle has recently been described, termed a Φ body. The catalase-positive spindle-shaped particles are limited by a single membrane and contain a crystalloid rod situated in the long axis of the spindle. The long axis of the organelle is 0.5–5 μm long. Thus the particles are readily identifiable under the light microscope upon cytochemical DAB-

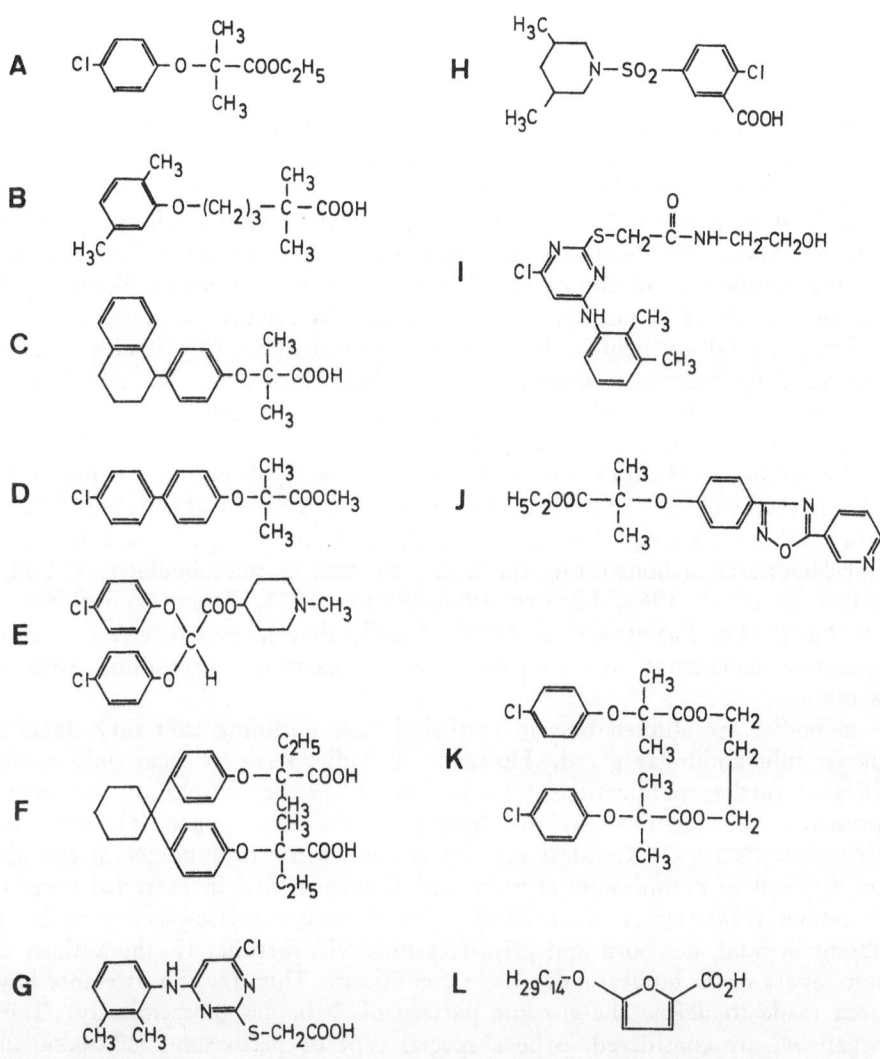

Fig. 33 A–L. Chemical structures of peroxisome proliferators. *A* Ethyl-α-*p*-chlorophenoxy-isobutyrate (clofibrate, CPIB). *B* 5-(2,5-dimethylphenoxy)-2,2-dimethyl-pentanoic acid (gem-fibrozil). *C* 2-Methyl-2-[*p*-(1,2,3,4-tetrahydro-1-naphthyl)phenoxy]propionic acid (nafenopin). *D* Methyl-2-[4-(*p*-chlorophenyl)phenoxy]2-methyl propionate (methyl clofenapate). *E* 1-Me-thyl-4-piperidyl-bis(*p*-chlorophenoxy)acetate (SaH-42,348). *F* 1,1-bis[4′-(1″-carboxy-1″-me-thylpropoxy)phenyl]cyclohexane (S-8527). *G* [4-Chloro-6-(2,3-xylidino)-2-pyrimidinylthio]-acetic acid (Wy-14,643). *H* 2-Chloro-5-(3,5-dimethylpiperidinosulfonyl)benzoic acid (tibric acid, CP-18,524). *I* 4-Chloro-6-(2,2 xylidino)-2-pyrimidinylthio (*N*-β-hydroxyethyl)-acet-amide (BR-931). *J* 3-[4-(1-Ethoxycarbonyl-1-methylethoxy)phenyl]-5-(3-pyridyl)-1,2,4-oxa-diazole (AT-308). *K* 1,3-Propyl-bis(2-*p*-chlorophenoxy)2-methyl propionate (simfibrate, CLY-503). *L* 5-Tetradecyloxy-2-furan-carboxylic acid (RMI 14,514).

staining for catalase (HANKER and ROMANOVICZ 1977). Crystalline rods seen free in the cytoplasm have been interpreted to be extruded from Φ bodies.

Under the electron microscope several transitional profiles between micro-bodies with crystalline cores and Φ bodies (as described above) have been observed. These profiles suggest a gradual transformation of microbodies to Φ bodies. During this transformation the crystalline core grows and becomes longer (the microbody transforms from spherical to elliptical, and thereafter to a Φ-like profile); finally, the core is extruded into the cytoplasm (HANKER and ROMANOVICZ 1977). Examination under the electron microscope shows that the crystalline inclusions of Φ bodies as well as the crystalline rods free in the cytoplasm are composed of parallel tubules 13 nm in diameter; the tubules are about 16 nm apart and are linked by lateral processes. In mouse salivary ductal epithelium, HANKER and ROMANOVICZ (1977) observed iso-citrate dehydrogenase activity in Φ bodies. However, the cytochemical reaction for isocitrate dehydrogenase activity is significantly weaker than the staining for peroxidatic activities of catalase.

According to HANKER and ROMANOVICZ (1977), Φ bodies are found "in certain epithelial cells of mice, clofibrate-fed rats, and in leukemic leukocytes". The authors interpret fine-structural descriptions of specialized forms of microbodies as demonstrating the transformation of microbodies to Φ bodies (HRUBAN et al. 1966, LANGER 1968, REDDY 1973, REDDY and SVOBODA 1972 a, 1973 a, TANDLER et al. 1969). Finally they suggest that Auer rods of leukemic leukocytes are conglomerates of extruded crystalline rods of Φ bodies.

Φ bodies are studied best in epithelial cells outlining excretory ducts of mouse submandibular gland. However, Φ bodies seem to occur only rarely. In two further publications (HANKER et al. 1977 a, b), 13 of the figures presented are identical to the figures in the first paper (HANKER and ROMANOVICZ 1977). Catalase activity is not limited to Φ bodies; it can also be detected in cytoplasmic crystals and filaments, i.e., in extruded cores of Φ bodies (HANKER et al. 1977 b). Cytoplasmic catalase-positive rods are absent in fetal, newborn and germ-free mice. Unfortunately, the authors do not report on Φ bodies under these conditions. Thus far, no attempts have been made to define the enzyme pattern of Φ bodies biochemically. These organelles are considered to be a special type of peroxisome (HANKER and ROMANOVICZ 1977).

Recently, Φ bodies were also identified in human granulocyte precursors in cases of acute myelogenous leukemia, and HANKER et al. (1978) claim that this finding is pathognomonic for this disease. They also give method-ological suggestions for optimum fixation, staining, and counterstaining of marrow and peripheral blood films.

III.G. Experimental Modification of Peroxisomes

III.G.1. Hypolipidemic Agents

Experiments using these agents to influence biochemical and morphological parameters of peroxisomes have provided valuable evidence as to their mor-

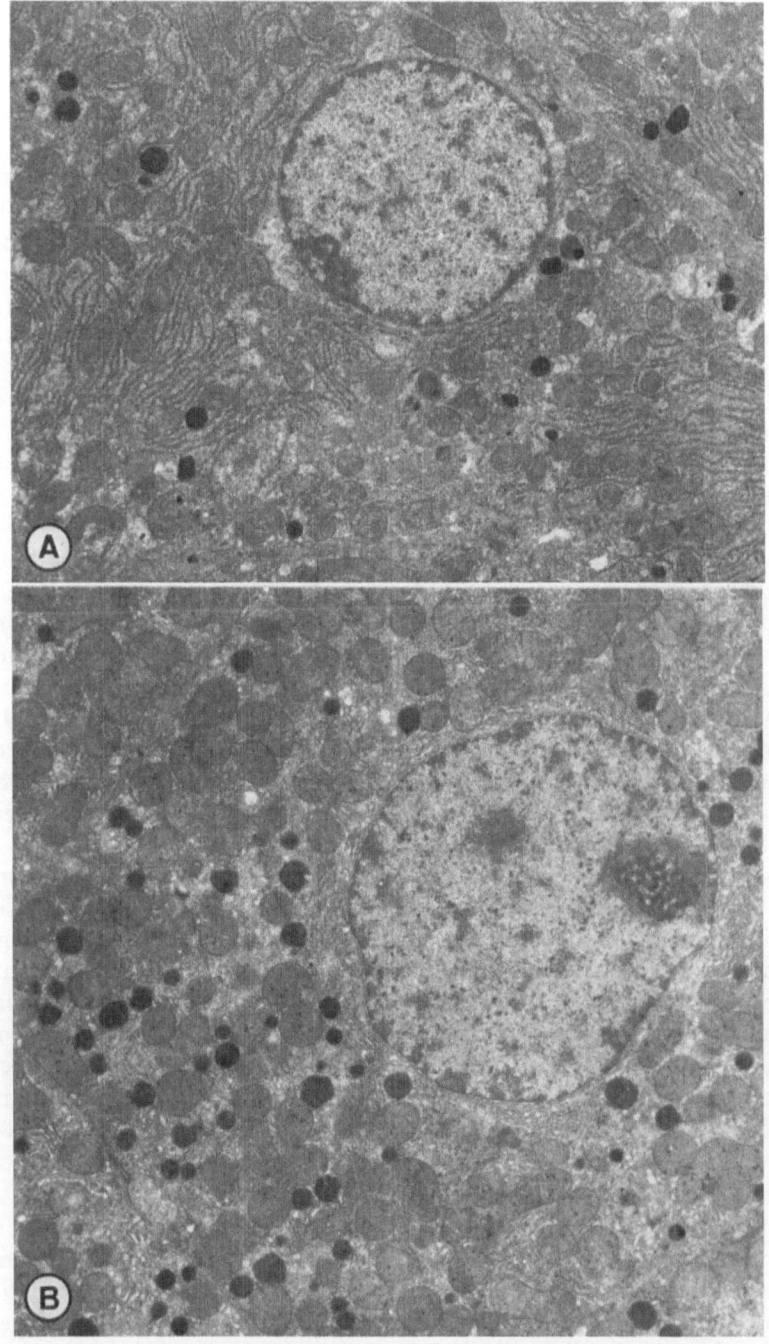

Fig. 34 A, B. Representative cytoplasmic areas of rat hepatocytes. *A* Normal. *B* After clofibrate treatment. Catalase staining with alkaline DAB medium. ×5,400. From: GOLDEN-BERG *et al.* (1976): Histochemistry 46, 192.

phology, enzymatic equipment, and origin, although the exact mechanism of these drugs has not as yet been clarified.

III.G.1.a. Clofibrate (Ethyl-α-p-chlorophenoxyisobutyrate, CPIB)

Clofibrate (Fig. 33 A), a compound that reduces serum cholesterol and triglyceride levels in various animal species and humans (ANTHONY et al. 1978, BERKOWITZ 1965, BEST and DUNCAN 1964, HAVEL and KANE 1973, HELLMAN et al. 1963, HOWARD et al. 1963, OLIVER 1963, THORP and WARING 1962), was the first hypolipidemic agent noted to have an effect on peroxisomes. An early publication on this subject by THORP and WARING (1962) described renal hypertrophy in chickens treated with clofibrate, and PAGET (1963) reported an increased number of mitochondria and dense bodies in rodent hepatocytes after clofibrate application.

In 1965 HESS et al. reported for the first time on effects of clofibrate treatment on hepatic peroxisomes. Administration of clofibrate (0.25–0.3% in the diet) to male rats caused hepatomegaly, an increase in smooth endoplasmic reticulum, and an increase in the number of peroxisomes. Subsequently, similar findings were published (ANTHONY et al. 1978, AVOY et al. 1965, AZARNOFF and SVOBODA 1966, AZARNOFF and TUCKER 1965, AZARNOFF et al. 1965, GOLDENBERG et al. 1976, HRUBAN et al. 1974 a, b, KOLDE et al. 1976, LEGG and WOOD 1970 a, PLATT and THORP 1966, REDDY 1973, SVOBODA and AZARNOFF 1966, 1971, SVOBODA et al. 1967) (Fig. 34). Clofibrate-induced alterations of the fine structure of hepatocytes in male rats and mice can be summarized as follows: After an initial reduction (RIEDE and ROHR 1974), the number of microbodies (peroxisomes) increases markedly; smooth endoplasmic reticulum proliferates; many microbodies lack a characteristic nucleoid; their matrix becomes flocculent; and they are frequently seen to be continuous with endoplasmic reticulum.

The proximity of microbodies and endoplasmic reticulum has been emphasized. Microbodies are often closely enshrouded with cisternae of the endoplasmic reticulum, the membranes adjacent to microbodies usually being free of ribosomes (LEGG and WOOD 1970 a). In liver tissue of untreated animals microbodies are randomly distributed in regions with rough endoplasmic reticulum; in early periods of clofibrate treatment the organelles are primarily seen in the boundary region between rough and smooth endoplasmic reticulum or near glycogen. In later periods of treatment they are predominantly observed in areas of smooth endoplasmic reticulum or glycogen (LEGG and WOOD 1970 a). An increased number of mitochondria (ANTHONY et al. 1978, KURUP et al. 1970) and morphological changes in mitochondria have been noted (LEE et al. 1974); pleomorphism (SVOBODA et al. 1967, KOLDE et al. 1976), membrane-limited matrical inclusions (SVOBODA et al. 1967), and an increased number of intramitochondrial granules (LEGG and WOOD 1970 a) have been reported. Clofibrate also causes increase in the number of peroxisomes in kidney epithelial cells (AZARNOFF et al. 1965) and in the number of microperoxisomes in intestinal epithelial cells of male rats (SVOBODA 1976).

The increase in the number of liver cell microbodies seems to reach a plateau after 2 weeks of treatment (SVOBODA and AZARNOFF 1971), and no further increase has been observed up to 30 days despite continuation of the drug (SVOBODA *et al.* 1967). By 30 days of treatment the shape of many micro-bodies is bizarre and matrical inclusions with linear striations or lamellae are observed (SVOBODA and AZARNOFF 1966, SVOBODA *et al.* 1967). Membrane connections with the smooth endoplasmic reticulum are still present (SVOBODA and AZARNOFF 1966, SVOBODA *et al.* 1967).

Table 21. *Effects of clofibrate on number and size of microbodies and on catalase activity*

Variables	Normal liver		Hepatoma 9618 A
	Adult rat	Newborn rat	
Number of microbodies/ 1,000 μm² cytoplasm			
Control	68.5 ± 5.46 (7) [a]	34.7 ± 3.98 (5)	40.1 ± 4.00 (7)
Treated	246.3 ± 22.0 [b] (8)	58.6 ± 2.42 [b] (5)	57.3 ± 6.83 (8)
Average diameter of microbodies (nm)			
Control	547 ± 11.7 (5)	522 ± 11.8 (5)	506 ± 7.9 (7)
Treated	692 ± 15.1 [b] (7)	670 ± 9.5 [b] (5)	560 ± 17.5 [c] (8)
Catalase activity [d]			
Control	114.5 ± 4.94 (10)	28.5 ± 0.51 (7)	82.4 ± 2.91 (13)
Treated	184.7 ± 5.53 [b] (14)	71.2 ± 2.43 [e] (9)	95.2 ± 3.00 [c] (13)

From: TSUKADA *et al.* 1975 b.

[a] Values are means ± SE. Figures in parentheses are numbers of examinations.

[b] $p < 0.001$.

[c] $p < 0.01$.

[d] Catalase activity is expressed in terms of $k/100$ mg dry tissue, where k is the mean of the observed first-order velocity constant at 1, 2, and 3 minutes.

[e] $p < 0.05$.

Linear striations and lamellae, termed "matrical plates", were first described by MAUNSBACH (1966) in renal proximal tubules of normal rats and were later found as characteristic peroxisomal inclusions after treatment with acetylsalicylic acid, clofibrate, or dimethrin (HRUBAN *et al.* 1974 a). Intra-peroxisomal tubular structures (characteristic after dimethrin treatment) are induced by clofibrate only in hypophysectomized rats, in rats bearing liver tumors (REDDY and SVOBODA 1973 a), and in hyperplastic preneoplastic hepatocytes (TSUKADA *et al.* 1975 a). Matrical fibrils and tubules have been considered to represent regularly arranged enzyme molecules (HRUBAN *et al.* 1974 a, REDDY and SVOBODA 1973 a). DAB staining of microbodies in animals treated with acetylsalicylic acid, clofibrate, and dimethrin is stronger in matrix regions between the fibrils and tubular walls (HRUBAN *et al.* 1974 a). Therefore, it is unlikely that the fibrils and tubules are formed by catalase.

Other enzymes (*e.g.*, isocitrate dehydrogenase) have also been discussed in this context (HRUBAN *et al.* 1974 a).

Morphometrical analysis of clofibrate-induced ultrastructural alterations of hepatocytes (ANTHONY *et al.* 1978, KOLDE *et al.* 1976, MOODY and REDDY 1975, STÄUBLI and HESS 1966, TSUKADA *et al.* 1975 b) revealed a significant increase in the volume of individual hepatocytes (ANTHONY *et al.* 1978, KOLDE *et al.* 1976), and in the numerical density and volume density of microbodies (ANTHONY *et al.* 1978, KOLDE *et al.* 1976, STÄUBLI and HESS 1966). The average microbody diameter increases (TSUKADA *et al.* 1975 b) and the number of microbodies which contain nucleoids is markedly decreased (TSUKADA *et al.* 1975 b; Table 21). Furthermore, the surface density of the smooth endoplasmic reticulum is significantly enhanced, whereas the surface of the rough endoplasmic reticulum is decreased. Increased smooth endoplasmic reticulum has been reported in hepatocytes of all animals treated with any drug that induces peroxisome proliferation.

Several facts (for instance, continuities between peroxisomes and endoplasmic reticulum or similarities between the membranes of these compartments) indicate a relationship between the proliferation of smooth endoplasmic reticulum and peroxisomes. Microsomal cytochrome P-450 concentration, NADPH-cytochrome c reductase, and the rate of ethylmorphine *N*-demethylation are all increased after clofibrate treatment (ANTHONY *et al.* 1978, KANEKO *et al.* 1969, LEWIS *et al.* 1974). A possible involvement of the smooth endoplasmic reticulum in the mechanisms responsible for the hypocholesterolemic effect of clofibrate has been suggested (ANTHONY *et al.* 1978). On the other hand, a nonspecific pharmacological response to clofibrate must also be considered.

KOLDE *et al.* (1976) found a slight increase in the numerical density of mitochondria after 8 days of clofibrate treatment, and a conspicuous decrease after 21 days; the volume density of mitochondria did not change during this time (KOLDE *et al.* 1976). ANTHONY *et al.* (1978) reported a significant increase in mitochondrial volume density after 7 days of clofibrate treatment. This finding is of particular interest with respect to the clofibrate-induced increase in carnitine acetyltransferase (KÄHÖNEN 1976, MARKWELL *et al.* 1977) and fatty acid oxidation (MANNAERTS *et al.* 1978).

After withdrawal of clofibrate the number of microbodies in hepatocytes and kidney epithelia is reduced to normal within 3 weeks (SVOBODA and AZARNOFF 1971, SVOBODA *et al.* 1967). During the first weeks after withdrawal of the drug the electron opacity of the matrix is markedly decreased in some of the microbodies, whereas other organelles appear normal. One or more interruptions of the limiting membrane concomitantly occur (SVOBODA and REDDY 1972). Some of the organelles are incorporated into autophagolysosomes (SVOBODA *et al.* 1967), but the mode of removal or destruction of the major part is still uncertain (SVOBODA and AZARNOFF 1971) (see Section I.B.5.).

Characteristic changes in biochemical parameters attend the appearance of hepatomegaly and morphological alterations. Microbody proliferation is paralleled by a significant increase in liver catalase activity (CHIGA *et al.*

1971, HAYASHI et al. 1975, KOLDE et al. 1976, KRISHNAKANTHA and KURUP 1972, REDDY et al. 1971, SVOBODA and AZARNOFF 1966, SVOBODA et al. 1967, TSUKADA et al. 1975 b). Generally, 3 types of peroxisomal enzyme response to clofibrate may be distinguished (GOLDENBERG et al. 1976):

1. The synthesis of the first group of enzymes (e.g., urate oxidase) is not enhanced by the proliferative stimulus of clofibrate; these enzymes exhibit lower specific activities in the peroxisomal fraction.

Table 22. *Effect of hypolipidemic drugs on hepatic carnitine acyltransferase and acyl-CoA hydrolase activities in male Swiss-Webster mice*

Treatment	Carnitine acetyltransferases (units/mg protein)			Acyl-CoA hydrolases (units/mg protein)		
	Acetyl-CoA	Octanoyl-CoA	Palmitoyl-CoA	Acetyl-CoA	Octanoyl-CoA	Palmitoyl-CoA
Control	4.0 ± 0.5 [a]	17.0 ± 1.4	8.5 ± 1.4	16.4 ± 0.9	38.0 ± 1.3	37.8 ± 1.4
0.25% clofibrate (28 days)	27.6 ± 3.5 [b]	75.8 ± 7.2 [b]	16.8 ± 2.8	18.2 ± 2.2	39.8 ± 0.9	43.9 ± 1.1
0.125% nafenopin (28 days)	105.7 ± 10.2 [b]	177.6 ± 13.0 [b]	37.6 ± 3.0 [b]	34.7 ± 2.5 [b]	73.2 ± 3.6 [b]	77.6 ± 3.4 [b]
0.125% tibric acid (28 days)	58.5 ± 2.9 [b]	164.2 ± 26.0 [b]	27.7 ± 2.0 [b]	23.1 ± 0.1 [b]	58.1 ± 6.2 [c]	52.6 ± 4.1 [c]
0.125% Wy-14,643 (28 days)	70.4 ± 3.0 [b]	196.6 ± 22.1 [b]	33.9 ± 4.7 [c]	25.7 ± 2.0 [b]	65.3 ± 3.0 [b]	73.4 ± 11.1 [c]

From: MOODY and REDDY 1978 a.

[a] Values are means ± SE.

[b] $p < 0.001$.

[c] $p < 0.01$.

2. The synthesis of the second group of enzymes (e.g., catalase) is proportionally increased as the formation of total peroxisomal protein is increased. Their specific activities remain constant; the rise of activity per gram liver reflects the increased volume of peroxisomes.

3. The synthesis of the third group of enzymes (e.g., carnitine acetyltransferase) is strongly and selectively induced by clofibrate.

Clofibrate application markedly increases liver carnitine acetyltransferase activity in wild-type and acatalasemic mice (MOODY and REDDY 1974) and in male rats (GOLDENBERG et al. 1976, KÄHÖNEN 1976, KÄHÖNEN and YLIKARI 1974, MARKWELL et al. 1977, SOLBERG et al. 1972). Recently, MOODY and REDDY (1978 a) showed that in mice treated with clofibrate or other hypolipidemic agents the most dramatic increase occurs in the hepatic short-chain (8–26-fold) and medium-chain (4–11-fold) carnitine acyltransferases, whereas the rise in long-chain carnitine acyltransferases is only 2–4-fold (Table 22).

The activities of acyl-CoA-hydrolases are increased slightly; the extent of increase is independent of the substrate used (MOODY and REDDY 1978 a).

Furthermore, carnitine acetyltransferase and carnitine octanoyltransferase are the first enzymes whose peroxisomal specific activities have been reported to be enhanced by clofibrate (GOLDENBERG et al. 1976, KÄHÖNEN 1976, MARKWELL et al. 1977). Unlike the stimulation of mitochondrial glycerol-3-phosphate dehydrogenase (WESTERFIELD et al. 1968), the increase in carnitine acetyltransferase is not mediated by thyroxine (KÄHÖNEN and YLIKARI 1974). Peroxisomal as well as mitochondrial and microsomal activities of carnitine acetyltransferase are enhanced (KÄHÖNEN 1976, MARKWELL et al. 1977); however, the increase is significantly higher in the mitochondrial fraction than in the two other components. This causes a conspicuous change in the distribution of carnitine acetyltransferase within the hepatic cell after clofibrate treatment (KÄHÖNEN 1976, MARKWELL et al. 1977) (Table 23).

Phenobarbital, which is known to induce the proliferation of smooth endoplasmatic reticulum, does not influence the specific activities of peroxisomal, mitochondrial, and microsomal carnitine acetyltransferase (MARKWELL et al. 1977) (Table 24).

LAZAROW and DE DUVE (1976) demonstrated that rat liver peroxisomes possess an activity for the oxidation of palmitoyl-CoA. The peroxisomal activity for oxidation of palmitoyl-CoA is greatly increased in animals treated with clofibrate, tibric acid, and Wy 14,643 (LAZAROW 1977). LAZAROW (1978) also demonstrated that in clofibrate-fed animals peroxisomal fatty acid oxidation is a β-oxidative process, suggesting that the peroxisomes represent a major site for the β-oxidation of long-chain fatty acids. PALTAUF and MAGNET (1979) showed induction of fatty acid oxidation by magnesium-4-chlorophenoxyisobutyrate in peroxisomal fraction of rat liver. The stimulation was found to be about twice that induced by clofibrate. Further studies (MANNAERTS et al. 1978, 1979) provided evidence that the mitochondrial pathway represents the major route of oleate oxidation in the liver of clofibrate-treated animals. It appears that the increase in β-oxidation also traces back to an increased biosynthesis of CoA induced by clofibrate (SAVOLAINEN et al. 1977, SKREDE and HALVORSEN 1979).

CHRISTIANSEN (1978) observed stimulated oxidation and esterification, and simultaneously, an enhanced capacity for chain shortening in isolated hepatocytes of clofibrate-treated rats. It is suggested that the oxidation and esterification of very long chain fatty acids are limited by the capacity of an extramitochondrial chain-shortening system, which most probably is localized in peroxisomes (CHRISTIANSEN 1978).

Furthermore, clofibrate and other hypolipidemic agents cause a marked increase of an 80,000-dalton polypeptide in subcellular liver fractions (REDDY and KUMAR 1977). Enhancement of this polypeptide is also observed in the absence of catalase synthesis, which indicates that it does not represent a catalase subunit. No increase is noted in postmicrosomal supernatants. It is not solubilized by 30 mM potassium chloride or 20 mM EDTA treatment and therefore, most probably, is membrane bound (REDDY and KUMAR 1977). A comparable finding was reported recently by HÜTTINGER et al. (1979).

Table 23. *Subcellular distribution of carnitine acetyltransferase in liver of normal and clofibrate-treated rats*

	Normal	Clofibrate
Peroxisomes	0.91 ± 0.22 (47.8)	5.8 ± 1.2 [a] (9.4)
Mitochondria	0.69 ± 0.37 (36.2)	48.7 ± 10.1 [a] (79.3)
Membranous fraction	0.31 ± 0.11 (16.0)	4.4 ± 1.7 [a] (7.1)
Soluble fraction	? [b]	2.6 ± 1.7 (4.2)
Total	1.91 ± 0.55	61.4 ± 9.3 [a]

From: KÄHÖNEN 1976.

Results are given as nmol/(min · mg total protein in gradient). Values are means ± SE; five animals in clofibrate group and seven in control group. Values in parentheses are percentages of total.

[a] Significant values: $p < 0.001$.

[b] Activity could not be measured because of the high acetyl-CoA hydrolase activity.

Table 24. *Effect of clofibrate and phenobarbital on the activity of liver enzymes*

	Control	Clofibrate	Phenobarbital
Total activity [μmol/(min · g liver)]			
Catalase	57,000 ± 3,000	122,000 ± 23,000 [a]	61,000
Urate oxidase	0.351 ± 0.068	0.480 ± 0.092	
NADPH-cytochrome c reductase	1.508 ± 0.167	1.979 ± 0.449	3.1
Carnitine acetyltransferase	0.420 ± 0.047	5.405 ± 1.225 [b]	
Carnitine octanoyltransferase	0.699 ± 0.058	3.362 ± 0.818 [a]	
Carnitine palmitoyltransferase	0.195 ± 0.021	1.018 ± 0.275 [a]	
Specific activity [nmol/(min · mg protein)]			
Peroxisomal peak			
Catalase	(5.93 ± 2.61) × 10⁶	(5.04 ± 0.46) × 10⁶	7.8 × 10⁶
Urate oxidase	6.3 ± 18.7	43.4 ± 7.3	63.5
Carnitine acetyltransferase	14.0 ± 1.0	24.6 ± 3.8 [a]	14.0
Carnitine octanoyltransferase	11.1 ± 1.0	37.4 ± 7.7 [a]	12.4
Mitochondrial peak			
Carnitine acetyltransferase	14.1 ± 0.6	137.1 ± 46.2 [a]	15.0
Carnitine octanoyltransferase	26.8 ± 2.9	84.9 ± 9.2 [a]	28.0
Carnitine palmitoyltransferase	12.9 ± 0.9	36.3 ± 4.7 [b]	12.6
Microsomal peak			
NADPH-cytochrome c reductase	109 ± 8	160 ± 22	232 [a]
Carnitine acetyltransferase	7.0 ± 0.3	10.6 ± 1.0 [a]	6.4
Carnitine octanoyltransferase	8.0 ± 0.5	20.9 ± 3.9 [a]	7.6

From: MARKWELL et al. 1977.

Mean values ± SD are from three experiments for control and clofibrate, and one experiment for phenobarbital.

[a] Statistically different from control, $p < 0.01$.

[b] Statistically different from control, $p < 0.001$.

After clofibrate treatment of rats two proteins of the peroxisomal (but not of the microsomal) membrane with molecular weights of 40,000 and 80,000 dalton are largely increased.

Microbody-proliferating effects of clofibrate and other hypolipidemic agents repeatedly have been used to elucidate the question as to whether microbodies exist as individual entities or represent specialized parts of the endoplasmic reticulum (see Section I.B.5.). In their study on the synthesis and turnover of rat liver microbodies, POOLE et al. (1970) showed that enzymes are homogeneously distributed in peroxisomes of different classes of size. Based on these studies, the authors discussed two alternatives: a) that micro-bodies exist as individual particles; or b) that microbodies are connected with one another and microbody proteins form a common pool in which they exchange material. Morphologically demonstrable continuities of microbodies with the endoplasmic reticulum and continuities between two or more micro-bodies, along with morphological irregularities after clofibrate application, are compartible with this "pool" hypothesis. Irregularities in size and shape after treatment with clofibrate or other agents seem to reflect varying quantities of microbody protein that accumulate in endoplasmic reticulum channels (REDDY and SVOBODA 1971 a, 1973 b). Additional support for the pool hypothesis comes from the observations that the degradation of microbodies after with-drawal of clofibrate occurs without a significant increase in autophagy (SVOBODA and REDDY 1972).

Contrary to these observations, LEGG and WOOD (1970 a) could not find any continuities between microbodies and endoplasmic reticulum in rat liver after clofibrate treatment. These authors refer to small particles near normal size microbodies, that in some instances are connected with microbodies. Such situations are especially frequent during short periods of clofibrate applica-tion. The small particles positively react when incubated for catalase activity. Similar structures were described by GOLDENBERG et al. (1976). These groups of small particles near normal size microbodies have been interpreted to reflect a process of fragmentation or budding off from preexisting microbodies (LEGG and WOOD 1970 a). It has been considered as an alternative mechanism of proliferation.

In a series of studies, attempts were made to elucidate the connection be-tween microbody proliferation and the hypolipidemic effect of clofibrate. Up to a dietary level of 0.25% microbody proliferation and increased catalase activity are evoked only in male animals (with the exception of acatalasemic mice), although the hypolipidemic effect is also present in females. Therefore, the influence of various hormones is of particular interest. SVOBODA et al. (1969) and SVOBODA and AZARNOFF (1971) demonstrated that the microbody-catalase response significantly depends on male sex hormone, but is largely independent of thyroid and adrenal hormones. Thyreodectomy abolishes the hypolipidemic effect of clofibrate but does not influence microbody proliferation (AZARNOFF and SVOBODA 1969). Comparison of the clofibrate effect in young adult and in retired breeder rats (representing a model for hyperlipidemia; ANTHONY et al. 1978) showed that microbody proliferation is more pronounced in young adult rats although the hypocholesterolemic effect was more marked in retired breeder rats. These results deny a positive correlation between the hypocholesterolemic effect and the extent of microbody proliferation. Similar observations have been made in experiments with ubiquinone suggesting that the serum lipid-lowering effect and peroxisome proliferation are two mechanisms independent of each other (AZARNOFF and SVOBODA 1969, SVOBODA and AZARNOFF 1971, SVOBODA et al. 1969).

Clofibrate causes increased concentrations of ubiquinone in rat liver by blocking the catabolism of ubiquinone (LAKSHMANAN et al. 1968). As ubiquinone exhibits effects on hepatic cholesterol synthesis and serum cholesterol levels similar to those of clofibrate, the clofibrate effect was considered to be due to ubiquinone (KRISHNAIAH et al. 1967); however, ubiquinone administered to male rats for 3 weeks has no effect on the number of microbodies and the catalase levels (REDDY and SVOBODA 1971 b). Therefore, ubiquinone per se does not appear to be involved in the microbody-catalase response to clofibrate (REDDY and SVOBODA 1971 b).

Because of the coincidence of microbody proliferation and increased catalase activity, microbody formation has been considered to be an expression of increased catalase levels (SVOBODA and AZARNOFF 1966). However, comparison of morphometric and biochemical results shows that catalase activity per volume unit of microbodies decreases continually during treatment with clofibrate (KOLDE et al. 1976). Microbody proliferation also occurs when liver catalase activity is negligible, as has been shown in experiments with acatalasemic mice (REDDY et al. 1969 a, b) and with the catalase inhibitors AIA and aminotriazole (LEGG and WOOD 1970 b, REDDY et al. 1970, 1977 a). These observations strongly suggest that synthesis of protein(s) other than catalase is induced by clofibrate (REDDY 1973). This suggestion is supported by recent results concerning enhanced carnitine acyltransferase activity (GOLDENBERG et al. 1976, KÄHÖNEN 1976, MARKWELL et al. 1977, MOODY and REDDY 1974, 1978 a) and increased fatty acid oxidation (LAZAROW 1977, 1978, MANNAERTS et al. 1978) after clofibrate treatment. Thus, the most prominent increase in peroxisomal enzymes is related to activities that are involved in lipid metabolism. Contrary to former suggestions, these findings indicate that the hypolipidemic effect and proliferation of hepatic peroxisomes are related events (MOODY and REDDY 1978 a).

REDDY and KUMAR (1979) studied the effect of clofibrate at dietary concentrations of 0.25%, 0.5%, 1%, and 2% on hepatic peroxisomes of female F-344 rats. They demonstrated that the previously observed sex difference in the inducibility of enzymes associated with peroxisomes is obviated by increased doses. Although at the 0.25% dose level clofibrate has only a minimal effect on hepatic peroxisomes and catalase in female rats, the increase in hepatic catalase, carnitine acetyltransferase, peroxisome number, and an 80,000-dalton polypeptide in postnuclear liver pellets in female rats treated with higher doses of clofibrate are similar to the changes observed in male rats fed clofibrate at the 0.25% or 1% level.

Gemfibrozil [5-(2,5-dimethylphenoxy)-2,2-dimethyl-pentanoic acid] is a newly developed hypolipidemic agent (Fig. 33 B). Its effect on the ultrastructure of hepatic microbodies of rats has been described by FUKUDA and co-workers (1978). Gemfibrozil, similar to clofibrate, causes a marked increase in relative liver weight and in liver catalase activity and a reduction of serum triglyceride levels. The number of hepatic microbodies is increased following gemfibrozil treatment, and microbodies frequently contain characteristic tubular substructures with an inner diameter of approximately 65 nm and an outer diameter of 100 nm. Similar substructures have been observed in hepatic microbodies following the application of dimethrin (see Section III.G.5.) and after the administration of clofibrate under several experimental conditions (see Section III.G.1.a.).

KÄHÖNEN and YLIKAHRI (1979) described the effects of clofibrate and gemfibrozil on the activities of mitochondrial carnitine acyltransferases in rat liver. Gemfibrozil does not enhance the acylcarnitine oxidation in mitochondria, as does clofibrate; the effects of gemfibrozil on mitochondrial α-glycerophosphate dehydrogenase and carnitine acyltransferases are not as marked as those of clofibrate. In contrast to other results (see Section III.G.1.a.), these data do not support a connection between the hypolipidemic effect and the effect on β-oxidation.

III.G.1.b. Methyl Clofenapate (Methyl-2-[4-(p-chlorophenyl) phenoxy] 2-methyl propionate; ICI 55, 695)

Methyl clofenapate (Fig. 33 D) is a hypolipidemic drug closely related to clofibrate (CRAIG 1972, THORP 1970). It also resembles this agent in that it increases hepatic catalase, glycerophosphate dehydrogenase, and carnitine acetyltransferase activities and induces microbody proliferation in hepatocytes of rats and mice and in kidney epithelial cells of mice (KRISHNAKANTHA and KURUP 1972, MOODY and REDDY 1974, REDDY 1974). Methyl clofenapate is 5–16 times more effective than clofibrate in stimulating microbody proliferation and in envincing hypolipidemia (REDDY 1974). Microbodies of treated animals are enlarged and continuities with the endoplasmic reticulum are frequently seen.

III.G.1.c. Nafenopin (2-Methyl-2-[p-(1,2,3,4-tetrahydro-1-naphthyl)- phenoxy] propionic acid; SU-13,437)

Nafenopin (Fig. 33 C), a hypolipidemic agent structurally related to clofibrate (BERKOWITZ 1969, BEST and DUNCAN 1968, 1969, 1970, HESS and BENCZE 1968), is approximately 5 times as active as clofibrate (BEST and DUNCAN 1970, HESS and BENCZE 1968). Its effects on liver and intestinal peroxisomes (PŠENIČNIK and PIPAN 1977) and peroxisomal enzymes (MOODY and REDDY 1974) are similar to those of clofibrate and methyl clofenapate (LEIGHTON et al. 1975, REDDY et al. 1973 b, 1974 a, TUCHWEBER et al. 1976) in either sex (REDDY et al. 1974 a). Administration at dietary levels of 0.125 or 0.25% to male and female rats or mice significantly increases the number of liver peroxisomes, causes proliferation of smooth endoplasmic reticulum, and increases liver catalase activity. Variations in the size and shape of peroxisomes, altered nucleoids, and continuities with smooth endoplasmic reticulum are observed (REDDY et al. 1974 a). Morphological changes are apparent after 1.5 days of treatment; they become more conspicuous 2 and 5 days later.

Nucleoids are absent after 3.5 days of treatment (LEIGHTON et al. 1975). Prolonged treatment (up to 8 weeks) causes a sustained increase in the number of peroxisomes combined with severe morphological changes (REDDY et al. 1974 a). Continuities between peroxisomes and between peroxisomes and smooth endoplasmic reticulum are clearly seen (REDDY et al. 1974 a). The DAB method for catalase staining frequently reveals varying reaction intensities in proliferated microbodies (LEIGHTON et al. 1975).

Morphological and biochemical changes are reversed within 14 days after withdrawal of the drug. The electron density of the peroxisomal matrix decreases markedly during this time; nucleoids seem to be present in dilated channels of the endoplasmic reticulum and free in the hyaloplasm. There is no evidence for increased cellular autophagy. Nucleoids free in endoplasmic reticulum channels devoid of matrix material have been interpreted to indicate a gradual depletion of matrix proteins as a mechanism to reduce the peroxisome compartment (REDDY et al. 1974 a).

LEIGHTON et al. (1975) have shown that proliferated peroxisomes behave differently than normal peroxisomes. The total activity (per gram liver) of each of the oxidases gradually decreases, whereas catalase activity increases. This change results from a fivefold increase in catalase in the supernatant, the particle-bound fraction being reduced. As the average diameter of peroxisomes increases from 0.58 ± 0.15 µm in controls to 0.73 ± 0.25 µm in treated animals, the measured peroxisomal enzymes are highly diluted in proliferated particles (LEIGHTON et al. 1975). Approximately one-half of the normal peroxisomes and all proliferated peroxisomes loose matrix material during tissue fractionation and ghosts are formed. Submission of isolated particles to mechanical stress does not reveal increased fragility of proliferated peroxisomes. These results indicate that matrix extraction and increased enzyme activities in the supernatant result from transmembrane passage of peroxisomal proteins.

In a study on nafenopin-activated peroxisomes in neonatal rat liver, STÄUBLI et al. (1977) found approximately a sixfold increase in specific volume and approximately a twofold increase in the specific number of peroxisomes. Corresponding biochemical and autoradiographic data led to the assumption that the peroxisomal population is composed of at least two subpopulations: a) one exhibits control patterns of size distribution (its population is diminished during nafenopin treatment; b) the other is stimulated by nafenopin to proliferate, and these particles are enlarged. Variability in DAB staining of peroxisomes has been observed in hepatocytes of animals treated with clofibrate (LEGG and WOOD 1970 a, HAYASHI et al. 1975) or nafenopin (LEIGHTON et al. 1975). These cytochemical results agree well with the findings mentioned above.

Long-term treatment with nafenopin causes malignant tumors in acatalasemic mice (hepatocellular carcinomas; REDDY et al. 1976 b) and in rats (hepatocellular carcinomas and pancreatic acinar cell tumors; REDDY and RAO 1977). Observations of increased total liver DNA after acute or chronic nafenopin treatment (BECKETT et al. 1972, MOODY et al. 1977) and reports of increased mitotic indices in wild-type and acatalasemic mice (MOODY et al.

1977) are noteworthy in this context. Other hepatocarcinogenic agents have also been found to possess mitogenic effects.

III.G.1.d. SaH-42,348 [1-Methyl-4-piperidyl-bis(p-chlorophenoxy) acetate]

SaH-42,348 (Fig. 33 E) is a hypolipidemic agent (TIMMS et al. 1969, KRITCHEVSKY 1971) that is one of the aryloxyisobutyrate proliferators of peroxisomes (REDDY et al. 1975, MOODY and REDDY 1976 b). It induces increased liver catalase and carnitine acetyltransferase activities (MOODY and REDDY 1976 b). Detailed morphometric analysis (MOODY and REDDY 1976 a) of ultrastructural changes after treatment with SaH-42,348 revealed a conspicuous increase in surface area and volume density of peroxisomes, but only a moderate increase in peroxisome number, average peroxisome volume, and average peroxisome diameter. The increase in relative peroxisome volume and surface area is more prominent in centrilobular than in periportal regions. The volume and surface area of smooth endoplasmic reticulum are also increased, while these values for mitochondria, Golgi areas, and lysosomes are not significantly changed by SH-42,348 treatment. After withdrawal of the drug most of the increased values return to normal within 8 days; the values for smooth endoplasmic reticulum, however, decrease only partially (Fig. 35).

In regard to mechanisms of peroxisome degradation, two morphometric results are of particular interest: First, lysosomal volume density is markedly increased 4 days after withdrawal of the drug, and second, the increased values for smooth endoplasmic reticulum do not return to normal, as do the values for peroxisomes. The increase in lysosomal volume density suggests that the lysosomal system is significantly involved in the removal of SaH-42,348-stimulated peroxisomes. This effect is not observed during reconstitution after withdrawal of nafenopin (STÄUBLI et al. 1977). However, there is no increase in autophagic vacuoles that contain peroxisomes or smooth endoplasmic reticulum. Rapid dissolution of the peroxisomal matrix or incorporation of peroxisomes into autophagic vacuoles by fusion might account for this observation (MOODY and REDDY 1976 a). Persistently elevated morphometric values for smooth endoplasmic reticulum lend support to the hypothesis that peroxisomal matrix proteins are retracted into the endoplasmic reticulum, and empty limiting membranes might contribute in part to the increased smooth endoplasmic reticulum values (MOODY and REDDY 1976 a) (Fig. 35).

III.G.1.e. Simfibrate [CLY-503, 1,3-Propyl-bis(2-p-chlorophenoxy)2-methyl propionate]

Simfibrate (Fig. 33 K) is similar to clofibrate in its molecular structure and its pharmacological properties (HIRAI and OGAWA 1975). It causes proliferation of mouse liver peroxisomes to "giant" peroxisomes (measuring up to 2 µm in diameter) when given intraperitoneally in a daily dose of 1,000 mg/kg for 3–14 days (HIRAI and OGAWA 1973, 1975). Frequent loop- or hook-shaped profiles of smooth endoplasmic reticulum as well as continuities between peroxisomes and smooth or rough endoplasmic reticulum have been noted after simfibrate treatment. It has been suggested that these loop- or hook-shaped segments of endoplasmic reticulum may play an important role in the initial stage of peroxisome formation (HIRAI and OGAWA 1975) (see Section I.B.5.).

III.G.1.f. Wy-14,643 ([4-Chloro-6-(2,3-xylidino)-2-pyrimidinylthio] acetic acid) and Tibric Acid [2-Chloro-5-(3,5-dimethylpiperidinosulfonyl) benzoic acid]

The increased frequency of peroxisomes after treatment with the hypolipidemic compounds Wy-14,643 (Fig. 33 G) and tibric acid (Fig. 33 H) (PEREIRA *et al*. 1975, SANTILLI *et al*. 1974) is of particular importance. In

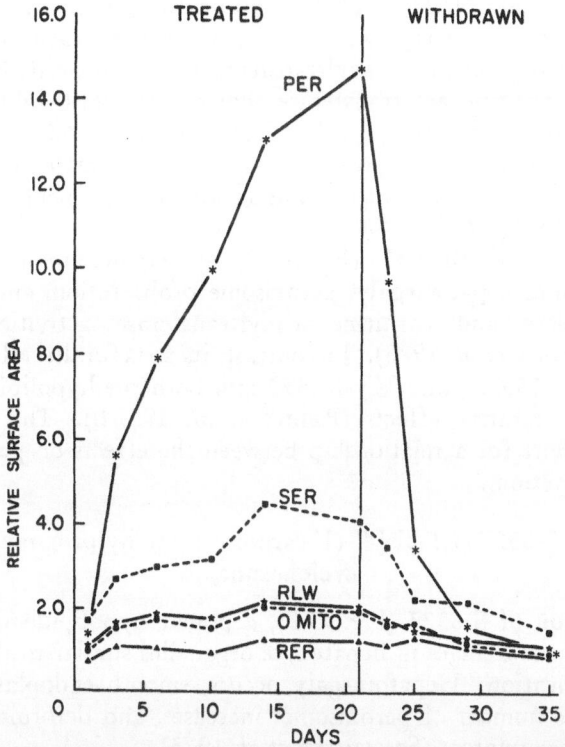

Fig. 35. Total liver values for the surface area of hepatocyte organelles in F-344 rats receiving 0.10% SaH 42-348, expressed as ratio to control values (relative surface area) and compared to the ratio of liver weight in treated animals relative to that in control animals. *PER* peroxisomes; *SER* smooth endoplasmic reticulum; *RER* rough endoplasmic reticulum; *OMITO* outer mitochondrial membrane; *RLW* relative liver weight. From: MOODY and REDDY (1976 a): J. Cell Biol. 71, 771.

contrast to many other peroxisome-proliferating agents, these drugs are not structurally related to clofibrate. Using peroxisome-proliferating agents that are structurally related to clofibrate, it is not possible to decide definitely whether or not a relationship exists between the proliferation of peroxisomes and the hypolipidemic effect; similar properties of structurally related compounds may cause these findings. However, the fact that hypolipidemic compounds other than clofibrate and its analogues also stimulate the proliferation of hepatic peroxisomes is additional evidence for a causal relationship between peroxisome proliferation and the hypolipidemic effect.

Wy-14,643 and tibric acid, administered to male rats and mice at a dietetic level of 0.125–0.25% for 1–4 weeks, cause a significant increase in liver weight, an enlarged volume of smooth endoplasmic reticulum, an increased number of peroxisomes, and increased activities of peroxisomal enzymes (Moody and Reddy 1976 b, 1978 a, Reddy and Krishnakantha 1975). The increase in total liver DNA (Moody and Reddy 1978 a) suggests (similar to results with nafenopin; Beckett et al. 1972, Moody et al. 1977) that liver growth results from a combination of cellular hypertrophy and hyperplasia. Liver catalase and NAD$^+$-α-glycerophosphate dehydrogenase display a 2–3-fold increase in activity, whereas the activity of carnitine acetyltransferase (short-chain carnitine acyltransferase) is enhanced 8–26-fold; the medium-chain carnitine acyltransferase shows a 4.5–11.0-fold increase, and long-chain carnitine acyltransferase is enhanced only 2.0–4.0-fold.

Lazarow (1977) recently reported an 11–18-fold increase in the capacity of male rat liver to oxidize palmitoyl-CoA following treatment with clofibrate, tribric acid, or Wy-14,643.

Administration of BR-931 (Fig. 33 I), an ethanolamine derivative of Wy-14,643, causes hepatomegaly, peroxisome proliferation, and enhancement of liver catalase and carnitine acetyltransferase activities, similar to Wy-14,643 (Reddy et al. 1978). In contrast, its structurally related analogues Wy-15,690, Wy-15,647, and Wy-15,490 lack both the hypolipidemic and the peroxisome-proliferative effects (Reddy et al. 1977 b). This finding lends additional support for a relationship between the effects on peroxisomes and hypolipidemic action.

III.G.1.g. S-8527 (1,1-bis[4'-(1"carboxy-1"methylpropoxy)-phenyl] cyclohexane)

Administration of S-8527 (Fig. 33 F), a potent hypolipidemic agent (Toki et al. 1973), causes changes in hepatocyte organelles similar to those seen after clofibrate application. Hepatomegaly occurs, smooth endoplasmic reticulum proliferates, the number of peroxisomes increases, and deformations of mitochondria become apparent (Sakamoto et al. 1973).

III.G.1.h. Di(2-ethylhexyl)phthalate (DEHP) and Related Compounds

Oral administration of DEHP (Fig. 36 B), a plasticizer extensively used in the manufacture of PVP plastics (Autian 1973), has been shown to cause hepatomegaly (Carpenter et al. 1953, Lake et al. 1974, 1975, Nikonorow et al. 1973) and to decrease the activities of glucose-6-phosphatase, aniline-4-hydroxylase, and succinate dehydrogenase (Lake et al. 1975). Serum cholesterol and triglyceride levels are substantially reduced when DEHP is administered at dietetic levels of 0.5, 2, and 4%. The hypolipidemic effect is most prominent after 1 week of treatment with 4% DEHP (Lake et al. 1975). The concomitant increase in the number of peroxisomes is paralleled by increased activities of hepatic catalase (2-fold) and carnitine acetyltransferase

(10–20-fold) (REDDY et al. 1976 a). Furthermore, enhancement of fatty acyl-CoA oxidizing activity in rat liver peroxisomes has been reported to be induced by DEHP (OSUMI and HASHIMOTO 1978 a). Similar effects are obtained with the plasticizers di(2-ethylhexyl)adipate (Fig. 36 A) and di(2-ethylhexyl)-sebacate (Fig. 36 D), and with the related compounds 2-ethyl-n-hexanol

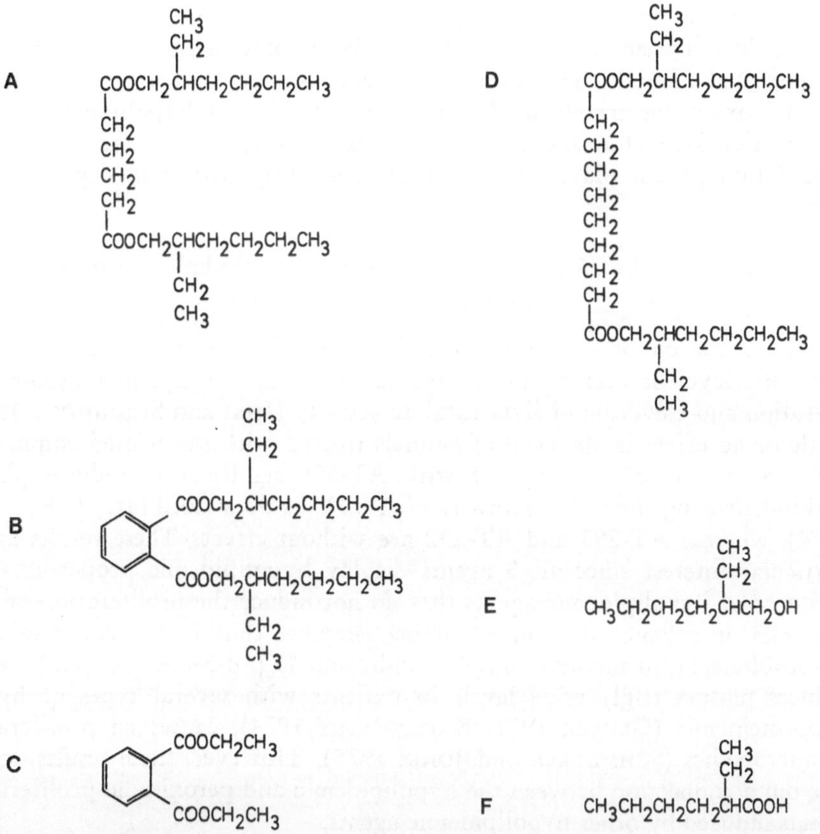

Fig. 36 A–F. Chemical structures of plasticizers and structurally related compounds. A Di-(2-ethylhexyl)adipate. B Di(2-ethylhexyl)phthalate (DEHP). C Diethylphthalate. D Di(2-ethylhexyl)sebacate. E 2-Ethyl-n-hexanol. F 2-Ethylhexanoic acid.

(Fig. 36 E) and 2-ethylhexanoic acid (Fig. 36 F); 2-ethylhexylaldehyde induces a moderate increase in the number of peroxisomes (MOODY and REDDY 1978 b). The respective straight-chain analogues to 2-ethyl-n-hexanol and 2-ethylhexanoic acid—2-hexanol and 2-hexanoic acid—as well as the plasticizers adipic acid and diethylphthalate (Fig. 36 C) have neither hypolipidemic nor peroxisome-proliferating effects (MOODY and REDDY 1978 b, MOODY et al. 1976). These results indicate that 2-ethyl-n-hexanol is the essential part of the molecule that is responsible for peroxisome proliferation (MOODY and REDDY 1978 b).

The subcellular distribution of the enzymes of the fatty acid β-oxidation system and their induction by DEHP in rat liver has been described by OSUMI and HASHIMOTO (1979 a). Their data indicate that all the enzymes of fatty acyl-CoA oxidation are located in peroxisomes and are induced by DEHP. In the DEHP-treated animals the activities of the enzymes of the peroxisomal β-oxidation system are at the same level as or higher than those of the mitochondrial enzymes. On the other hand, the activities of the β-oxidation enzymes in peroxisomes are extremely low in the control group; thus, it seems to be unlikely that the peroxisomal system is important in untreated animals. However, as the activities of the enzymes of peroxisomal β-oxidation are elevated after administration of hypolipidemic agents (see Section III.G.1.a. and III.G.1.f.), it is conceivable that the peroxisomal β-oxidation system plays an important role when fatty acid degradation is enhanced.

III.G.1.i. AT-308 (3-[4-(1-ethoxycarbonyl-1-methylethoxy)-phenyl]-5-(3-pyridyl)-1,2,4-oxadiazole)

Administration of the hypolipidemic agent AT-308 (Fig. 33 J) to rats at a dietetic level of 0.25% for 7 days causes hepatomegaly, peroxisome proliferation and elevation of liver catalase activity (IMAI and SHIMAMOTO 1973). Little or no effect is observed in animals treated with the related compounds AT-293 or AT-232. Treatment with AT-308 significantly reduces phenobarbital sleeping time (ARAKAWA et al. 1978) (as does clofibrate; LEWIS et al. 1974), whereas AT-293 and AT-232 are without effect. These results are of particular interest since all 3 agents display hypolipidemic properties. The existence of hypolipidemic agents that do not induce the proliferation of peroxisomes or enhance drug-metabolizing enzymes thus is proved. Similarly, oxandrolone (17-α-methyl-2 oxa-5 α-androstan-17 β-d-3-one, Anavas), which reduces plasma triglyceride levels in patients with several types of hyperlipoproteinemia (GLUECK 1971, KRITCHEVSKY 1974), causes no proliferation of microbodies (SCHMUCKER and JONES 1975). However, these results do not rule out a connection between the hypolipidemic and peroxisome-proliferating effects induced by other hypolipidemic agents.

The peroxisome-proliferating effect of AT-308 is considered to be due to the phenoxyisobutyrate moiety (ARAKAWA et al. 1978) of the drug molecule.

III.G.1.j. RMI 14,514 (5-Tetradecyloxy-2-furan-carboxylic acid)

Unusual responses of rat hepatic and renal peroxisomes to the hypolipidemic compound RMI 14,514 (Fig. 33 L) (KARIYA et al. 1975, PARKER et al. 1975, 1977) have recently been reported (SVOBODA 1978). The morphological (liver) and biochemical effects (liver and kidney) on peroxisomes induced by this agent are different from those induced by other hypolipidemic agents. Effects are seen in both sexes. Treatment with RMI 14,514 causes an increase in the number of peroxisomes in liver, kidney, and intestinal mucosa. Ultrastructural abnormalities are observed in liver peroxisomes only; the altered organelles frequently display cavitations and compartmentation of the matrix. Additionally, dense bands, canaliculi, or cisternae extend from the periphery of the peroxisomes into the cytoplasm. In addition to increased catalase levels in the liver and kidney of male and female animals, activities of urate oxidase (liver, males and females), α-hydroxy acid oxidase (liver and kidney, males and females) and D-amino acid oxidase (liver, males and females; kidney, males) are enhanced.

III.G.2. Aminotriazole (3-Amino-1,2,4-triazole)

Aminotriazole inhibits the enzymatic activity of catalase without affecting its synthesis. The effect of aminotriazole on liver catalase was initially studied by HEIM et al. (1955, 1956). They found that 3 hours after intraperitoneal injection of aminotriazole, catalase activity is reduced to approximately 10⁰/o of control values. Later on, PRICE et al. (1961, 1962) demonstrated that 1 hour after aminotriazole administration only 2⁰/o of the normal enzyme activity remains, and it can be attributed to red blood cell catalase; the activity returns to a normal level approximately 5 hours after aminotriazole administration. HEIM et al. (1955, 1956) studied the binding mechanism of aminotriazole to catalase and found evidence of both reversible and irreversible inhibition. In vitro catalase is inhibited only by a high concentration of aminotriazole (FEINSTEIN et al. 1958, HEIM et al. 1956), and this inhibition is partially reversible by dilution. Significantly smaller quantities of aminotriazole are necessary for complete inhibition in vivo, and this inhibition is not reversible; aminotriazole forms an irreversible complex with catalase by binding to its protein moiety (MARGOLIASH and NOVOGRODSKI 1958, MARGOLIASH et al. 1960, PRICE et al. 1961, 1962).

The return to normal catalase activity after the administration of aminotriazole is accompanied by a corresponding uptake of radioactive iron (PRICE et al. 1961, 1962). This indicates that regeneration of catalase activity is due to the synthesis of new catalase. Cytochemical findings in rat hepatocytes (WOOD and LEGG 1970) correspond to biochemical results: within 1 hour after the intraperitoneal injection of aminotriazole, catalase staining of peroxisomes is abolished and the return of reactivity coincides with the return of catalase activity.

Aminotriazole does not affect the positive DAB reaction of red blood cells and of azurophilic granules in leukocytes. No changes can be observed in the morphology or number of peroxisomes. The absence of evidence for destruction of peroxisomes or for formation of new peroxisomes after aminotriazole administration, together with the rapid return of enzyme activity, indicates that newly synthesized catalase enters preexisting peroxisomes (WOOD and LEGG 1970).

III.G.3. Allylisopropylacetamide (AIA)

AIA inhibits the synthesis of liver catalase (PRICE et al. 1962, RECHCIGL and PRICE 1968, SCHMID et al. 1955); it also causes various metabolic abnormalities in hepatocytes, such as increased formation of porphyrins and related compounds (GOLDBERG and RIMINGTON 1955,

GRANICK 1965, 1966, LABBE et al. 1961, ROSE et al. 1961, SCHMID 1963), increased lipid synthesis (BIEMPICA et al. 1967, 1971, LABBE et al. 1961, LOTTS-FELD and LABBE 1965), and increased protein synthesis (KAUFMANN et al. 1966, LOTTSFELD and LABBE 1965, MARVER et al. 1966). Studies on the incorporation of radioactive glycine (SCHMID et al. 1955) and radioactive iron (RECHCIGL and PRICE 1968) into catalase have shown that AIA selectively prevents the synthesis of catalase, but not of other heme compounds (LABBE et al. 1961). AIA most probably inhibits the synthesis of catalase apoprotein or prevents the association of the heme prosthetic group with the protein moiety (LEGG and WOOD 1972; cf., KAWAMATA et al. 1975 a, b). PRICE et al. (1962) found that administration of both aminotriazole and AIA caused catalase activity to rapidly rise to a low plateau of approximately 8% of the normal level; this indicates the existence of a second catalase whose activity is inhibited by aminotriazole but whose synthesis is not blocked by AIA (PRICE et al. 1962).

Significant morphological and cytochemical changes are seen in hepatocytes after AIA application (BIEMPICA et al. 1967, 1971, MOSES et al. 1970, POSOLAKI and BARKA 1968, ROHEIM et al. 1971). Initial alterations occur 3 hours after a single injection of AIA and become most prominent after 12 hours. At this time lipid droplets occupy large areas of the cytoplasm, smooth endoplasmic reticulum is increased, rough endoplasmic reticulum is diminished, and ribosomal aggregates are decreased. The changes revert to normal at 72 hours (BIEMPICA et al. 1967). After three or more injections of AIA the accumulation of lipids decreases and a prominent increase of smooth endoplasmic reticulum is noted (BIEMPICA et al. 1971).

AIA-induced morphological and cytochemical changes in peroxisomes (LEGG and WOOD 1970 b, 1972, REDDY et al. 1977 a) are visible 48 hours after administration of the drug and are more obvious 3 days later: peroxisomes are decreased in size and the matrix is reduced; an irregular annular zone of low electron density appears between the matrix and the limiting membrane. The number of peroxisomes remains constant, but the organelles are gathered in relatively small areas of the cytoplasm. Cytochemical studies reveal diminished DAB reactivity 24 hours after AIA injection and the staining is completely abolished after 3–5 days. Morphological reconstitution to normal occurs within 24 hours after cessation of AIA administration. Cytochemically, only traces of peroxidatic activity are present at this time; the staining reaction returns to normal 5–10 days later. Combined administration of AIA, aminotriazole, and lipid-lowering, peroxisome-proliferating agents has shown that the hypolipidemic and peroxisome-proliferating effects are also evoked when catalase activity is negligible (LEGG and WOOD 1970 b, REDDY et al. 1970, 1977 a). This finding suggests that peroxisomal catalase is not necessary for the induction of peroxisome proliferation.

III.G.4. Salicylates

Administration of acetylsalicylic acid to adult rats at a dietary level of 1% causes an increased number, variation in size, and the occurrence of fibrillar matrix material in hepatic and renal microbodies (HRUBAN et al. 1966, 1974 a,

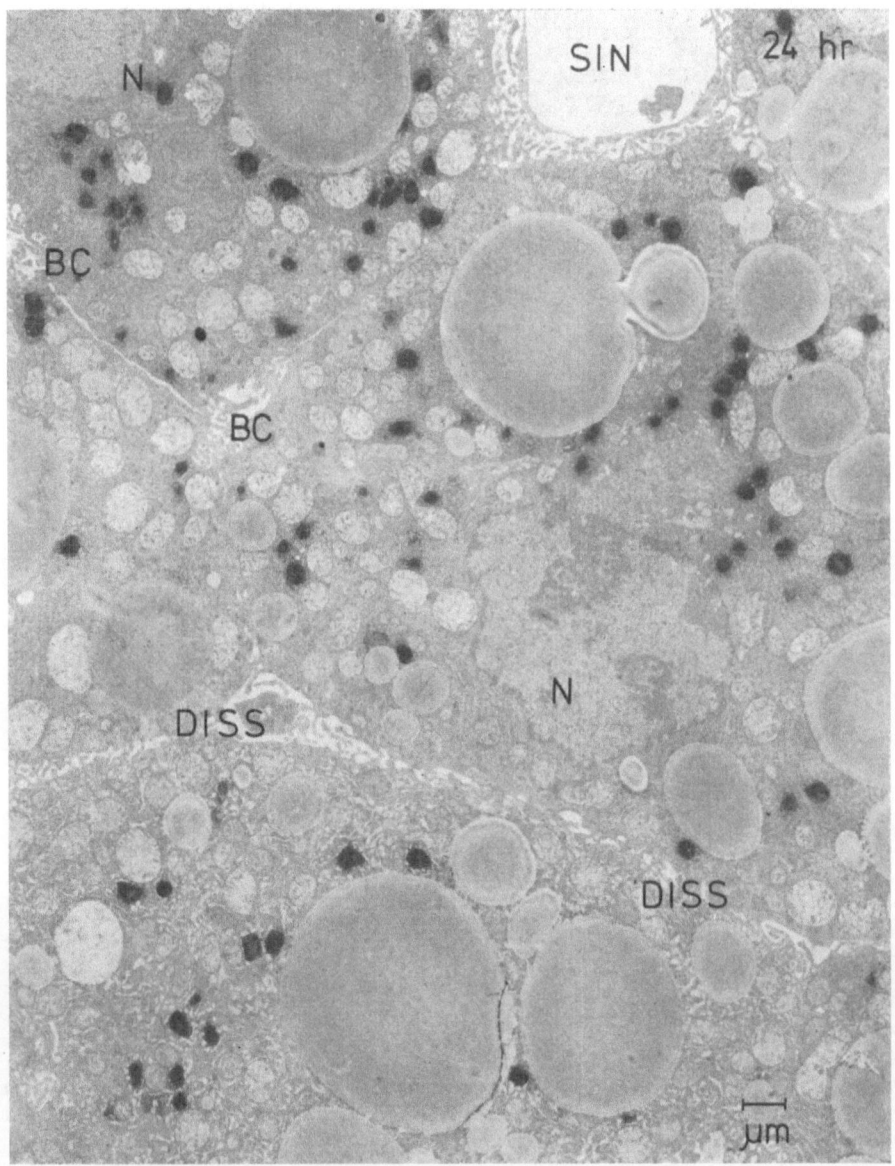

Fig. 37. Low-power electron micrograph of regenerating rat liver 24 hours after subtotal hepatectomy demonstrates early structural changes: complete loss of glycogen, appearance of numerous lipid vacuoles, and disaggregation of the endoplasmic reticulum. The number of microbodies stained by the cytochemical catalase reaction is not reduced. *N* nucleus; *SIN* sinusoid; *BC* bile canaliculus; *DISS* Disse's space. ×5,500. From: GOLDENBERG *et al.* (1975 a): Histochemistry 44, 49.

RIEDE and ROHR 1974). The size of microbodies varies markedly; small particles and very large organelles are present within the same cell. Although crystalloids appear to be unaffected, additional structures, termed "matrical plates" are seen in the matrix (HRUBAN *et al.* 1974 a). The formation of

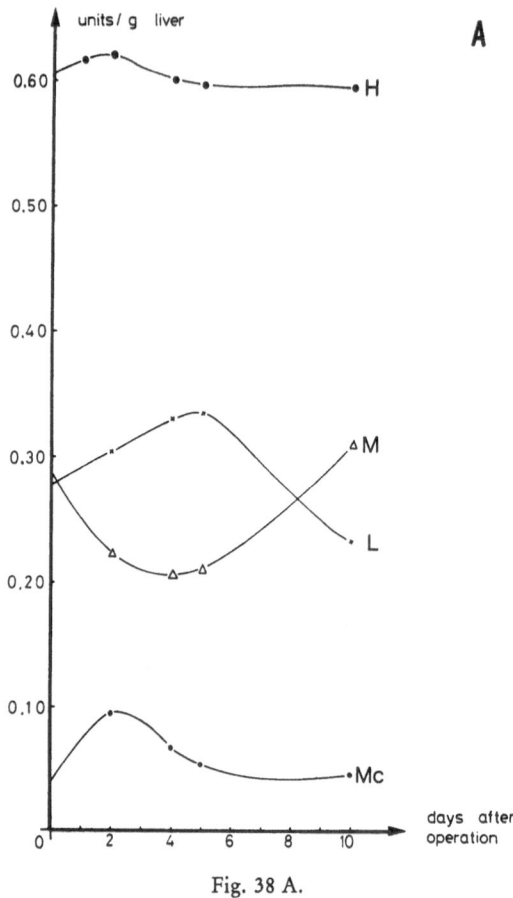

Fig. 38 A.

Fig. 38. Enzyme activities in rat liver homogenate (*H*) and in heavy (*M*) and light (*L*), microsomal (*Mc*), and soluble (*S*) cell fractions after subtotal hepatectomy. *A* Urate oxidase. *B* Catalase. *C* α-Hydroxy acid oxidase. Enzyme activities are given in terms of the first-order rate constant *k* (catalase) or in micromoles of substrate turned over per minute (urate oxidase and α-hydroxy acid oxidase). From: GOLDENBERG *et al.* (1975 a): Histochemistry 44, 51—53.

matrical plates can also be induced by the administration of sodium citrate, propionate, or aminotriazole (GOTOH *et al.* 1975).

"Gastruloid cisternae" are frequently formed in hepatocytes of acetylsalicylic acid-treated animals. Gastruloid cisternae have been interpreted as transitional structures between rough endoplasmic reticulum and microbodies that are involved in the packaging of catalase in microbodies (HRUBAN *et al.*

1974 a). Similar structures have been reported to occur frequently after sim-fibrate treatment (HIRAI and OGAWA 1975) (Fig. 14). Microbodies are also abundant in hepatocytes of mice treated with acetylsalicylic acid, but in this species they are devoid of matrical plates (HRUBAN et al. 1966).

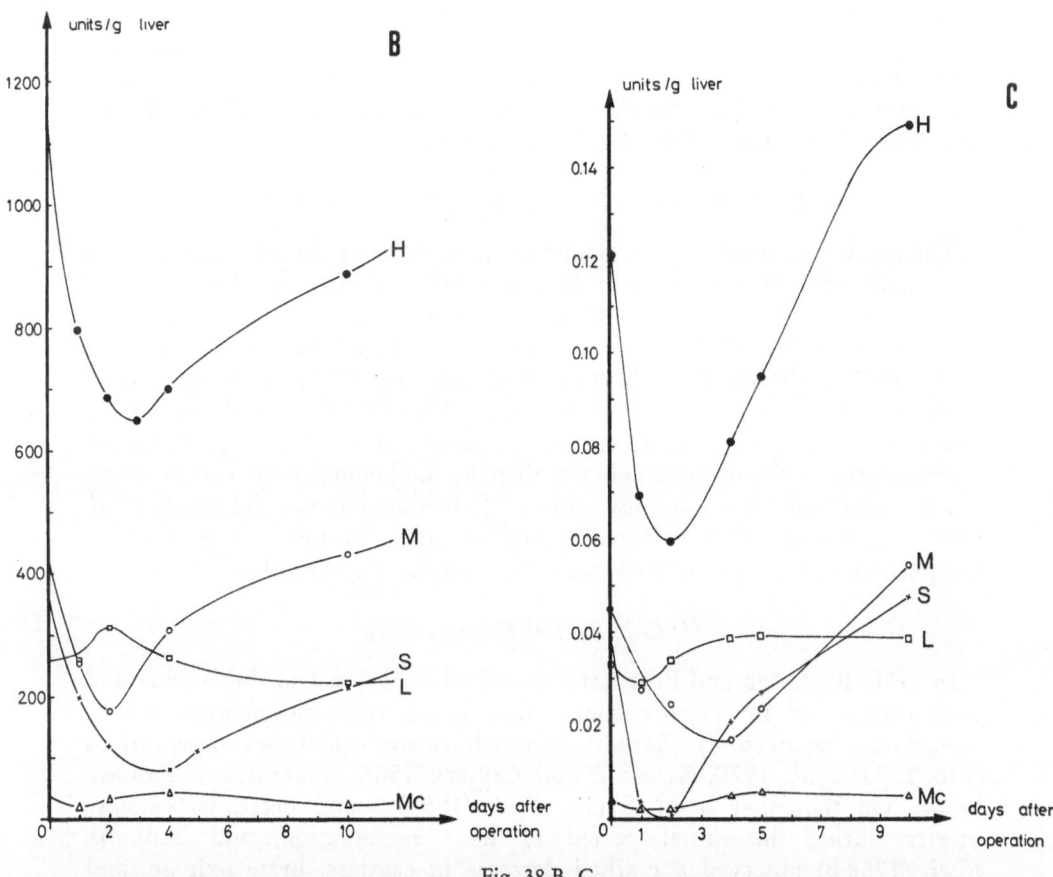

Fig. 38 B, C.

Similar results regarding variations in the size and shape of hepatic micro-bodies have been obtained after treatment with salicylic acid, salicyl anilide, and hexahydrosalicylic acid. The last drug causes structures that are inter-preted as extrusions of crystalloid cores from microbodies (HRUBAN et al. 1966).

III.G.5. Dimethrin

The insecticide dimethrin (2,4-dimethylbenzylchrysanthemumic acid) in-creases the number of microbodies and causes variations in their shape and size when administered to male rats for 2 months at a dietary level of 2%/o (HRUBAN et al. 1974 a). The ratio of liver weight to body weight is in-creased in dimethrin-fed animals (AMBROSE 1964, MARSI et al. 1964). Liver weights return to normal after withdrawal of the drug. Matrical plates induced by dimethrin have essentially the same structure as those induced by acetylsalicylic acid or clofibrate. Serum cholesterol levels change in different directions after administration of acetylsalicylic acid, clofibrate and dimethrin. It appears from these observations that there is no correlation between the presence of matrical plates and serum cholesterol levels (HRUBAN et al. 1974 a). Structures composed of helically arranged grooved fibrils characteristically appear within microbodies after treatment with dimethrin (HRUBAN et al. 1974 a). Similar structures have been rarely observed after meclizine administration (HRUBAN et al. 1974 b).

III.G.6. Various Other Compounds and Treatments

The number of hepatic peroxisomes is increased after the administration of taurocholic acid (SCHAFFNER and JAVITT 1966) and ethionine (WOOD 1965, 1967). Chronic intoxication with thioacetamide (SALOMON 1962) or halothane (ROSS and CARDELL 1972), and treatment with lidoflavine (VERHEYEN et al. 1975) or phenobarbital (BURGER and HERDSON 1966) causes an increase in the number of hepatic peroxisomes. Retrorsine (AFZELIUS and SCHOENTAL 1967) or cortisone (WIENER et al. 1968) reduces the number of peroxisomes.

Irregularly outlined microbodies with pale, nonhomogeneous matrix material are observed after administration of β-3-thionylalanine (HRUBAN et al. 1963). Azaserine causes enlarged crystalline cores (HRUBAN et al. 1965 b); this phenomenon may be related to an effect on the synthesis of uric acid.

III.G.7. Partial Hepatectomy

In 1956, ROUILLER and BERNHARD described an increase in the number and polymorphism of rat hepatic microbodies during regeneration after partial (two thirds) hepatectomy. Several authors have confirmed these observations (RIGATUSO et al. 1970, STENGER and CONFER 1966, VIRÁGH and BARTÓK 1966). On the other hand, STEIN et al. (1951) found decreased catalase ictivity during the initial period of liver regeneration, and TSUKADA et al. (1968 b) observed a marked decrease in catalase, urate oxidase, and microbody frequency both in partially hepatectomized and sham-operated rats. Morphometric analysis (ROHR et al. 1970) did not reveal altered peroxisomal volume densities during the regeneration period.

SAITO et al. (1973) noted a slight decrease in the number of microbodies 6 hours after partial hepatectomy, a moderate decrease after 12 and 18 hours, and extremely low numbers 24 hours after operation. Small peroxisomes were rarely seen 24 hours after hepatectomy. This suggests that the decrease in the number of peroxisomes is mainly due to a decreasing number of smaller peroxisomes. Gradual recovery of peroxisome numbers began 28 hours after the operation. GOLDENBERG et al. (1975 a) observed an early increase in the number of peroxisomes during the first day after partial hepatectomy, and a marked decrease (to about 50%) after 24–36 hours (Fig. 37). Later on, the number and size of peroxisomes increased again. The remarkable decrease in number 24–36 hours after partial hepatectomy correlates with the first mitotic wave during this time of regeneration (BECKER 1970). These findings disagree with the results of RIGATUSO et al. (1970), who found increasing numbers of peroxisomes up to 36 hours after partial hepatectomy.

Activities of peroxisomal enzymes after minimal hepatectomy have been measured by LAMY et al. (1970, 1973). Catalase, α-hydroxy acid oxidase, and D-amino acid oxidase are reduced, whereas changes in urate oxidase activities are not significant.

The behavior of the enzymes catalase, α-hydroxy acid oxidase, and urate oxidase in total liver homogenate and in heavy and light mitochondrial, microsomal, and soluble fractions 10 days after partial hepatectomy are summarized in Fig. 38. Catalase and α-hydroxy acid oxidase activities are depressed in the total liver homogenate as well as in the mitochondrial and soluble fractions. They do not change appreciably in the microsomal and lysosomal fractions. In contrast, urate oxidase activity remains unchanged in the total liver homogenate, decreases in the mitochondrial fraction, and increases in the lysosomal and microsomal fractions. The symmetrical behavior of urate oxidase activity in the mitochondrial and lysosomal fractions indicates that microbodies become smaller during liver regeneration; this finding has been interpreted to correspond to the fragmentation or budding of peroxisomes detected in morphological studies (FAHIMI and VENKATACHALAM 1970, RIGATUSO et al. 1970). The variations in enzyme activities may be caused by either different populations of peroxisomes or different turnover rates of oxidases and catalase (GOLDENBERG et al. 1975 a).

III.H. Peroxisomes Under Pathological Conditions

III.H.1. Acatalasemia

Five strains of mutant mice with low levels of catalase activity in the blood and solid tissue, including one acatalasemic strain (Csb), have been described by FEINSTEIN et al. (1966, 1967). AEBI et al. (1968) demonstrated that the catalase molecule of the acatalasemic strain is unstable. The enzyme is rapidly inactivated by heat and becomes inactivated during incubation when assayed. Wild and mutant strains exhibit similar levels of catalase activity when measured at 20 °C or when assayed at 37 °C and incubated for 15 seconds, whereas the enzyme of the mutant strain is rapidly inactivated when incubated for 5 minutes at 37 °C. The half-life of catalase in acatalasemic mice is half that in wild-type mice (JONES and MASTERS 1976). Serum glycerol-glyceride levels are considerably lower in acatalasemic than in wild-type mice (GOLD-FISCHER et al. 1971, REDDY et al. 1974 a). Hepatic peroxisomes are few in number in acatalasemic mice of both sexes. The organelles display high variability in shape and contain a variable, electron-dense central core (REDDY et al. 1969 a).

Cytochemical studies with benzidine and DAB reveal a more rapid and more intense staining of hepatic and renal peroxisomes in acatalasemic mice than in wild-type mice (GOLDFISCHER and ESSNER 1970). Whereas the catalase inhibitor aminotriazole markedly decreases the reaction in wild strains at

37 °C, there is little effect on peroxisomal staining in the mutant strain. Disregarding the effects of the pH and temperature of the incubation medium, it seems that DAB and benzidine staining is enhanced in peroxisomes of the acatalasemic strain, which are relatively resistant to aminotriazole. These results suggest that the cytochemical reaction is due to a peroxidase and not simply to peroxidatic activity of catalase, and further that the "inactivated" catalase of the acatalasemic animals actually is an active peroxidase that oxidizes DAB and benzidine (GOLDFISCHER and ESSNER 1970). Biochemical (TANFORD and LOVRIEN 1962) and cytochemical (GOLDFISCHER and ESSNER 1969) studies have shown that catalase is degraded into peroxidatically active subunits at alkaline pH. The assumption that a peroxisomal peroxidase exists is interesting with respect to the observation that the injection of a subunit of hepatic catalase with peroxidatic activity causes hypolipidemia (CARAVACA and MAY 1964, CARAVACA et al. 1967). Whether or not a relationship exists between the presence of a peroxisomal peroxidase and hypolipidemia in acatalasemic mice (GOLDFISCHER et al. 1971, REDDY et al. 1974 a) cannot be determined at present; it is unlikely (REDDY et al. 1974 a), since hepatic lipogenesis in acatalasemic mice has been shown to be abnormal (CUADRADO and BRICKER 1973).

Treatment of acatalasemic mice with clofibrate (REDDY and SVOBODA 1971 a, REDDY et al. 1969 a, b), nafenopin (REDDY et al. 1974 a), or methyl clofenapate (REDDY 1974)—three hypolipidemic agents that induce the proliferation of peroxisomes and enhance catalase activity in normal mice—causes an increase in the number of peroxisomes in acatalasemic mice without a measurable increase in catalase activity. The peroxisomal response to hypolipidemic agents in acatalasemic mice shows that the proliferation of peroxisomes can be induced in the absence of catalase activity. Similar results have been obtained with aminotriazole and AIA (LEGG and WOOD 1970 b, REDDY et al. 1970, 1977 a).

A hereditary deficiency in blood and tissue catalase in humans has been reported by AEBI et al. (1962), SZEINBERG et al. (1963), and TAKAHARA (1952) (for a review see HRUBAN and RECHCIGL 1969). Investigations presented by OGATA et al. (1974) indicate that human acatalasemia may be classified into two types: one ("Swiss type") type exhibits minute catalase activity in the whole body (AEBI et al. 1964, NAKAMURA et al. 1952, OGATA et al. 1974); the other type displays minute catalase activity only in the blood (OGATA et al. 1974).

Contrary to the results of FEINSTEIN et al. (1966), SHAPIRA et al. (1974), using specific anticatalase antibody techniques, have shown that there is only a partial immunological identity between catalase from normal red cells and the residual enzyme from individuals homozygous for Swiss-type acatalasemia. Decreased stability against heat and urea suggests that the low levels of catalase activity (2% of normal) in hemolysates are due to an instability of the mutant enzyme. This finding resembles observations in one of the mutants (Csb) of acatalasemic mice, which exhibits about the same level of residual activity as individuals homozygous for Swiss-type acatalasemia. Thus, the mouse mutant may represent a model counterpart for Swiss-type acatalasemia.

The findings in humans as well as in mice present evidence that the lack of red cell catalase is due to the synthesis of a relatively unstable mutant enzyme protein (SHAPIRA et al. 1974), causing a gradual decrease in enzyme activity in erythrocytes; this is masked in tissues with higher turnover rates, such as liver tissue (AEBI et al. 1968).

Leukocyte catalase from individuals heterozygous and homozygous for Swiss-type acatalasemia has been reported to differ from normal enzyme; it varies interindividually with regard to heat stability and electrophoretic mobility (WYSS and AEBI 1975). Hybridization between normal and unstable subunits has been suggested as a possible explanation.

III.H.2. Hepatomas and Other Malignant Tumors

Catalase activity has been reported to be low or absent in a number of tumors such as Novikoff's hepatoma, hepatoma LC 18, and hepatoma 3683 of rats, and various transplantable hepatomas of mice and hamsters (GREEN-STEIN 1954, 1955, KAMPSCHMIDT 1965, RECHCIGL et al. 1962); on the other hand, investigations on slowly growing hepatomas have revealed normal or even high catalase activities (MORRIS et al. 1964, ONO 1966, RECHCIGL et al. 1962, 1969). DALTON (1964) found that the size of microbodies is generally related to the growth rate of hepatomas; this result has been confirmed by HRUBAN et al. (1965 a) and RECHCIGL et al. (1969), who found micro-bodies only in slowly growing hepatomas; in fast growing hepatomas micro-bodies and peroxisomal enzymes were reported to be absent. Obviously, this must be seen in the context of a general loss of morphological organization in hepatomas (MALICK 1972).

In their detailed study on peroxisomal behavior in transplantable hepatomas of different growth rates, MOCHIZUKI et al. (1971) showed that microbodies are usually smaller in hepatoma cells than in hepatocytes, and demonstrated that microbody number, size, and internal structure generally reflect the growth rate of the tumor. Only a few, small microbodies that are mostly devoid of nucleoids are seen in fast growing hepatomas, whereas microbodies of hepatomas with intermediate growth rates are larger and more abundant, and are usually provided with nucleoids. Slowly growing hepatomas display numerous large microbodies. Furthermore, enzyme activities are frequently in good agreement with morphological differentiation: hepatomas with numerous large microbodies exhibit high catalase and D-amino acid oxidase activities, and hepatomas with microbodies with nucleoids show urate oxidase activity. Catalase activity roughly correlates inversely with hepatoma growth rates, but D-amino acid oxidase activity and urate oxidase activity do not necessarily; high urate oxidase activity is not uncommon in hepatomas with intermediate growth rates (MOCHIZUKI et al. 1971).

Studies on rat liver tumors induced by 3'-methyl-4-(dimethylamino)-azo-benzene revealed similar features (ITABASHI et al. 1975). Peroxisomes in poorly differentiated carcinomas and adenocarcinomas are less numerous than in well-differentiated carcinomas. Peroxisomes in well-differentiated carci-nomas resemble those in hepatocytes. Microbodies in poorly differentiated carcinomas are small in size; the matrix is mostly scanty and they usually lack a nucleoid. In well-differentiated hepatomas catalase values are depressed to one-third of those found in nontumorous portions of the liver. In poorly differentiated hepatomas and in adenocarcinomas catalase activity is lowered to one-tenth of control values.

In addition to these changes in the tumor itself, the effect of spontaneous, transplanted, or induced tumors on catalase activity in the liver and kidney of the host should be considered. It has been found that, without exception,

Table 25. *Catalase activity in male F-344 rats bearing transplanted tumors and treated with clofibrate (CPIB)* [a]

Treatment	No. of animals	Age of tumor (weeks) [b]	Catalase activity (units/mg protein) (mean ± SE)
Normal rats			
Control	5	—	40 ± 2.3
CPIB treated	6	—	85 ± 4.2
Transplanted tumors (subcutaneous)			
Ethionine hepatoma	5	8	19 ± 4.7
Ethionine hepatoma + CPIB	8	10	63 ± 6.1
Aflatoxin hepatoma	5	9	16 ± 3.9
Aflatoxin hepatoma + CPIB	4	12	59 ± 4.1
Actinomycin tumor (mesothelioma)	4	13	13 ± 5.1
Actinomycin tumor + CPIB	6	14	70 ± 5.6

From: REDDY and SVOBODA 1971 c.

[a] Clofibrate was administered in the diet for 4 weeks before sacrifice.

[b] Age of tumor at sacrifice.

catalase activity is significantly reduced in the liver and kidneys of individuals bearing a malignant tumor (Table 26). The effect is strongly related to the size of the tumor (RECHCIGL et al. 1969). The mechanism of this long-distance action is not understood. It has been suggested that it might be due to a so-called "toxohormone"—a substance (probably a protein or a lipoprotein; DELMON et al. 1958) extracted from human cancer tissue that depresses liver catalase when injected into normal mice (NAKAHARA and FUKUOKA 1948, 1961). However, depression of catalase is also observed in non-tumor-bearing animals after subcutaneous injury (NISHIMURA et al. 1963) and following the administration of various chemicals (DAY et al. 1954, HRUBAN and RECHCIGL 1969, KAMPSCHMIDT 1965). On the other hand, BHAWAN et al. (1975) reported an increase in the number of hepatic microbodies in rats 4–12 weeks after transplantation of mammary carcinomas.

Treatment with clofibrate (see Section III.G.1.a.) causes a slight increase in peroxisomal size but does not significantly alter the number of peroxisomes in Morris hepatoma 9618 A, a type representative of highly differentiated hepatomas (Tsukada et al. 1975 b) (Table 21); similar results have been obtained in both well and poorly differentiated 3′-methyl-4-(dimethylamino)-azobenzene-induced carcinomas (Itabashi et al. 1975). After clofibrate treatment matrical plates occur much more frequently in peroxisomes of Morris hepatoma 9618 A than in those of normal hepatocytes; catalase activity is only slightly increased (Tsukada et al. 1975 b) (Table 21). This finding indicates that even in highly differentiated hepatomas the control mechanisms

Table 26. *Liver and kidney catalase activity of hepatoma-bearing rats*

Hepatoma	Tumor size (percent body weight)	No. of obser- vations	Liver catalase (mean ± SE)		Kidney catalase [a] (mean ± SE)	
			units/g	units/mg N	units/g	units/mg N
Control		6	164 ± 6	4.9 ± 0.1	62	2.0
3,683	19	6	90 ± 3	3.2 ± 0.2	44	
Control		3	177 ± 9	5.5 ± 0.9	43	
LC 18	15	6	46 ± 5	2.0 ± 0.3	32	1.3
Control		2	163 ± 1	5.4 ± 0.01	50	2.2
Novikoff	14	4	77 ± 12	2.7 ± 0.5	51	1.9
Control		4	206 ± 8	6.7 ± 0.4	57 ± 2	2.1 ± 0.1
5,123	3	7	160 ± 10	5.8 ± 0.3	54 ± 1	2.0 ± 0.1
5,123	18	13	90 ± 8	2.9 ± 0.3	31 ± 4	1.2 ± 0.05

From: Rechcigl et al. 1962.

[a] All kidney samples, with the exception of the hepatoma 5,123 group, were pooled.

for the formation of peroxisomes are significantly impaired (Tsukada et al. 1975 b). On the other hand, in the liver of male rats that bear subcutaneously transplanted tumors (in these animals the initial level of catalase activity is low), clofibrate treatment significantly increases the number of peroxisomes and the catalase activity (Reddy and Svoboda 1971 c) (Table 25). This might mean that the proliferative stimulus of clofibrate acts directly on liver tissue itself, whereas the vector arising from the tumor can be transferred humorally according to the concept of a toxohormone (Nakahara and Fukuoka 1961). The effect of clofibrate and other peroxisome-proliferating agents on hepatocytes might be mediated by the peroxisome proliferation-associated polypeptide (see Section III.G.1.a.), which is increased in liver cell fractions after treatment with the proliferator (Reddy and Kumar 1977). This direct action on liver tissue might compensate for the influence of the tumor. Clofibrate neither restores catalase activity nor increases the number of peroxisomes in hepatomas or in irreparably altered cells of hyperplastic nodules induced by 3′-methyl-4-(dimethylamino)-azobenzene (Itabashi et al. 1975, 1977) and α-benzene hexachloride (Tsukada et al. 1979).

III.H.3. Zellweger Syndrome

Zellweger syndrome is a rarely occurring familial malady of autosomal recessive heredity. It involves cerebral, renal, and skeletal abnormalities, liver disease with iron storage and increased serum iron content, severe hypotonia, and death within 6 months (Bowen et al. 1964, Danks et al. 1975, Gilchrist et al. 1976, Passarge and McAdams 1967, Smith et al. 1964, Vitale et al. 1969). Mitochondria in hepatocytes and cortical astrocytes are of distorted appearance, and the glycogen content of these cells is increased. Brain and liver mitochondrial preparations show diminished oxygen consumption. This defect has been attributed to a disturbance within the nonheme iron protein region of the respiratory chain (Goldfischer et al. 1973). Furthermore, hepatocytes and renal proximal tubule cells lack peroxisomes, and smooth endoplasmic reticulum is sparse. As peroxisomes are postulated to originate from the endoplasmic reticulum (see Section I.B.3.), a relationship between the scarcity of smooth endoplasmic reticulum and the lack of peroxisomes has been suggested (Goldfischer et al. 1973).

Recently, Versmold and co-workers (1977) reported a case similar to Zellweger syndrome, but differing in some details. Contrary to Zellweger syndrome, patients survive into the second year of life. They show slight muscular hypertonia, and a functional abnormality of cytochrome b seems to occur. Although smooth endoplasmic reticulum is present in hepatocytes these cells are devoid of peroxisomes and catalase. Sometimes concentric membrane glycogen arrays are formed; a similar observation has been made in human and rat hepatoma cells (Flaks 1968, Ghadially and Parry 1966).

III.H.4. Various Other Pathological Conditions

The number and size of hepatic peroxisomes are increased in cases of viral hepatitis (Schaffner 1966), chronic passive congestion (Safran and Schaffner 1967), and essential fatty acid deficiency (Wilson and Leduc 1963).

The number of liver peroxisomes and the activities of catalase and urate oxidase are decreased during pneumococcal sepsis in rats (Canonico et al. 1975). Similarly, turpentine-induced inflammation results in a depression of peroxisomal enzyme activities. Measurements of the rates of degradation and synthesis of catalase have revealed that the decrease in catalase activity comes from a 60% decrease in its rate of synthesis (Canonico et al. 1977).

"Irreversible" shock causes increase in the number of hepatic peroxisomes in dogs (Blair et al. 1968); the number of hepatic peroxisomes is also increased in rats and mice with endotoxemia (de Palma et al. 1967, Levy et al. 1968).

Peroxisomes of human hepatocytes under pathological conditions are discussed further in Section VI.B.7.a.

III.I. Hypotheses on the Biogenesis of Peroxisomes

Older models of peroxisome biogenesis were based on the assumption that peroxisomes are exclusively derived from the endoplasmic reticulum; the peroxisomal proteins (matrical, membrane-bound, and core-bound) were considered without exception to stem from membrane-bound ribosomes and to be transferred through the cisternae of the endoplasmic reticulum or by membrane flow (FRANKE *et al.* 1971) to developing peroxisomes. The strongest support for this view clearly came from morphological observations: NOVIKOFF and SHIN (1964), and later ESSNER (1967) and REDDY and SVOBODA (1971 a), described extensive continuities between endoplasmic reticulum and microbodies and considered the organelles as dilatations of the endoplasmic reticulum that eventually bud off (see Section I.B.3. and I.B.4.).

As further evidence in support of this theory, HIGASHI and PETERS (1963 a, b) reported that a pulse of ^{3}H-leucine results in a strong peak of labeled catalase (detected by immunoprecipitation) in the rough microsomal fraction. In the mitochondrial fraction containing peroxisomes the labeling is delayed, indicating that catalase (and presumably other peroxisomal proteins) originates in the rough endoplasmic reticulum and is transferred into the peroxisomes. As most of the activity found in the endoplasmic reticulum might be due to absorbed secretory proteins such as albumin, this interpretation has been called into question (DE DUVE 1973).

In these models the continuity of peroxisomes with the endoplasmic reticulum, at least at early stages of their development, was taken into account. More recent studies—including pulse labeling followed by rate sedimentation of peroxisomes, which renders feasible the calculation of size distribution—have shown that the peroxisomes sprouting from the endoplasmic reticulum do not grow continually at a constant rate for a fixed life span (4.4 days), but rather are destroyed randomly irrespective of their age (POOLE *et al.* 1970).

According to the older models, for a short time after a radioactive pulse the specific radioactivity of the proteins should be higher in the small particles, whereas later on due to the growth of the particles a shift from the smaller to the larger particles should occur. As no difference in the specific radioactivity of catalase has been detected among peroxisomes of different size, it must be concluded that the period of formation of the particles is very short compared to the duration of the experiment (from 1 hour up to 1 week). Of course, this conclusion is only valid if the peroxisomes are individual particles growing independently of each other. These experiments led POOLE *et al.* (1970) to an alternative model in which clusters of peroxisomes remain connected to each other by parts of the endoplasmic reticulum channels so that they form a single pool and are able to exchange material by direct communication (*cf.*, Figs. 9 and 10).

All of these models rely on the supposition that the peroxisome as a whole stems from the endoplasmic reticulum. The plausible alternative that peroxisomal proteins might be formed both on free and bound ribosomes was introduced to interpret results suggesting the synthesis of catalase on free polysomes. SAKAMOTO and HIGASHI (1973) reported that catalase peptides arise not only from membrane-bound but also from free polysomes. LAZAROW and DE DUVE (1973 a, b) showed that after a pulse of ^{3}H-leucine and ^{3}H-δ-aminolevulinate an immunoprecipitable heme-free precursor of catalase arises in the cytosol and is transferred into the peroxisomes, where presumably the active tetrameric catalase is assembled. LAZAROW and DE DUVE speculate that this precursor is synthesized on free polysomes and note some contradictions to the "budding" hypothesis of peroxisome biogenesis. However, there is no evidence contrary to the alternative hypothesis that some of the peroxisomal proteins (especially the matrical ones) are synthesized on free polysomes, whereas others (especially the intrinsic membrane proteins) are formed on ribosomes bound to the endoplasmic reticulum.

The latter mechanism has recently been studied in some detail (CHANG et al. 1978, KATZ et al. 1977, ROTHERMAN and LENARD 1977, TONEGUZZO and GOSH 1978). In a process mediated by amino-terminal signal sequences, such proteins are transferred through the membrane during their synthesis by ribosomes bound to the endoplasmic reticulum membrane by specific junctions. In this "cotranslational transfer" (BLOBEL and DOBBERSTEIN 1975 a, b) secretory proteins, beginning with their amino-terminals, slide into the lumen of the endoplasmic reticulum. If this cotranslational transfer can be interrupted at a certain stage of protein synthesis, the carboxy-terminal stays on the cytoplasmic side of the endoplasmic reticulum and the protein spans the membrane. The interrupted form of cotranslational transport becomes effective during the fixing of intrinsic transmembrane proteins.

GOLDMAN and BLOBEL (1978) incubated free or membrane-bound polysomes from rat liver in cell-free rabbit reticulocyte lysates. When assayed by immunoprecipitation, catalase and urate oxidase are detected among the translation products only in the free-polysome system or in a system containing phenol-extracted RNA from free polysomes. Whereas serum albumin (when formed in the presence of pancreas microsomes) is not susceptible to proteolytic degradation, urate oxidase and catalase (not segregated into microsomes) are digested. Thus, both these enzymes are not segregated by cotranslational transport but presumably pass the peroxisomal membrane by posttranslational transfer. Although the latter should be mediated by signal sequences of the enzyme precursors which are recognized by receptors in the peroxisomal membrane, no larger precursors of urate oxidase and catalase (containing signal sequences) have been detected. Furthermore, the electrophoretic mobilities of urate oxidase and catalase formed in the cell-free system are identical to those of urate oxidase and catalase isolated from liver. ROBBI and LAZAROW (1978) found that the catalase apomonomer translated in cellfree systems from rat liver polysomal RNA exhibits (on SDS electrophoresis) a molecular weight of about 4,000 daltons higher than that of the monomer of purified catalase. On the other hand, after *in vivo* labeling both the apomonomer precursor and the completed enzyme from peroxisomes also have the higher molecular weight. The authors therefore suspect that (due to a contaminating protease) a smaller artifact is formed from native catalase during the purification procedure. In any case, the question as to whether the signal hypothesis is valid for catalase and uricase remains open.

GOLDMAN and BLOBEL (1978) drew a preliminary scheme of the biogenesis of peroxisomes: The intrinsic membrane proteins enter the rough endoplasmic reticulum membrane by cotranslational insertion; the peroxisome proliferationassociated polypeptide, which increases after application of clofibrate or nafenopin (REDDY and KUMAR 1977), might be such a characteristic intrinsic protein. Thanks to a common sorting sequence, the peroxisomal proteins become separated from other membrane proteins; this movement corresponds to the morphologically observable budding of peroxisomes from the endoplasmic reticulum. The characteristic enzyme proteins of the peroxisomal matrix and core formed on the free polysomes are then towed through a hydrophilic pore, which is opened with the help of a specific signal sequence on the amino-terminal of the enzyme; this is the process of posttranslational transfer.

In conclusion, different peroxisomal proteins may be assumed to arise from different sources; in particular, intrinsic membrane proteins might stem exclusively from ribosomes bound to the endoplasmic reticulum, whereas other proteins (such as the enzymes of the peroxisomal matrix) may be synthesized by free polysomes.

IV. Peroxisomes in Nonmammalian Animal Tissues

IV.A. Birds

Hepatic and renal peroxisomes have been studied in the developing chicken (ESSNER 1970). Peroxisomes with needle- or diamond-shaped crystalloids (so-called avian type) have been seen in hepatocytes at all stages of development, but most frequently at the time of hatching (STEPHENS and BILS 1967). Peroxisomes with crystalloids are relatively rare in hepatocytes of adult animals. Only microbodies devoid of cores have been observed by AFZELIUS (1965) and SHNITKA (1966). Peroxisomes with crystalloids, as well as those devoid of cores, are stained by the cytochemical catalase reaction (ESSNER 1970). In kidney tissue peroxisomes are frequently seen in the proximal convolute, whereas they are absent in epithelia of the distal convolute. The number of kidney peroxisomes that contain a crystalloid increases during development and reaches a maximum in adult animals.

IV.B. Amphibians

Microbodies in the toad (*Bufo marinus*) have been studied at the fine structural level by ROELS et al. (1970). Hepatic peroxisomes measure 0.3–0.5 µm in diameter. They contain a finely granular matrix and never show a core or density. The organelles are fairly reactive when stained with the cytochemical DAB procedure for catalase. Microbodies in kidney tissue measure about 0.6 µm in diameter. They react strongly with the alkaline DAB medium. Contrary to liver peroxisomes, a thin, fibrillar nucleoid is occasionally identified within the finely granular matrix. Kidney peroxisomes are found in the proximal convolute, and are most frequent in the lower part of the proximal tubule. Continuities between microbodies and smooth endoplasmic reticulum have been observed (ROELS et al. 1970).

Microbodies from frog (*Rana pipiens*) hepatocytes have been isolated by
VISENTIN and ALLEN (1969). These particles (or at least some of them) were
shown to contain a nucleoid. The enzymes urate oxidase and allantoinase
follow the distribution of D-amino acid oxidase, α-hydroxy acid oxidase, and
catalase. These observations led to the assumption that amphibian hepatic
peroxisomes fulfill an uricolytic function.

Peroxisomes or microperoxisomes have been identified in the thyroid and
pancreas of the urodele *Pleurodeles waltlii* (PICHERAL 1972). Microperoxisomes are also present in interrenal cells of the urodeles *Triturus cristatus* and
Salamandra salamandra (BERCHTOLD 1975 a, b). In *Triturus* the number of
microperoxisomes in interrenal cells has been shown to be related to the
steroidogenic activity of these cells. Microperoxisomes are only rarely seen
in hypophsectomized animals, whereas their number is drastically increased
after corticotropic stimulation (BERCHTOLD 1975 b). Continuities between
microperoxisomes and smooth endoplasmic reticulum have been demonstrated.
Strikingly elongated organelles are often seen, reaching a length of up to
1.0 μm (with diameters of about 0.1 μm). Microperoxisomes are never seen
in autophagic vacuoles (BERCHTOLD 1975 a).

IV.C. Fish

Hepatic microbodies of the carp (*Cyprinus carpio*) have been studied cytochemically and biochemically. The spherical organelles contain a finely
granular or flocculent matrix, are devoid of cores, and are often gathered in
clusters (Fig. 55). Membrane continuities with smooth endoplasmic reticulum
have been demonstrated (KRAMAR et al. 1974) (Fig. 11 C). Carp liver peroxisomes are 0.2–0.75 μm in diameter and readily stain with the cytochemical
catalase medium. The organelles have been prepared by various centrifugation
procedures (GOLDENBERG et al. 1978 a, KRAMAR et al. 1974); these results are
discussed in detail in Section VI.B.7. Urate oxidase of carp hepatic peroxisomes is bound to the microbody membrane; the organelles are involved in
purine degradation (GOLDENBERG 1977 a, b).

Comparable organelles devoid of nucleoids have been studied in the liver of
Lebistes reticulatus (HEUSEQUIN 1973). These particles display a finely
granular matrix and stain with alkaline DAB media. However, histochemical
methods do not allow a distinction between peroxidase staining and catalase
staining. Two types of peroxisomes can be discerned according to their size:
a) smaller ones with a mean diameter of 0.5 μm; and b) larger ones with
diameters of about 1 μm. In *Lebistes reticulatus* peroxisomes are either isolated
or gathered in clusters (HEUSEQUIN 1973).

In the renal tubules of sticklebacks (*Gasterosteus aculeatus trachurus*), peroxisomes have been shown to contain catalase and D-amino acid oxidase by
means of cytochemical methods (VEENHUIS and WENDELAAR BONGA 1977).
Microbodies with diameters of 0.2–0.5 μm have been found in epithelial cells
of all kidney tubules, but most frequently in the second segment of proximal
tubules. The matrix of these organelles is homogeneously electron dense without crystalline cores. No membrane continuities with the endoplasmic

reticulum have been observed. The organelles react strongly for catalase and D-amino acid oxidase.

The volume density of peroxisomes does not differ between fresh-water and sea-water fishes. Since fresh-water and sea-water fishes differ significantly in secretion and reabsorption of ions, there is no evidence that peroxisomes are involved in this part of kidney function (VEENHUIS and WENDELAAR BONGA 1977). Continuities between microperoxisomes and smooth endoplasmic reticulum have been observed in kidney epithelium of *Cyprinus carpio* (Fig. 11 C).

The yolk and the *zona radiata* of fertilized Atlantic salmon (*Salmo salar*) eggs have been studied for urate oxidase activity (HAMOR and GARSIDE 1973). Positive histochemical staining is restricted to small granules and vesicles, suggesting the presence of peroxisomes.

IV.D. Arthropods

IV.D.1. Insects

Microbodies have been studied during the development of the fat body of *Calpodes etlius* and catalase and urate oxidase activity have been demonstrated with cytochemical methods (LOCKE and McMAHON 1971). Insect peroxisomes are ovoid, measuring about 1.1×0.9 µm (mature microbodies). They display a granular matrix and a core that is made up of coiled tubules. Microbodies decrease in size during the fifth larval stage and are destroyed by autophagy at the end of this stage. Another process of "atrophy" occurs during the fourth larval stage. Atrophy is not coupled to or followed by autophagy. During atrophy the core is reduced and myelin figures appear in microbodies. The amount of myelin increases during this time and the small vesicles that remain are finally lost (LOCKE and McMAHON 1971). Continuities between microbodies and endoplasmic reticulum are frequently observed during the formation and growth of the organelles.

It has been suggested that the function of microbodies in the insect fat body (which can be compared to the liver of vertebrates) is to convert urate to allantoic acid (LOCKE and McMAHON 1971).

Morphological studies have confirmed the presence of microbodies in oenocytes of *Gryllus bimaculatus* (ROMER 1974). The organelles are spherical or ovoid, measuring 0.1–1.0 µm in diameter. The matrix is granular and occasionally tubular inclusions are seen. The number of peroxisomes in *Gryllus* oenocytes depends on the moulting stage. Microbodies increase continually in size and number after moulting. As reported in *Calpodes* (LOCKE 1969), microbodies in *Gryllus* oenocytes seem to originate from the endoplasmic reticulum (ROMER 1974). Comparable observations have been reported by GNATZY (1970) in *Culex*, and by ROMER (1973) in oenocytes of *Blaptica*, *Heliothrips*, and *Gerris*. Using a pH 9.0 DAB medium, CASSIER and FAIN-MAUREL (1972) succeeded in demonstrating catalase in irregularly formed microbodies of *Locusta* oenocytes.

Abundant peroxisomes have been identified with DAB methods in the firefly photocytes (*Photuris*, mainly in *Photuris versicolor*; HANNA et al.

1976). The organelles are 1–2 μm in diameter, display a granular matrix, and show a dense core. It seems that each microbody contains a core (OERTEL et al. 1976). HANNA et al. (1976) reported possible continuities with the endoplasmic reticulum. They also suggested a possible involvement of peroxisomes in the process of chemiluminiscence, probably by being responsible for recycling of oxidized D-luciferin.

Microperoxisomes have been identified in oocytes of *Drosophila melanogaster* during the latest stages of oogenesis (GIORGI and DERI 1976). The organelles have been shown to contain catalase by means of cytochemical methods. They are continuous with the endoplasmic reticulum. It is interesting to note that GIORGIO and DERI (1976) also observed a positive catalase reaction in those segments of the smooth endoplasmic reticulum connected to the peroxisomes. This is the only reported observation of this nature.

IV. D. 2. Crustacea

Peroxisomes have been identified both biochemically and cytochemically in the digestive ceca of two species of *Porcellio* (*P. dilatus* and *P. laevis*; DONADEY 1975).

IV.E. Molluscs

Epithelial cells in the digestive diverticulum of the protobranch bivalve *Nucula sulcata* have been shown to contain peroxisomes (OWEN 1972 a, b). The organelles include crystalline cores and stain with DAB in the presence of hydrogen peroxide. Peroxisomes in bivalves are relatively large, measuring up to 1.5 μm in length and 0.5–0.8 μm in cross section. DAB reaction product seems to be evenly distributed throughout the organelles; the regions of the cores, however, are somewhat more reactive than adjacent areas (Fig. 5 b from OWEN 1972). Peroxisomes are most frequently seen near lipid inclusions. Microbodies have also been observed in digestive tubules of *Mya arenaria* (PAL 1971, 1972).

DANNEN and BEARD (1977) studied peroxisomes in the cells of the kidney sac in *Arion ater* and *Ariolimax columbianus* (pulmonate gastropodes). The organelles are 0.25 μm in diameter when they appear as circular profiles; sometimes they are up to 0.8 μm long. Morphologically they appear as normal microbodies that sometimes bear a nucleoid (in *Ariolimax* only). Peroxisomes in *Arion* are reactive with alkaline DAB media for catalase staining, but those in *Ariolimax* are not (DANNEN and BEARD 1977).

IV.F. Worms

Biochemical studies of AVERON and ROTHSTEIN (1974) provided evidence for the presence of peroxisomes in *Turbatrix aceti*. These particles lack urate oxidase, but contain catalase, D-amino acid oxidase, α-hydroxy acid oxidase, and isocitrate lyase. Their sedimentation characteristics are different from those of rat liver peroxisomes. It is not possible to separate nematode peroxisomes from mitochondria by means of gradient centrifugations methods (sucrose).

IV.G. Coelenterates

Microperoxisomes have been identified in two species of *Hydra*, *H. littoralis* and *H. pseudoligactis* (HAND 1976), by cytochemical methods for catalase and α-hydroxy acid oxidase activity. Microperoxisomes of characteristic size and shape are present in epitheliomuscular, digestive, and gland cells. The organelles are round or oval in shape with diameters of 0.13–0.4 μm. Elongated profiles reach up to 0.8 μm in length. The organelles readily stain with alkaline DAB media for catalase. The amount of reaction product after incubation for α-hydroxy acid oxidase depends on the substrate employed (see Section II.B.4.). Microperoxisomes in *Hydra* are seen near lipid droplets and glycogen areas (HAND 1976).

V. Peroxisomes and Related Particles in Protozoa

V.A. Morphology and Cytochemical Identification

V.A.1. Aerobic Protozoa

Peroxisomes of *Tetrahymena pyriformis* are round or oval bodies 0.2–0.7 μm in diameter (average 0.4 μm). They are bounded by a tripartite unit membrane 8 nm thick. The matrix material is amorphous or finely granular and displays moderate electron density in glutaraldehyde/osmium tetroxide-fixed specimens (FOK and ALLEN 1975, HIRAI 1974, WILLIAMS and LUFT 1968). Continuities between peroxisomes and endoplasmic reticulum have not been observed. Nucleoids are absent in the GL strain of *Tetrahymena* (FOK and ALLEN 1975, but see below).

Prolonged incubation (up to 4 hours at 37 °C) in alkaline DAB media in the presence of hydrogen peroxide causes a dense reaction product that is evenly distributed in the peroxisome matrix. The staining intensity increases with incubation time. No variation in staining intensity is observed within the peroxisome population of a given cell, but variations between different cells have been observed (FOK and ALLEN 1975). The reaction is inhibited in the presence of 20 mM aminotriazole, 100 mM sodium azide, or 100 mM potassium cyanide.

In contrast, HIRAI (1974) could not find any DAB-reactivity in *Tetrahymena* microbodies. Considering the reliability of DAB methods, this is a surprising and important fact. It may be of particular significance that HIRAI used a 1–2 weeks old stationary culture of *Tetrahymena* (no catalase staining), whereas FOK and ALLEN (1975) studied exponentially growing cultures (positive catalase reaction).

STELLY et al. (1975) found crystalline cores in the peroxisomes of the GL strain of *Tetrahymena*. This serves as an example of the fact that crystalloids may bee seen in peroxisomes that are devoid of urate oxidase (MÜLLER et al. 1968). It is disappointing that STELLY et al. (1975) found that *Tetrahymena* peroxisomes react with DAB media at pH 6.0, but not at pH 9.0. No explanation could be given by the authors.

Abundant microbody-like organelles have been identified in acetate- and ethanol-grown *Euglena gracilis* cells (GRAVES *et al.* 1971), whereas these organelles were only rarely seen in glucose-grown *Euglena* cells. These microbodies measure 0.4–0.8 μm in diameter and contain a finely granular matrix. Some of the organelles include a core; others contain membranous material of whorl-like configuration.

Morphologically comparable organelles have been found in *Paramecium* (diameter 0.3–1.0 μm). These microbodies are devoid of cores but contain a few tubular inclusions reminiscent of short mitochondrial cristae (WILLIAMS and LUFT 1968). They stain with DAB in the presence of hydrogen peroxide at pH 9.0 (STELLY *et al.* 1975). Comparable morphological and histochemical observations have been reported by CHILDS (1973 a), who studied peroxisomes in the amoeba *Hartmanella culbertsoni*. Cell fractionation studies in this amoeba showed that catalase and urate oxidase are associated with an identical particle population that is different from lysosomes (CHILDS 1973 b). Similar results have been obtained with *Acanthamoeba*, the peroxisomes of which also include catalase and urate oxidase (MÜLLER and MØLLER 1969). In CHILDS' study (1973 b) the various fractions were morphologically controlled. Isolated peroxisomes appeared less homogeneously filled with matrix material and also DAB-reactive proteins behave in the same manner. Most probably this is caused by the leaking of microbody contents during the preparation steps.

Studies of intact protozoa have shown a positive DAB-reaction in *Paramecium aurelia* (AUDETTE 1975). The histochemical catalase reaction is negative in *Volvocida* and in 13 dinoflagellata (MÜLLER 1975). A positive reaction for oxidases has been obtained in *Chlamydomonas reinhardtii* (GIRAUD and CZANINSKI 1971).

Microbodies of usual morphology have been observed in *Trypanosoma brucei* and in *T. equiperdum* (STEIGER 1973). The organelles contain a crystalline core and for some time were erroneously believed to contain α-glycerophosphate oxidase ("α-glycerophosphate oxidase bodies"; but see Section V.B.4.). Microbodies of *T. brucei* and *T. cruzei* are also positive for peroxidatic activities (BAYNE *et al.* 1969, WARTON 1975, WARTON and MODLINSKA 1976); however, catalase and the usual peroxisomal oxidases are not present in *T. brucei*. Microbodies in lower trypanosomes stain with the histochemical catalase medium (*Crithidia fasciculata;* MUSE and ROBERTS 1973).

V.A.2. Anaerobic Protozoa

In *Trichomonas suis* (a flagellate that has no mitochondria) paraxostylar and paracostal granules have been identified as organelles of terminal oxidation, comparable to peroxisomes (CORNFORD and OUTKA 1970). These bodies are about 0.5 μm in diameter and display a homogeneous matrix with an amorphous density. This nucleoid stains with cytochemical catalase media. The membrane of the organelles is provided with a tetrazolium-reducing flavoprotein oxidase that also reacts with oxygen (CORNFORD and OUTKA 1970). CORNFORD and OUTKA (1970) propose to classify these organelles as type of peroxisomes; they are now termed "hydrogenosomes". Hydrogenosomes contain hydrogenase and pyruvate synthase activities instead of catalase or other peroxisomal enzymes (LINDMARK and MÜLLER 1973).

V.B. Biochemical Characterization of Microbodies and Microbody-Like Particles

The specialized types of microbodies described in protozoa thus far reflect the intermediate position of this phylum between plants and animals and its physiological plasticity. Evolutionary adaption (*e.g.*, to the conditions of parasitic life) might also have contributed to the development of an impressive variability in metabolic patterns. Only some basal metabolic types of protozoan microbodies will be discussed in this section.

V.B.1. Peroxisomes

Most microbodies studied in protozoa seem to contain catalase. The presence of hydrogen peroxide-producing oxidase has been described in some cases and urate oxidase has been found (*e.g.*, in the isolated microbodies of *Acanthamoeba*), but D-amino acid oxidase and α-hydroxy acid oxidase have not been demonstrated (MÜLLER and MØLLER 1969). It has already been mentioned (Section I.E.2.) that urate oxidase and catalase are the only peroxisomal enzymes in peroxisomes of *Dictyostelium discoideum* (a slime mold that develops myxamoebae). On the other hand, the peroxisomes of the ciliate *Tetrahymena pyriformis* (the example of a preparation of protozoan peroxisomes; BAUDHUIN *et al.* (1965 b), contain α-hydroxy acid, glyoxylate, and D-amino acid oxidases, whereas urate oxidase is not detected (BAUDHUIN *et al.* 1965 b, MÜLLER *et al.* 1968).

V.B.2. Glyoxysomal Particles and "Leaf" Peroxisomes of Phytoflagellates

In *Euglena gracilis* the microbodies contain enzymes of the glyoxylate cycle (WHITE and BRODY 1974) and β-oxidation (GRAVES and BECKER 1974). Whereas levels of the peroxisomal marker enzymes (enzymes of the glycolate pathway, *i.e.*, hydroxypyruvate reductase and glycolate dehydrogenase) rise dramatically when the acetate-supplemented culture is exposed to light for 24 hours ("greening") or longer periods, specific activities of the glyoxylate cycle markers malate synthase and isocitrate lyase remain essentially unchanged; catalase is increased (1.4-fold). From these observations and from the appearance of multilobed microbodies during greening, WHITE and BRODY (1974) conclude that *Euglena* contains two classes of microbodies: a) particles with glyoxylate cycle enzymes and catalase (glyoxysomes), whose number is not changed appreciably during greening, and b) particles with peroxisomal enzymes (corresponding to the leaf peroxisomes of higher plants), which are multiplied during greening (presumably by division).

The inverse adaptation has also been described in *Euglena:* The transition from photoautotrophic growth to heterotrophic growth on acetate induces an increase in the specific activities of malate synthase and isocitrate lyase, whereas the specific activity of hydroxypyruvate reductase remains constant.

Heterotrophic growth causes a switch to the glyoxylate cycle by derepression of the corresponding enzymes (COLLINS and MERRETT 1975).

As mentioned above, the peroxisomes of *Euglena* possess glycolate dehydrogenase instead of glycolate oxidase (which is present in leaf peroxisomes of higher plants). The peroxisomal glycolate dehydrogenase is found in *Euglena* only under conditions favorable for photorespiration (illumination in air without carbon dioxide); cells transferred to darkness or a nitrogen atmosphere or heterotrophically grown cells are devoid of peroxisomal glycolate dehydrogenase. However, the mitochondrial moiety of this enzyme which delivers reducing equivalents for coupled oxidative phosphorylation (COLLINS et al. 1975, YOKOTA et al. 1978 b), is found under all growth conditions (YOKOTA et al. 1978 a).

V.B.3. Hydrogenosomes of Trichomonads

The microbody profiles detected in anaerobic protozoa, *e.g.*, aerotolerant trichomonads (see Section V.A.1.) seem to be devoid of catalase and other peroxisomal markers. MÜLLER (1973 a, b) isolated these particles from *Tritrichomonas foetus* by isopycnic sucrose gradient centrifugation (buoyant density of the particles $1.24–1.26 \text{ g/cm}^3$). Catalase in the nonsedimentable portion of the cytoplasm was not derived from microbody-like structures, which were obviously undamaged, as proved by latency studies.

Trichomonads do not contain mitochondria, and the microbody-like structures are the only redox organelles active in the disposal of reducing equivalents arising from glucose degradation. In *Tritrichomonas* the glycolytic accumulation of lactate is avoided by reduction of hydrogen ions to molecular hydrogen due to reducing equivalents originating from oxidative decarboxylation of pyruvate (RYLEY 1955). Thus, an additional energy-conserving step becomes possible, namely the generation of ATP at the expense of acetyl-CoA formed from pyruvate.

LINDMARK and MÜLLER (1973), who proved this pathway to be localized in the microbody-like particles, coined the term "hydrogenosomes". The first step that takes place in the particles, the decarboxylation of pyruvate, is coupled to a transfer of reducing equivalents to some primary electron acceptor, presumably ferredoxin (as in hydrogen-producing bacteria such as *Hydrogenomonas*). This acceptor acts as coenzyme of the hydrogenase and hydrogen ions are reduced to molecular hydrogen:

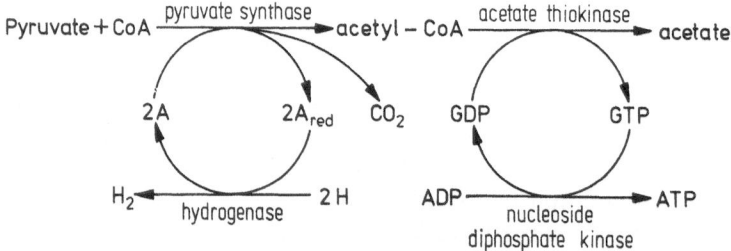

where A is an electron acceptor and A_red is the reduced form.

Glycerol-3-phosphate dehydrogenase and NAD(P)-linked decarboxylating malate dehydrogenase have also been detected in isolated hydrogenosomes (MÜLLER 1973 a, b, LINDMARK and MÜLLER 1973). The latter enzyme converts pyruvate into malate, and as such initiates the formation of succinate; succinate, in addition to acetate, is found as a product of aerobic and anaerobic carbohydrate degradation in *Tritrichomonas* (RYLEY 1955).

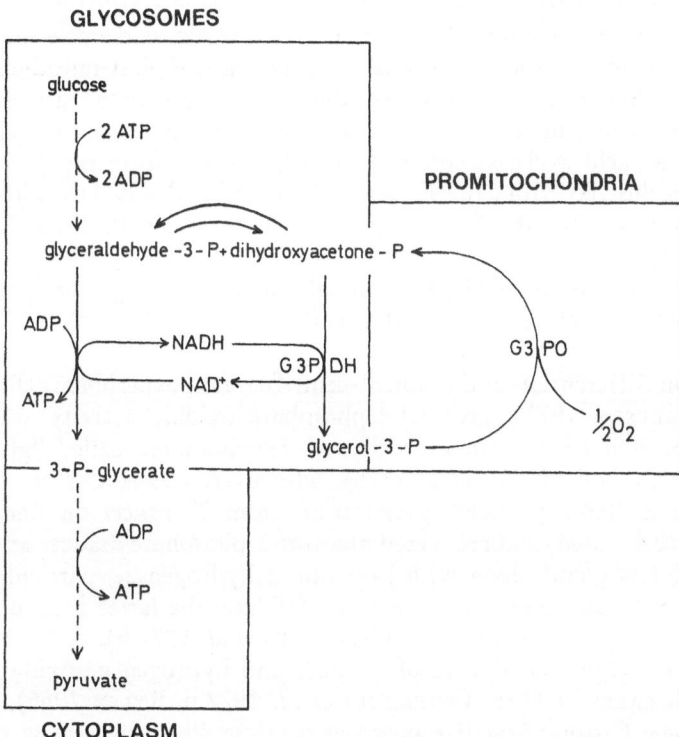

Fig. 39. Degradation of glucose in *Trypanosoma brucei* by division of labor between glycosomes, promitochondria, and cytoplasm. *G 3 PDH* glycerol-3-phosphate dehydrogenase; *G 3 PO* glycerol-3-phosphate oxidase (other enzymes omitted); *glycerol-3-P* glycerol-3-phosphate, etc.

LINDMARK and MÜLLER (1974) demonstrated the presence of cyanide-insensitive superoxide dismutase activity in *T. foetus* and *Monocercomonas* sp. In *Tritrichomonas* about one-sixth of the total activity is found in the hydrogenosomes. It traces back to a form of the enzyme that is electrophoretically discernible from the cytoplasmic one but has a similar molecular weight of about 38,000. The occurrence of superoxide dismutase in hydrogenosomes is surprising. This enzyme has been reported to be absent from microbodies in liver (PEETERS-JORIS et al. 1975) and leaves (ASADA et al. 1973) and is located rather in the cytosol and in the mitochondria. Thus, in the anaerobic trichomonads—in which the hydrogenosomes are the only redox organelles—hydrogenosomal superoxide dismutase may compensate for mitochondria. One might speculate that in trichomonads superoxide dismutase is the most important prerequisite for aerotolerance (LINDMARK and MÜLLER 1974).

V.B.4. Glycosomes of Trypanosomes

The African trypanosomes (trypanozoa, e.g., Trypanosoma rhodesiense, T. gambiense, T. brucei) cause sleeping sickness in man, wild game, and cattle. They appear in two forms with different carbohydrate catabolism. When living in the tsetse fly, the insect vector of the sickness, trypanosomes are able to degrade glucose completely via the tricarboxylic acid cycle and terminal oxidation thanks to the presence of an intact kinetoplast-mitochondrion. In the animal blood stream, however, the parasite possesses only an inactive promitochondrion, in which the synthesis of cytochromes and enzymes of the tricarboxylic acid cycle is repressed. Thus the latter form supplies its energy demand exclusively from the Embden-Meyerhof-pathway but with pyruvate instead of lactate as the final product (OPPERDOES et al. 1976, VICKERMAN 1965, 1971).

The reoxidation of NADH is brought about by a glycerol-3-phosphate dehydrogenase/glycerol-3-phosphate-oxidase system (GRANT and SARGENT 1960, 1961, GRANT et al. 1961).

Based on differential- and gradient-centrifugation experiments (BAYNE et al. 1969, CLARKSON 1975), glycerol-3-phosphate oxidase activity was thought to be associated with the microbodies of Trypanosoma called "glycerophosphate oxidase bodies" (MÜLLER 1975). However, OPPERDOES et al. (1977 a) fractioned a "large particle" preparation from T. brucei on linear sucrose gradients (0.4–2.0 M) and recovered glycerol-3-phosphate oxidase at a buoyant density of 1.17 g/cm³ along with isocitrate dehydrogenase, particulate malate dehydrogenase, and oligomycin-sensitive ATPase; the latter is an unequivocal marker for the promitochondrion (OPPERDOES et al. 1977 b).

Whereas T. brucei is devoid of catalase and hydrogen peroxide-producing oxidases (CLARKSON 1975, OPPERDOES et al. 1977 b, RYLEY 1955), the insect trypanosome Crithidia luciliae possesses catalase which in sucrose gradient is enriched at the isopycnic position of NAD-linked glycerol-3-phosphate dehydrogenase (1.26 g/cm³; OPPERDOES et al. 1977 b). Therefore, it seems reasonable to consider the microbody-like profiles of all trypanosomes as organelles of the same metabolic type and to use glycerol-3-phosphate dehydrogenase as a marker for them when catalase is absent. In T. brucei NADH-linked glycerol-3-phosphate dehydrogenase is associated with particles banding at 1.23 g/cm³; in morphological controls only microbody profiles are seen, and the fraction contains less than 3% glycerol-3-phosphate oxidase (OPPERDOES et al. 1977 a). Therefore, in trypanosomes only glycerol-3-phosphate dehydrogenase may be ascribed to the microbodies, whereas glycerol-3-phosphate oxidase must be considered as a mitochondrial enzyme.

The concept of the glycerophosphate oxidase body seems to be invalidated and the question remains as to which enzymes besides glycerol-3-phosphate dehydrogenase are really housed by the microbodies of Trypanosoma. OPPERDOES and BORST (1977) localized glycerol kinase and seven glycolytic enzymes (hexokinase, glucosephosphate isomerase, 6-phosphofructokinase, aldolase, triosephosphate isomerase, glyceraldehydephosphate dehydrogenase, and phosphoglycerate kinase) in microbodies of *T. brucei*, thus providing a rather surprising answer to this question (Fig. 39). Obviously this compartmentation of latent glycolytic enzymes is an advantage for the efficiency of the Embden-Meyerhof pathway, the energy source of the blood stream form. DHAP, glycerol-3-phosphate, and 3-phosphoglycerate, the metabolites that glycosomes share with promitochondria and cytoplasm, must pass the glycosomal membrane (Fig. 39). OPPERDOES and BORST (1977) assume that this is done by specific translocators, whereas glucose and glycerol might permeate freely . The formation of glycerol via glycerol kinase (which has been found in the glycosomes) provides an additional possibility for energy conservation. It might become important when, under anaerobic conditions, glycerol-3-phosphate cannot be reoxidized by promitochondria and accumulates largely.

In summary, the result of purely glycosomal reactions is

$$\text{glucose} + \text{ATP} + \text{P}_i \rightarrow \text{3-P-glycerate} + \text{glycerol-3-P} + \text{ADP}$$

In spite of the unfavorable equilibrium, under anaerobic conditions ATP might be formed by glycerol kinase:

$$\text{glycerol-3-P} + \text{ADP} \rightarrow \text{glycerol} + \text{ATP}$$

The sum of both equation is

$$\text{glucose} + \text{P}_i \rightarrow \text{3-P-glycerate} + \text{glycerol}.$$

By means of the reactions in the soluble compartment under anaerobic conditions glycerol and pyruvate are formed as end products:

$$\text{glucose} + \text{P}_i + \text{ADP} \rightarrow \text{glycerol} + \text{pyruvate} + \text{ATP} + \text{H}_2\text{O}$$

Aerobically the net balance of glucose degradation in the three compartments concerned explains the formation of pyruvate:

$$\text{glucose} + \text{O}_2 + 2\,\text{ADP} + 2\,\text{P}_i \rightarrow 2\,\text{pyruvate} + 2\,\text{ATP} + 4\,\text{H}_2\text{O}$$

VI. Microperoxisomes and Catalase-Positive Particles

VI.A. Distribution in Mammalian Tissues and Cells

VI.A.1. Adult Animals

Improved cytochemical methods for visualizing the peroxidatic activities of catalase (FAHIMI 1969) by means of strong alkaline DAB media (HERZOG and FAHIMI 1974 b, NOVIKOFF and GOLDFISCHER 1968, 1969, NOVIKOFF et al. 1972 a) enabled the identification of catalase-positive particles (CPs) in almost all the mammalian cell types investigated thus far. The organelles are bordered by a single unit membrane and lack a crystalline core within the matrix. The matrix is seen to have a moderate electron density, and a fine granularity becomes more prominent after incubation in DAB media. The size and shape of the organelles vary within a wide range; the smallest organelles are spherical with a diameter of about 0.1 μm (Fig. 40). Other particles are ovoid, and in some cases a rod- or worm-like shape is found. The latter CPs are up to 1 μm long, whereas the small diameter does not greatly exceed 0.1 μm (*e.g.*, in yolk sac visceral epithelium of the mouse, HRUBAN et al. 1972 a;

Harderian gland of the mouse and rat, Böck *et al.* 1975). In most cases the organelles are round to oval in shape with a diameter of 0.1–0.35 μm (Fig. 48).

NOVIKOFF and co-workers (NOVIKOFF and NOVIKOFF 1972, NOVIKOFF *et al.* 1973 a) claimed that the peculiar morphology of these organelles, combined with the histochemical demonstration of catalase, would satisfy the term

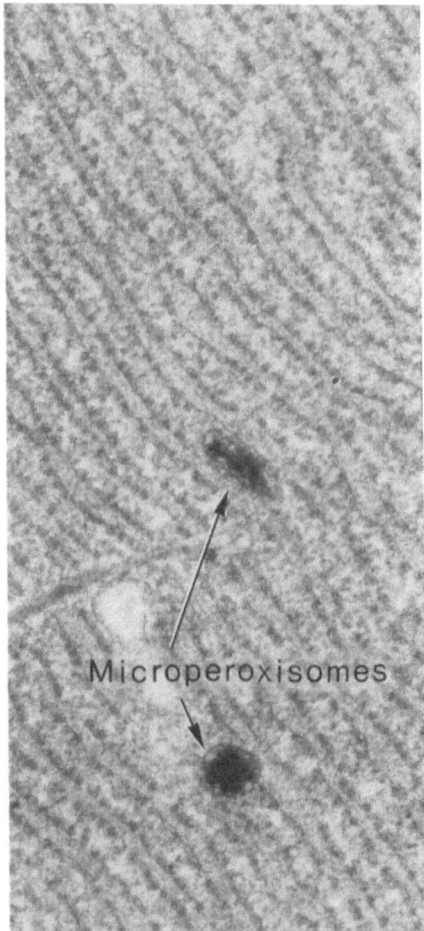

Fig. 40. Mouse exocrine pancreas incubated to demonstrate peroxidatic activity of catalase. Microperoxisomes are rarely seen in the ergastoplasm of acinar cells. ×25,000.

"microperoxisome". However, other authors prefer the term "microbody" (HRUBAN *et al.* 1972 a), which does not suggest the existence of a peroxisomal enzyme system (DE DUVE and BAUDHUIN 1966), or call these organelles CPs (BÖCK 1973 c, BÖCK *et al.* 1975). In fact, biochemical data indicate a remarkable variety in the enzyme patterns of isolated CPs. In the Harderian gland catalase sediments in a particulate fraction, whereas urate oxidase, D-amino acid oxidase, and α-hydroxy acid oxidase activities are not detectable in the same preparation (BÖCK *et al.* 1975). Thus, the particles resemble CPs,

but not (micro)peroxisomal enzyme systems. On the other hand, morphologically comparable organelles isolated from the bovine adrenal cortex
and the rat preputial gland (GOLDENBERG *et al.* 1975 b) contain catalase as
well as D-amino acid oxidase and α-hydroxy acid oxidase, but lack urate
oxidase. D-amino acid oxidase also sediments with catalase in a particulate
fraction prepared from guinea pig intestine (CONNOCK *et al.* 1974). These
organelles resemble true microperoxisomes (for details see Section VI.B.1.).

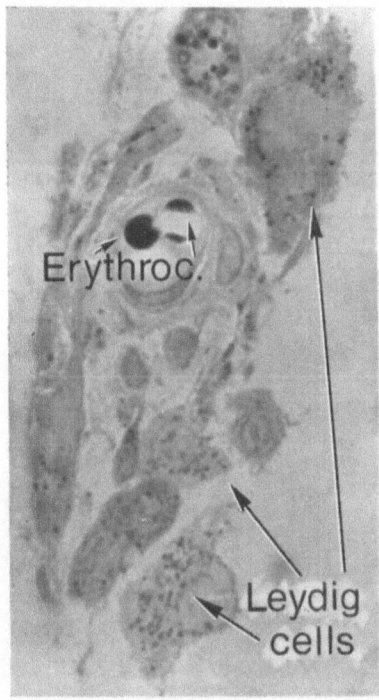

Fig. 41. Rat testicular interstitial tissue incubated to demonstrate peroxidatic activity of
catalase. Numerous microperoxisomes are clearly seen in Leydig cells. Erythrocytes are also
stained. ×500.

Therefore the term CPs is preferred until the peroxisomal nature of the
organelles under investigation can be established.

More recently, NOVIKOFF *et al.* (1973 a) used the term "microperoxisome"
fore coreless liver peroxisomes, such as human liver peroxisomes. These
organelles are up to 0.8 μm in diameter (mean 0.4 μm) and therefore are not
exactly *micro*peroxisomes. The authors claim that a large number of slender
connections exist between the smooth endoplasmic reticulum and this type
of organelle, which is thought to be a specific feature of microperoxisomes.
Although we do not agree with this view (as discussed in Sections I.B.3. and
VI.B.1.), this type of coreless microbodies will be referred to as a microperoxisome in order to avoid the introduction of additional new terms.

Microperoxisomes in human hepatocytes do not contain urate oxidase; the
correlation between the presence of crystalline cores and this enzyme (AFZELIUS

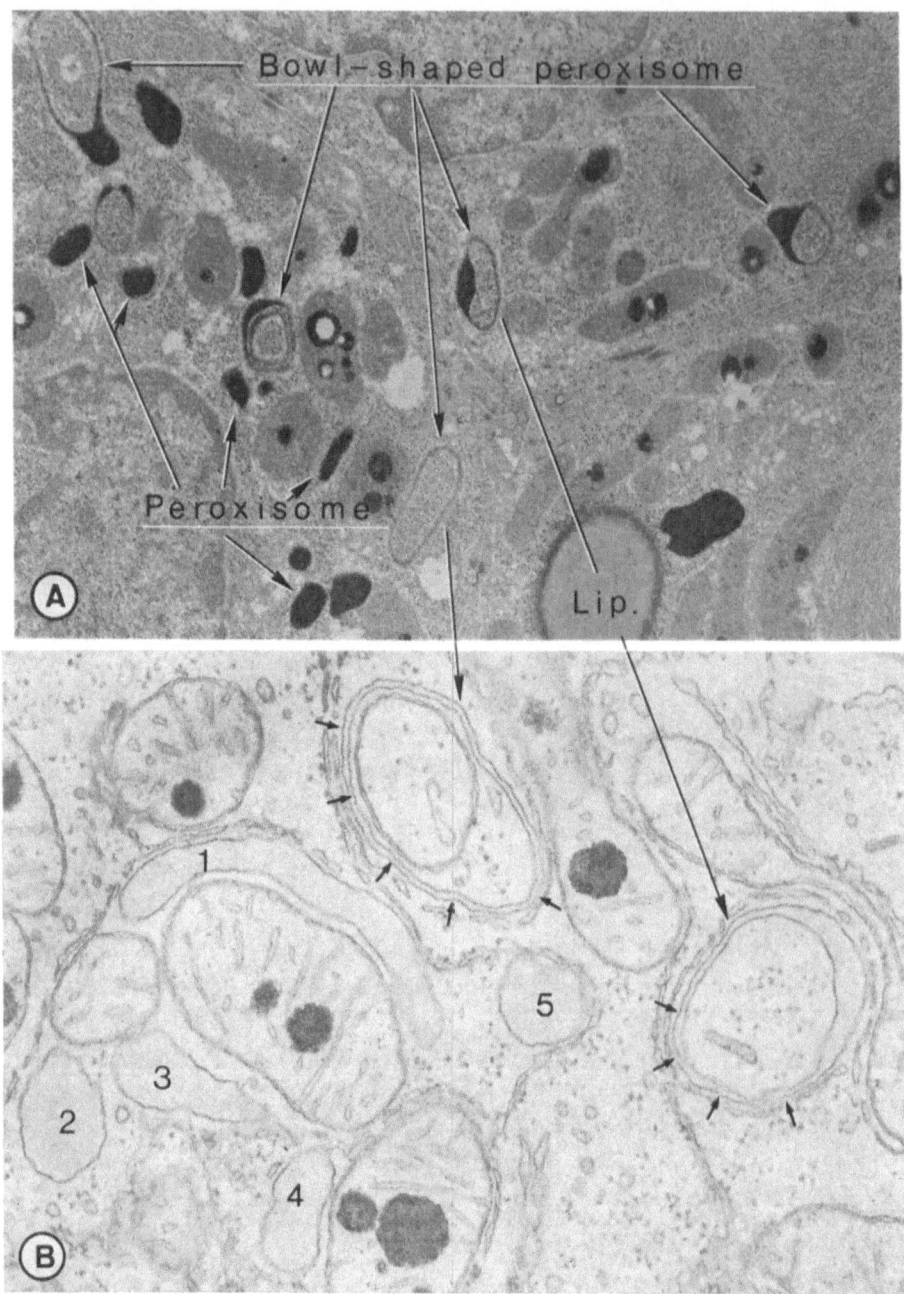

Fig. 42 A, B. Mouse preputial gland. *A* Cytochemical catalase reaction shows several irregularly formed peroxisomes, some of them bowl-shaped. *Arrows:* corresponding profiles in part *B*. *Lip* lipid droplet. Unstained. ×15,000. *B* Thin areas of bowl-shaped peroxisomes (*small arrows*) correspond to positively reacting organelles seen in part *A*. Conventional microbody profiles are numbered *1–5*. Serial sectioning and graphic reconstruction of these profiles reveals their continuity (see Fig. 43). Mitochondria contain large osmiophilic inclusion bodies. Glutaraldehyde/prereduced osmium fixation. ×60,000.

1965) is maintained. However, in nonmammalian hepatocytes, coreless microbodies (microperoxisomes) that contain urate oxidase have been observed.
Their existence was proved in combined morphological and biochemical
studies on carp hepatocytes (*Cyprinus carpio*; KRAMAR *et al.* 1974). In addition, SPORNITZ (1975) reported that hepatocytes of *Xenopus laevis*
(amphibia) occasionally contain microbody-like organelles. Since these
organelles are situated solely at the canalicular pole, and since they lack a
crystalline core or marginal plate, the author was not sure whether to term
them peroxisomes. The absence of a core is remarkable because urate oxidase

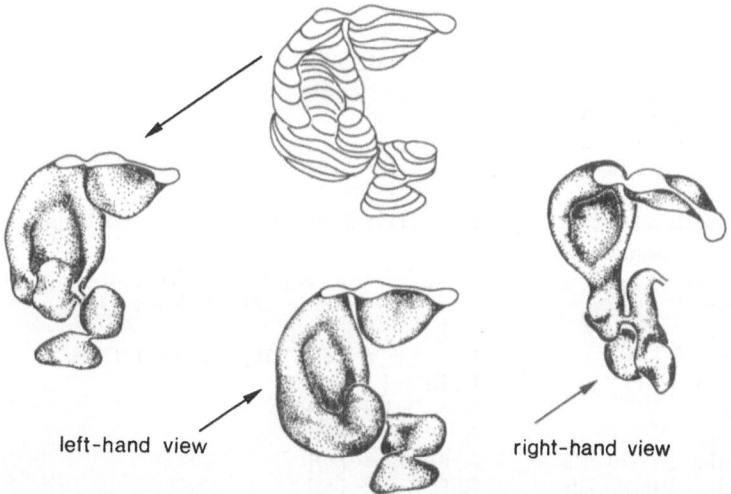

left-hand view   right-hand view

Fig. 43. Graphic reconstruction of serial sections of secretory epithelium of the mouse
preputial gland by means of the Perspektomat P-40 apparatus (Forster, Schaffhausen, Switzer-
land), showing four microperoxisomes connected by delicate channels. One microperoxisome
looks like an erythrocyte, another one is bowl-shaped. Courtesy of K. GORGAS and P. BÖCK.

occurs in other amphibia species (*Rana esculenta*, BROWN 1964; *Rana pipiens*,
VISENTIN and ALLEN 1969). In *Bufo marinus*, ROELS *et al.* (1970) identified
coreless microbodies in hepatocytes by means of the histochemical catalase
reaction.

No significant observations are available concerning a possible relationship
between the size and shape of microperoxisomes and certain cell types or
certain cellular functions. The greatest number of microperoxisomes, however, is regularly observed in lipid-metabolizing cells. This is true in sebaceous,
uropygial, Harderian (Fig. 48) and preputial glands (Fig. 42), brown fat, and
adrenocortical, lutein, and Leydig cells (Fig. 41). HRUBAN *et al.* (1972 a, b)
therefore speculated that certain enzymes involved in cholesterol and steroid
metabolism may be located in microperoxisomes. Several authors take the
widespread distribution of microperoxisomes as an indication that these
particles are constituent organelles and are ubiquitous in mammalian cells
(HRUBAN *et al.* 1972 a, b, NOVIKOFF *et al.* 1973 a). Data concerning their distribution number in various tissues and organs are compiled in Table 27.

Table 27. *Distribution of catalase-positive particles (microperoxisomes?) in mammalian tissues and cells*

Location	Amount present [a]
Respiratory tract	
Ciliated cells	M, ++ [10]
Clara cells	M, R, ++ [42, 49]
Pneumocytes type II	M, R, ++ [42]; G, ++ [30]; rodents, rabbit, monkey, +/++ [49]; pig, ++ [21]
Gastrointestinal tract	
Esophagus	
Keratinocytes	M, + [11]; G, +++ [40]
Stomach	
Keratinocytes	M, + [11]
Surface epithelial cells	M, + [11]
Mucous neck cells	M, + [11]
Oxyntic cells	M, + [11]; G, +++ [40]
Zymogenic cells	M, + [11]; G, ++++ [40]
Intestine, absorptive cells	
Duodenum	G, R, ++++ [38]; G, ++ [30]; M, +++ [43]
Jejunum	G, ++++ [16, 17, 38]; M, +++ [30]; H, ++++ [38]
Ileum	G, ++++ [16, 17, 38]; M, +++ [30]
Coecum	M, R, ++++ [40]
Colon	G, R, ++++ [40]
Rectum	G, ++++ [40]
Intestine, mucous cells	G, R, +++ [40]
Intestine, Paneth cells	G, R, +++ [40]
Male urogenital system	
Testis	
Leydig cells	R, +++ [42 a, 44, 45]; M, ++ [30]; R, + [30]; G, ++ [7 a]; Leydig cell tumor, ++++ [46]
Sertoli cells	Cat, ++ [1]
Seminal vesicle, epithelium	M, + [30]
Prostate, epithelium	M, + [30]
Epididymis, caput, epithelium	M, + [30]
Urinary bladder, epithelium	M, + [30]
Female urogenital system	
Ovary	
Lutein cells	M, ++++ [9]; G, ++ [30]; M, + [30]; dog, + [30]; *Macaca mulatta*, ++ [22, 23]
Granulosa cells	M, +++ [9]
Interstitial cells	M, +++ [9]
Fallopian tube, ciliated cells	M, ++ [10]
Uterus	
Epithelium	M, + [30]; R, ++ [30]
Decidua cells	M, ++ [30]
Smooth muscle cells	M, ++ [30]
Urinary bladder, epithelium	M, + [30]

Table 27 (*continued*)

Location	Amount present [a]
Exocrine glands [b]	
Albuminous glandular cells	
Parotid gland	R, ++ [24]
Pancreas	R, ++ [24]; G, +++ [40]
Submandibular gland	R, ++ [24]; M, + [42 a]
Lacrimal gland	R, ++ [24, 40]
Nasal mucosal glands	R, ++ [24]
Von Ebner glands	R, ++ [24]
Sebaceous glands	
Harderian gland	M, R, ++++ [13]
Preputial gland	R, ++++ [20]; M, +++ [this book]
Uropygial gland	Hen, +++ [30]; duck, ++++ [personal unpublished data]
Supracaudal gland	G, male, +++ [30]
Meibomian gland	Dog, + [30]
Mammary gland	M, ++ [30]
Perianal glands	Dog, ++ [33]
Endocrine cells [b]	
Leydig cells	see [3]
Ovary	see [4]
Suprarenal gland	
Interrenal cells	*Triturus cristatus*, ++ [6, 7]; *Salamandra salamandra*, ++ [6]
Cortex	R, +++ [5, 36, 42 a]; G, +++ [8, 30]; dog, + [30]; cow, ++ [20]
Medulla	R, + [3, 29]
Thyroid	R, + [42 a]
Gastrointestinal endocrine cells	G, R, +++ [40]
Nervous tissue	
Brain	
Neurons	Newborn M, + [30, 31]; newborn R, ++; adult R, + [4, 37]
Catecholaminergic neurons	R, ++/+++ [37]
Oligodendrocytes	R, ++/+++ [37]
Astrocytes	R, + [37]
Spinal cord, neurons	R, + [37]
Spinal ganglion	
Ganglion cells	G, + [30]; R, existent [15]
Satellite cells	R, + [15, 29, 37]
Schwann cells	G, + [30]; R, + [15, 29]
Sympathetic ganglion cells	R, ++ [3]
Adrenal medullary cells	R, + [3, 29]
Ependyma	R, ++ [37]
Eye, pigment epithelial cells	M, R, ++ [34, 35, 47]

Table 27 (continued)

Location	Amount present [a]
Connective tissue	
Fibroblasts	M, ++ [12, 30]; R, + [30]; G, R, ++ [40]
Brown fat tissue	R, +++ [2]; adult M, + [30]; 3-day-old M, +++ [30]; ferret, + [30]
Yellow fat tissue	M, R, ++ [30, 31]
Chondroblasts	M, + [30]
Lymphoid tissue	
Lymph node	M, R, existent [31]
Thymus	M, + [30]
Spleen	G, + [30]; M, R, existent [31]
Lamina propria, trachea	M, ++ [12]
Granulocytes	G, ++ [40]
Endothelial cells	G, R, H, ++ [40]; M, + [30]
Reticuloendothelial cells, spleen	G, + [30]; M, R, existent [31]
Muscular tissue	
Cardiac muscle	M, ++ [26, 27]; R, ++ [25, 28]; rabbit, gerbil, G, *Macaca java, Tupaya*, +/++ [28]
Sceletal muscle	R, ++ [25]
Smooth muscle cells	
Uterus	M, ++ [30]
Aorta	M, + [30]; rabbit, ++ [50]
Kidney [b]	
Proximal convolute	M, +++ [18, 43]
Distal convolute	G, ++++ [14, 39]
Thin limb of Henle's loop	R, ++ [14, 40]
Thick limb of Henle's loop	R, +++ [14, 40]
Collecting ducts	G, +++ [40]
Podocytes	R, +++ [40]; ferret, + [30]
Mesangial cells	R, +++ [40]
Liver [b]	
Hepatocytes	H, R, ++++ [40, 41]; *Cyprinus carpio*, +++ [32]; *Xenopus laevis*, ++ [51]; *Bufo marinus*, +++ [48]
Von Kupffer cells	R, +++ [40]
Bile duct epithelium	R, + [40]
Regenerating hepatocytes	R, ++++ [19]
Gall bladder epithelium	M, existent [17 a]

[a] The quantity of microperoxisomes is given in arbitrary units: +, rare; ++, few; +++, many; ++++, large number. Abbreviations: G guinea pig; H human; M mouse; R rat. The following references are cited in parenthesis.

[b] These data also include observations from lower animals that are of interest in connection with mammalian microperoxisomes.

[1] AHLABO, I., 1972: J. Submicrosc. Cytol. 4, 83—88.
[2] AHLABO, I., BARNARD, T., 1971: J. Histochem. Cytochem. 19, 670—675.
[3] ARNOLD, G., HOLTZMAN, E., 1975: Brain Res. 83, 509—515.
[4] ARNOLD, G., HOLTZMAN, E., 1978: Brain Res. 155, 1—17.
[5] BEARD, M. E., 1972: J. Histochem. Cytochem. 20, 173—179.
[6] BERCHTOLD, J.-P., 1975 a: Cell Tissue Res. 162, 349—356.
[7] BERCHTOLD, J.-P., 1975 b: C. R. Acad. Sci. (Paris) D 281, 543—545.

[7 a] BLACK, V. H., 1974: Anat. Rec. 178, 507.
[8] BLACK, V. H., BOGART, B. I., 1973: J. Cell Biol. 57, 345—358.
[9] BÖCK, P., 1972 b: Z. Zellforsch. 133, 131—140.
[10] BÖCK, P., 1973 a: Z. Anat. Entwicklungsgesch. 141, 103—114.
[11] BÖCK, P., 1973 b: Z. Anat. Entwicklungsgesch. 141, 265—273.
[12] BÖCK, P., 1973 c: Z. Zellforsch. 144, 539—547.
[13] BÖCK, P., GOLDENBERG, H., HÜTTINGER, M., KOLAR, M., KRAMAR, R., 1975: Exp. Cell Res. 90, 15—19.
[14] CHANG, C. H., SCHILLER, B., GOLDFISCHER, S., 1971: J. Histochem. Cytochem. 19, 56—62.
[15] CITKOWITZ, E., HOLTZMAN, E., 1973: J. Histochem. Cytochem. 21, 34—41.
[16] CONNOCK, M., KIRK, P. R., 1973: J. Histochem. Cytochem. 21, 502—503.
[17] CONNOCK, M., POVER, W., 1970: Histochem. J. 2, 371—373.
[17 a] FOMINA, W. A., ROGOVINE, V. V., PIRUZYAN, A., MURAVIEFF, R. A., PODOVINNIKOVA, E. A., 1975: Experientia 31, 1030—1031.
[18] GOECKERMANN, J. A., VIGIL, E. L., 1975: J. Histochem. Cytochem. 23, 957—973.
[19] GOLDENBERG, H., HÜTTINGER, M., BÖCK, P., KRAMAR, R., 1975 a: Histochemistry 44, 47—56.
[20] GOLDENBERG, H., HÜTTINGER, M., LUDWIG, R., KRAMAR, R., BÖCK, P., 1975 b: Exp. Cell Res. 93, 438—442.
[21] GOLDENBERG, H., HÜTTINGER, M., KOLLNER, U., KRAMAR, R., PAVELKA, M., 1978 a: Histochemistry 56, 253—264.
[22] GULYAS, B. J., YUAN, L. C., 1975: J. Histochem. Cytochem. 23, 359—368.
[23] GULYAS, B. J., YUAN, L. C., 1977: Cell Tissue Res. 179, 357—366.
[24] HAND, A. R., 1973: J. Histochem. Cytochem. 21, 131—141.
[25] HAND, A. R., 1974 a: J. Histochem. Cytochem. 22, 207—209.
[26] HERZOG, V., FAHIMI, H. D., 1974 a: Science 185, 271—273.
[27] HERZOG, V., FAHIMI, H. D., 1976: J. Mol. Cell. Cardiol. 8, 271—281.
[28] HICKS, L., FAHIMI, H. D., 1977: Cell Tissue Res. 175, 467—481.
[29] HOLTZMAN, E., TEICHBERG, S., ABRAHAM, S. J., CITKOWITZ, E., CRAIN, St. M., KAWAI, N., PETERSON, E. R., 1973: J. Histochem. Cytochem. 21, 349—385.
[30] HRUBAN, Z., VIGIL, E. L., SLESERS, A., HOPKINS, E., 1972 a: Lab. Invest. 27, 184—191.
[31] HRUBAN, Z., VIGIL, E. L., SLESERS, A., 1972 b: Fed. Proc. 31, 641 (A).
[32] KRAMAR, R., GOLDENBERG, H., BÖCK, P., KLOBUČAR, N., 1974: Histochemistry 40, 137—154.
[33] KUHN, C., 1968: Z. Zellforsch. 90, 554—562.
[34] LEUENBERGER, P. M., NOVIKOFF, A. B., 1973: Exp. Eye Res. 17, 399—409.
[35] LEUENBERGER, P. M., NOVIKOFF, A. B., 1975: J. Cell Biol. 65, 324—334.
[36] MAGALHÃES, M. M., MAGALHÃES, M. C., 1971: J. Ultrastruct. Res. 37, 563—573.
[37] McKENNA, O., ARNOLD, G., HOLTZMAN, E., 1976: Brain Res. 117, 181—194.
[38] NOVIKOFF, P. M., NOVIKOFF, A. B., 1972: J. Cell Biol. 53, 532—560.
[39] NOVIKOFF, A. B., NOVIKOFF, P. M., DAVIS, C., QUINTANA, N., 1972 a: J. Histochem. Cytochem. 20, 1006—1023.
[40] NOVIKOFF, A. B., NOVIKOFF, P. M., DAVIS, C., QUINTANA, N., 1973 a: J. Histochem. Cytochem. 21, 737—755.
[41] NOVIKOFF, P. M., NOVIKOFF, A. B., QUINTANA, N., DAVIS, C., 1973 b: J. Histochem. Cytochem. 21, 540—558.
[42] PETRIK, P., 1971: J. Histochem. Cytochem. 19, 339—348.
[42 a] PICHERAL, B., 1972: J. Microsc. 13, 247—262.
[43] PIPAN, N., PŠENIČNIK, M., 1975: Histochemistry 44, 13—21.
[44] REDDY, J., SVOBODA, D., 1972 a: J. Histochem. Cytochem. 20, 140—142.
[45] REDDY, J., SVOBODA, D., 1972 b: Lab. Invest. 26, 657—665.
[46] REDDY, J., SVOBODA, D., 1972 c: J. Histochem. Cytochem. 20, 793—803.
[47] ROBISON, W. G., JR., KUWABARA, T., 1975: Invest. Ophthalmol. 14, 866—872.
[48] ROELS, F., SCHILLER, B., GOLDFISCHER, S., 1970: Z. Zellforsch. 108, 135—149.
[49] SCHNEEBERGER, E. E., 1972 a: J. Histochem. Cytochem. 20, 180—191.
[50] SHIO, H., FARQUHAR, M. G., DE DUVE, C., 1974: Am. J. Pathol. 76, 1—16.
[51] SPORNITZ, U. M., 1975: Anat. Embryol. 146, 245—264.

The presence of microperoxisomes in the pigmented retinal epithelium and *tapetum lucidum* of the American oppossum has been confirmed by HAZLETT *et al.* (1978). Continuities between these organelles and endoplasmic reticulum have consistently been observed. HAZLETT *et al.* (1978) assume that these microperoxisomes are involved in the maintenance and renewal of the tapetal lipid.

Recent cytochemical studies confirmed the concomitant presence of catalase and D-amino acid oxidase in microperoxisomes. VEENHUIS and VENDELAAR BONGA (1977) identified both these enzymes in coreless peroxisomes of the stickleback kidney tubules, and they have been observed in microperoxisomes in the rat central nervous system (ARNOLD and HOLTZMAN 1978).

CPs of unusual shape are present in the mouse preputial gland (Figs. 42 and 43). Organelles of comparable shape develop in hepatocytes of rats treated with the hypolipidemic drug RMI 14,514 (SVOBODA 1978). As yet, there is no evidence as to the functional significance of such shapes. These organelles look like a bowl or cup with a thickened free margin. Irregularly formed "normal" CPs are also seen in the mouse preputial gland. All CPs are devoid of cores or densities and are connected by delicate canals (Fig. 43). Considering the findings in microperoxisomes from rat preputial gland, it is likely that Figs. 42 and 43 show true microperoxisomes of unusual shape. These organelles are seen very close to cisternae of smooth endoplasmic reticulum and secretory lipid droplets.

VI.A.2. Fetuses and Neonates

Cytochemical studies of CPs during fetal and early postnatal life have been done in the postnatal rat central nervous system (ARNOLD and HOLTZMAN 1978), guinea pig Leydig cells (BLACK 1974), guinea pig adrenal cortex (BLACK and BOGART 1973), mouse enterocytes (CALVERT and MENARD 1978), mouse kidney (GOECKERMANN and VIGIL 1975), rat submandibular gland (MOORIDAN and CUTLER 1978), mouse kidney and enterocytes (PIPAN and PŠENIČNIK 1975), and pneumocytes type II (SCHNEEBERGER 1972 b). Results of these publications are discussed in Section III.E.

In summary, it has been noted that the number of CPs significantly increases with the beginning of specific cellular functions. The decreasing number of CPs observed during the first postnatal weeks in the rat central nervous system seems to be an exception to this rule; however, this finding may be explained with respect to the delay of myelination during this period.

VI.B. Biochemical and Morphological Identification

VI.B.1. Preparation by Centrifugal Methods

Only a few attempts have been made at the preparation and purification of microperoxisomes. In fact, a number of drawbacks make this task quite difficult. First, the number of particles per cell is much lower than the number of peroxisomes per hepatocyte. Second, most of the tissues exhibiting considerable numbers of microperoxisomes are rich in lipid and contain a great deal of connective tissue. Thus, homogenization becomes a crucial step if good yields are required. In some instances this problem can be solved by using an electric knife for prehomogenization (Böck et al. 1975). Third, because microperoxisomes and larger microsomes are comparable in size, differential centrifugation has proved to be a poor method for the purification of these particles.

The usual 5-fraction centrifugation scheme (see Section II.A.1.b.; Figs. 26 and 44) applied to five selected tissues containing nucleoid-free CPs has shown that most of the catalase activity is found in the light mitochondrial

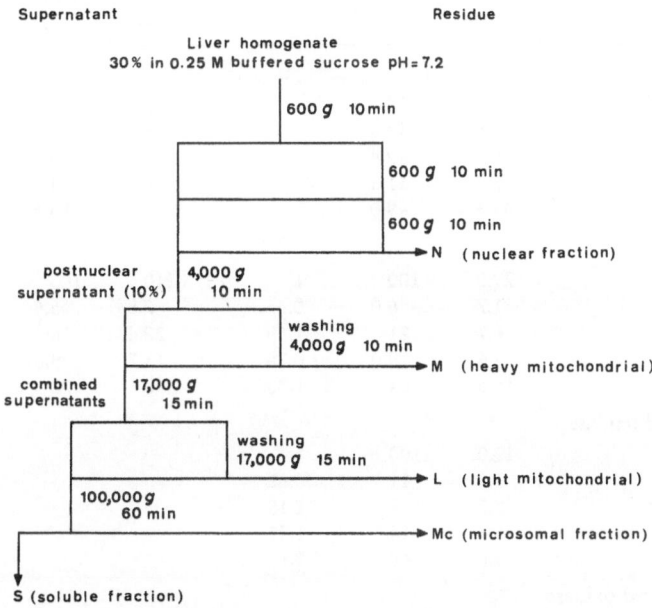

Fig. 44. Differential centrifugation scheme for carp liver tissue (according to GOLDENBERG 1977 b). Homogenizer: POTTER-ELVEHJEM (1,000 rpm, 1 stroke).

and microsomal fractions (Table 28). The microsomal fraction is sedimented at only 38,000 g (instead of the usual 100,000 g applied in the common scheme quoted above), leaving some microsomal material in the supernatant but without considerable loss of microperoxisomes. On the other hand, this situation seems to be more favorable than that encountered in the purification of liver peroxisomes, which overlap with the lysosomes in their key properties (size and density). Like liver peroxisomes, microperoxisomes are indeed more dense than mitochondria and microsomes [1], but, unlike liver particles, they are also much smaller than lysosomes. Thus there is one specific physical difference between microperoxisomes and lysosomes that makes pos-

---

[1] Microperoxisomes dealt with thus far seem to be of lower density than liver peroxisomes. Densities of about 1.20 g/cm³ have been found for particles from enterocytes, brown fat, and lung alveolar cells (CONNOCK et al. 1974, GOLDENBERG et al. 1978 a, PAVELKA et al. 1976).

sible their separation without unphysiological strategems (*e.g.*, the application of Triton WR-1339, used in order to effect the isopycnic banding of liver peroxisomes; LEIGHTON *et al.* 1968).

Rate sedimentation in zonal rotors takes advantage of the smaller size of microperoxisomes: in the zonal rotor loaded with a linear sucrose gradient

Table 28. *Distribution of protein and peroxisomal*

Cell fraction [a]	Bovine adrenal cortex			Pig lung		
	Units/g tissue	Recovery (%)	Rel. spec. act.	Units/g tissue	Recovery (%)	Rel. spec. act.
Protein						
PNS	70.0	100		85	100	
M	12.5	17.8		5.0	5.89	
L	8.7	12.4		2.86	3.36	
Mc	8.0	11.5		9.7	11.43	
S	41.3	49.0		71	83.5	
Catalase						
PNS	29.0	100	1	140	100	1
M	1.7	6.0	0.35	7.84	5.6	0.95
L	6.1	21	1.69	22.4	16	4.76
Mc	2.6	9.0	0.78	13.7	9.8	0.86
S	18.5	64	1.08	98	70	0.84
D-Amino acid oxidase						
PNS	10.0	100	1			
M	1.1	11	0.62			
L	2.7	27	2.18			
Mc	1.4	14	1.21			
S	4.8	48	0.81			
α-Hydroxy acid oxidase						
PNS	12.0	100	1			
M	2.0	18	1.1			
L	3.6	30	2.42			
Mc	2.4	20	1.73			
S	3.8	32	0.54			

Data from: BÖCK *et al.* (1975), GOLDENBERG *et al.* (1975 b), PAVELKA *et al.* (1976), and unpublished data from our laboratory.

Preparation: 2–5 g of tissue was minced with scissors in 10 volumes of 0.25 M sucrose; it was then homogenized for 5 minutes with a modified electric knife (equipped with razor blades) and for 1 minute in a Potter-Elvehjem homogenizer rotating at 1,000 rpm. The differential centrifugation scheme and designation of the fractions were as described for carp liver in Section VI.B.7.b., except that the microsomal fraction was pelleted at 38,000 g. The fractions were washed in 2 volumes of 0.25 M sucrose.

the particles migrate only a little faster than microsomes. Separation from mitochondria and the bulk of lysosomes seems to be fairly good, as seen by succinate dehydrogenase and acid phosphatase activities (CONNOCK 1973, CONNOCK and KIRK 1973).

This separation on the basis of size differences may be combined with a subsequent second separation based on density. It eliminates all particles of equal size but with a smaller density than microperoxisomes (mainly microsomes). CONNOCK et al. (1974) have developed this strategy in preparing microperoxisomes from guinea pig small intestine. Using a B-14 zonal rotor,

enzymes in cell fractions from five selected tissues

Rat preputial gland			Rat Harderian gland			Rat brown fat		
Units/g tissue	Recovery (%)	Rel. spec. act.	Units/g tissue	Recovery (%)	Rel. spec. act.	Units/g tissue	Recovery (%)	Rel. spec. act.
84.0	100		65.2	100		140	100	
10.0	12.0		5.5	8.7		18.2	13	
9.7	11.4		8.3	13.1		30.5	21.8	
11.3	13.4		9.3	14.7		15	10.7	
47.7	57.6		42.0	66.2		75	53.6	
200	100	1	226	100	1	300	100	1
8.0	4.0	0.33	14.6	6.5	0.75	10	3.33	0.256
68	34	2.98	113	50.0	3.81	100	33.3	1.53
99	49	3.77	67.4	29.8	2.20	25	8.33	0.78
21	10.5	0.17	37.4	16.5	0.25	162	54	1
9.0	100	1	Not detectable			Not detectable		
2.16	24	2.0						
1.35	15	1.3						
4.86	54	4.15						
0	0	0						
9.0	100	1	< 5			6.01 (with middle-chain		
1.44	16	1.33				α-hydroxy acids		
2.25	25	2.19				as substrate)		
5.31	59	4.54						
0	0	0						

[a] Protein, as measured by the Lowry method (LOWRY et al. 1957), is given in milligrams. A catalase unit corresponds to that amount of catalase which at 30 °C produces a first-order rate constant of $k = 1/min$. Activities of oxidases are expressed in nanomoles of substrate turned over per minute at 30 °C. Abbreviations: PNS postnuclear supernatant; M heavy mitochondrial fraction; L light mitochondrial fraction; Mc microsomal fraction; S soluble fraction.

they first perform the rate centrifugation procedure on a linear sucrose gradient (0.632–1.177 M, i.e., 20–35% w/w). Then the smaller particle fraction, rich in catalase and arylesterase, is purified by isopycnic banding on a denser linear sucrose gradient (0.988–1.796 M, i.e., 35–52% w/w; Figs. 45 and 46).

Remarkably enough, a smaller but significant fraction of the microsomal marker enzyme (arylesterase) always remains connected to the microperoxisomes. Presumably it corresponds to residual junctions between microperoxisomes and smooth endoplasmic reticulum cut up during homogenization. It has been possible to observe such "stumps" in pelleted peroxisomal preparations (Fig. 11 D–H). Owing to the close connection between microperoxisomes and the endoplasmic reticulum, the separation of CPs leads to great mor-

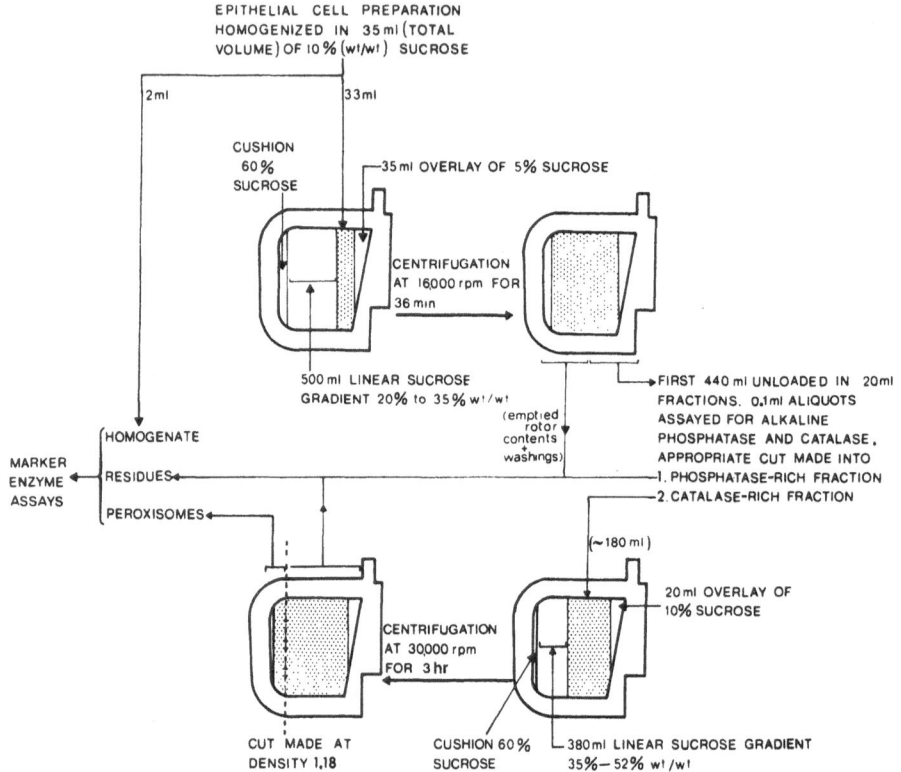

Fig. 45. Preparation of microperoxisomes from guinea pig enterocytes using a B-14 zonal rotor (rate sedimentation followed by isopycnic banding). From: CONNOCK et al. 1974.

phological changes in isolated particles. For example, isolated particles from Harderian glands never exhibit the characteristic rod-like shape found in the original tissue (BÖCK et al. 1975). This effect, which is obviously due to a morphological transformation during the preparation procedures, is discussed in detail in Section VI.B.3.

No method has yet been developed for the isolation of microperoxisomes that does not alter the particles in some respect. Furthermore, even the best preparations (CONNOCK et al. 1974) are not free of contaminations by other organelles. The removal of the endoplasmic reticulum seems to be especially difficult—which is not surprising considering the continuities between microperoxisomes and endoplasmic membranes.

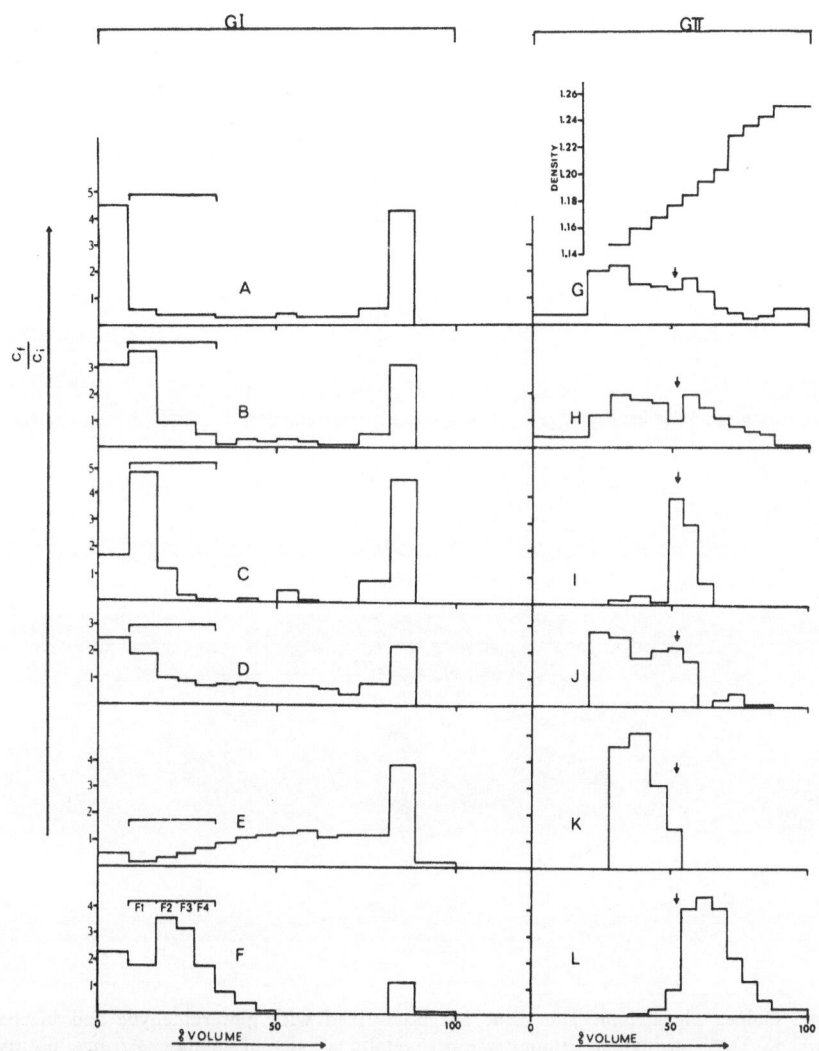

Fig. 46. Distribution of marker enzymes after rate zonal centrifugation (*GI*) of a homogenate from epithelial cells of guinea pig small intestine, followed by isopycnic centrifugation (*GII*) of the catalase-rich fractions from *GI* that contain the slowly sedimenting particles. Conditions as in Fig. 45, except that during the rate zonal step centrifugation time was shorter (22 minutes instead of 36 minutes). *A, G* Protein. *B, H* Aryl esterase. *C, I* Alkaline phosphatase. *D, J* Acid phosphatase. *E, K* Succinic dehydrogenase. *F, L* Catalase. Ordinate: Relative concentration or relative activity (see Section II.A.1.a.). From CONNOCK *et al.* 1974.

VI.B.2. Myocardium

Combined morphological and biochemical studies on microbodies in mouse cardiac muscle tissue have been carried out by HERZOG and FAHIMI (1974 a, c, d). Using an alkaline (pH 10.5) DAB medium, these authors found that peroxidatic activity of catalase is located within membrane-bound particles 0.2–0.5 μm in diameter (Fig. 47). The organelles are close to mitochondria and to the sarcoplasmic reticulum. The histochemical reaction is abolished in control preparations (incubation with 2×10^{-2} M aminotriazole, 10^{-1} M potassium cyanide, or 10^{-1} M azide in the medium). A fine granular matrix of moderate electron density is seen under these conditions. The particles appear round or oval, independent from the section plane (as seen from the

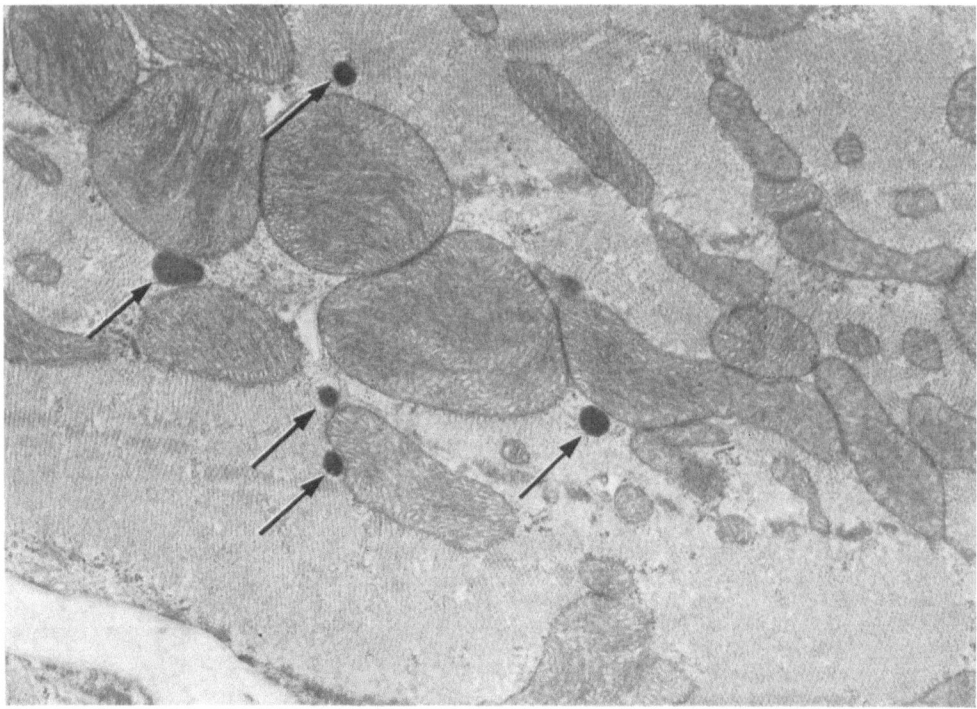

Fig. 47. Guinea pig atrium. The tissue has been fixed with glutaraldehyde and incubated in a pH 9.0 DAB medium to demonstrate peroxidatic activity of catalase. *Arrows:* positively reacting membrane-bound organelles. Note that the round or oval particles are intimately related to mitochondria, which do not react under these conditions. Brief counterstaining with lead citrate. ×17,000.

appearance of adjacent myofibrils). Generally the organelles lack a crystalline core or density (a few exceptions have recently been found; Hicks and Fahimi 1977). Thus they resemble microperoxisomes as described and defined by Novikoff et al. (1973 a, b).

Biochemical examinations were performed with mouse hearts perfused with saline (Herzog and Fahimi 1974 a). The tissue was homogenized and the differential centrifugation scheme of Stein and Stein (1968) was used. Catalase activity sedimented with both the mitochondrial and microsomal fraction. Catalase activity, however, was 164.0 units/mg protein in the microsomal fraction, whereas only 53.3 units/mg protein was measured in the mitochondrial fraction. This corresponds to the size ratio between mitochondria and CPs observed under the electron microscope. In a more recent study, HERZOG and FAHIMI (1975) succeeded in finding hydrogen peroxide-producing

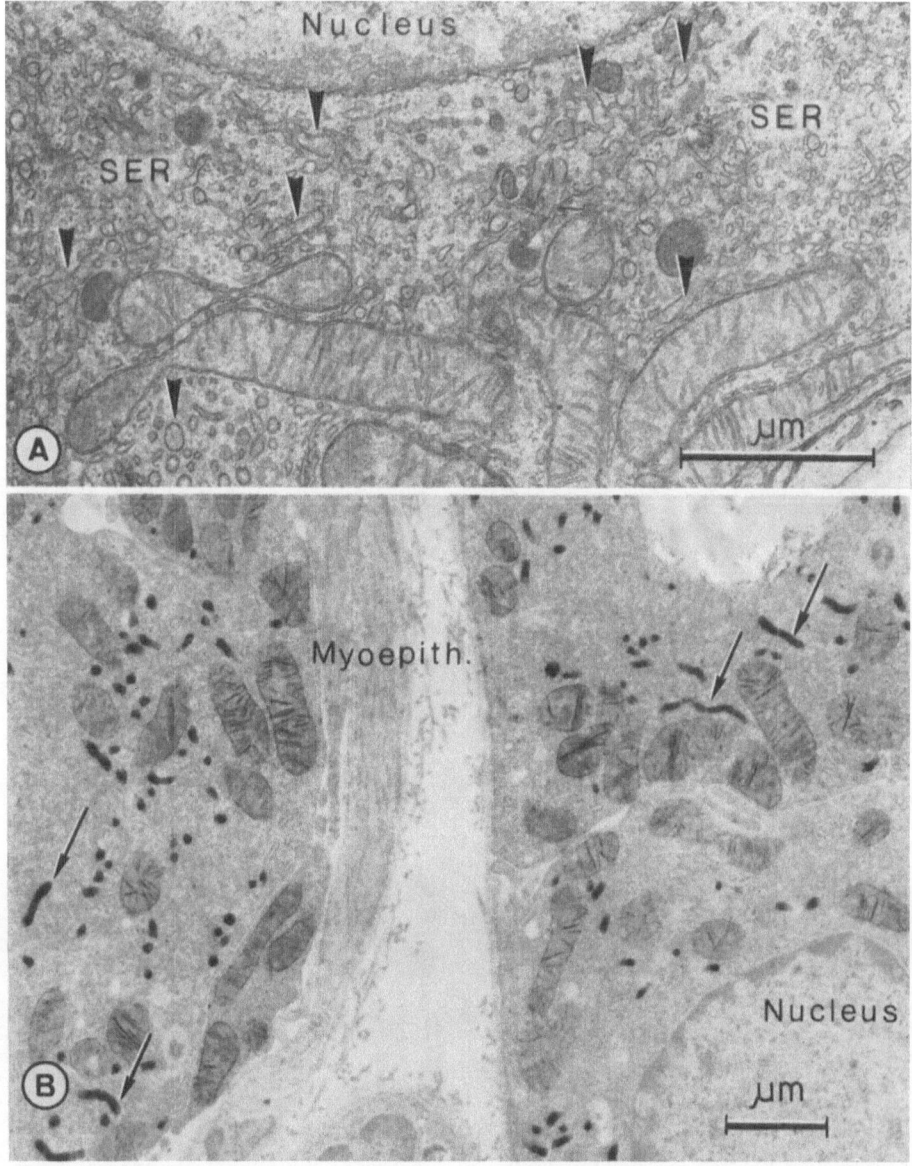

Fig. 48 A, B. Harderian gland of the rat. *A* A cytoplasmic region near the nucleus is seen. *Arrowheads:* some of the numerous profiles of microperoxisomes; the smooth endoplasmic reticulum *(SER)* is remarkably developed. Conventional glutaraldehyde/osmium tetroxide fixation. ×28,000. *B* Numerous microperoxisomes are stained after incubation to demonstrate peroxidatic activity of catalase. *Arrows:* rod- or worm-like organelles. *Myoepith.* myoepithelial cell. Unstained. ×14,000.

activities in this preparation: D-amino acid oxidase and α-hydroxy acid oxidase activities were found to be enriched in the microsomal fraction (0.13 and 0.77 milliunits/mg protein, respectively). The authors suggest that these organelles may be involved in myocardial lipid metabolism and note the fact that lipids act as fuel for the working myocardium (BING 1965).

Application of ethanol (36⁰/o of the dietary calories) over 5 weeks causes rat myocardial catalase to increase more than 100⁰/o, and also the number of

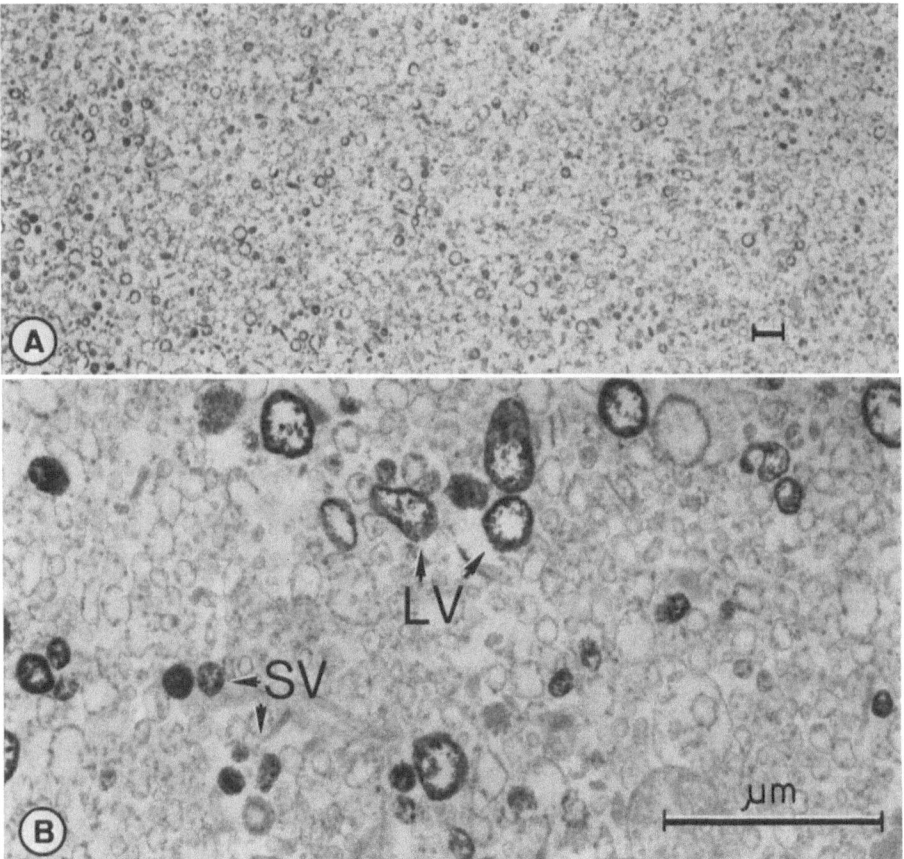

Fig. 49 A, B. Microsomal fraction prepared from Harderian gland of the rat by gradient centrifugation. The gradient was prepared by layering 4 portions of sucrose (2 ml 2 M, 5 ml 1.8 M, 5 ml 1.6 M, and 5 ml 1.4 M) and keeping them overnight. Centrifugation was performed for 4 hours at 96,000 g in the MSE 3 × 20 ml swing-out rotor. The highest catalase activity in the band with a density of 1.19 g/cm³ was pelleted and is shown after incubation to identify peroxidatic activity of catalase. *A* Low-power electron micrograph shows the homogeneity of the pellet. ×4,500 (Bar = 1 μm). *B* Higher magnification shows that the reaction product is restricted to certain smooth-surfaced vesicles. Two types of positively reacting vesicles can be discerned. Smaller vesicles (*SV*) are completely filled with reaction product, whereas in larger vesicles (*LV*) a rim of dense precipitates is seen attached to the inner surface of the membrane. Brief staining with lead citrate. ×30,000.

microperoxisomes in the myocytes is increased after this treatment. Hepatic catalase and peroxisomes, on the other hand, remain unaffected under these conditions (FAHIMI *et al.* 1979).

VI.B.3. Rat Harderian Gland

In mice and rats the Harderian gland is extremely rich in small organelles that react positively to the histochemical DAB method (BÖCK *et al.* 1975). The reaction has been interpreted as being specific for peroxidatic activities of catalase, since it is abolished by the addition of 2×10^{-2} M aminotriazole, whereas it is not markedly influenced by high concentrations of hydrogen peroxide in the incubation medium. In addition to round or oval

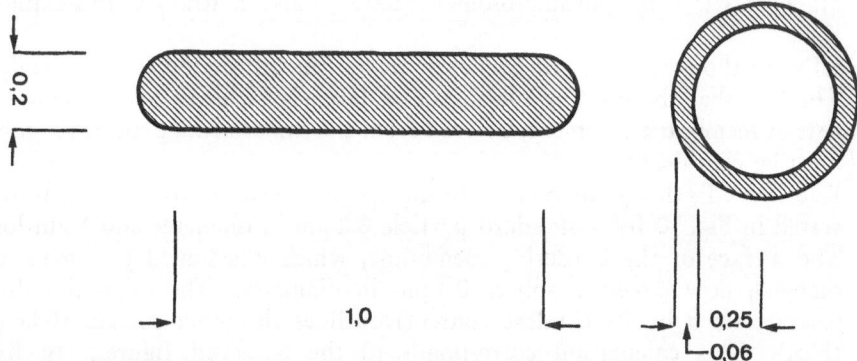

Fig. 50. Transformation of rod-shaped organelles to vesicles and the location of catalase within these vesicles. The surface of the rod-like organelle is identical to that of the sphere; the volume of the rod (*hatched*) is identical to the volume of the hatched zone at the inner surface of the vesicle (hatching symbolizes catalase activity). The indicated dimensions in micrometers fit well with the sizes of catalase-positive particles in the Harderian gland (rods) and in subcellular fractions prepared from this tissue (vesicles).

organelles 0.1–0.3 μm in diameter, rod- or worm-like particles are frequently seen (Fig. 48). They are also 0.1–0.3 μm in diameter and reach 1.5 μm in length. Obviously, the shape of these rod-like organelles depends on the plane of the section. All particles are bounded by a single unit membrane and homogeneously filled with reaction product. According to the criteria of NOVIKOFF and NOVIKOFF (1972), the particles are microperoxisomes.

Subcellular fractions from excised and pooled tissue (15–20 rats, 2–5 g tissue) were prepared as described in Section VI.B.1. and examined biochemically. The activities of catalase, glucose-6-phosphatase, acid phosphatase, and succinate dehydrogenase were measured in the five fractions. The postnuclear supernatant contained 96% of the total catalase activity. Specific catalase activities were 2.65 in the mitochondrial fraction, 13.70 in the lysosomal, and 7.22 in the microsomal; the postmicrosomal supernatant contained only 0.89. Thus the peak of main catalase activity is shifted to lighter fractions (as compared to liver), corresponding to the small size of CPs identified under the electron microscope.

Fragments of the pellets were also subjected to histochemical procedures. The peroxidatic activity of catalase was restricted to microsome-type vesicles, the bordering membranes of which were regularly found to be devoid of ribosomes. The diameters of these vesicles ranged from 0.1 to 0.4 μm. Rod-like particles were never found in the subcellular fractions. The reaction product, which showed faint granularity, was evenly distributed in smaller vesicles, but was attached to the inner surface of the limiting membrane in larger ones. This behavior was also observed when microsomes were prepared by gradient centrifugation (Fig. 49).

The unusual appearance of larger vesicles may be caused by the following:
1. Catalase may leak during the preparation process. However, it is not conceivable that larger vesicles only display this peculiarity, and the low cataalase content of the postmicrosomal fraction is also contrary to this explanation.
2. CPs are thought to be continuous with the smooth endoplasmic reticulum. The fact that the limiting membrane of larger vesicles to some extent consists of membrane fragments derived from the smooth endoplasmic reticulum must be considered.
3. Rod-like CPs change to vesicles during the preparation process. This is illustrated in Fig. 50 for a standard particle 0.2 μm in diameter and 1 μm long. The surface of the bordering membrane, which is assumed to remain unchanged, now covers a sphere 0.4 μm in diameter. The original volume (characterized by its catalase content) outlines the inner surface (0.06 μm thick). This calculation corresponds to the observed figures. In liver peroxisomes the remaining catalase is also found to be mostly attached to the inner surface of the peroxisome membrane and to the surface of the crystalline core when partial leakage has taken place (Fig. 20 B, C). This explanation seems to be the most likely, although point 2 above cannot be excluded.

Although Böck et al. (1975) demonstrated that catalase activity is present in a particulate form by means of biochemical and histochemical methods, they did not detect urate oxidase and D-amino acid oxidase activities. Furthermore, L-α-hydroxy acid oxidase activity was so low that the authors assumed that the recorded activities might represent non-specific dehydrogenation of the substrate by other enzymes.

These results indicate that considerable variation are to be expected in the enzyme patterns in microperoxisomes and, as an extreme, the existence of CPs that do not belong to the peroxisome family should be considered.

VI.B.4. Guinea Pig Enterocytes

Catalase has been seen within subcellular particles in the goldfish intestine (CONNOCK 1973), as well as in guinea pig small intestine (CONNOCK and KIRK 1973, CONNOCK et al. 1974). Histochemical studies by NOVIKOFF and NOVI-KOFF (1972) also revealed catalase located within microperoxisomes (Fig. 51). The concomitant occurrence of catalase and D-amino acid oxidase in subcellular particles of comparable sedimentation properties was shown more recently by CONNOCK et al. (1974). Therefore, it seems to be established that these particles represent a peroxisomal system and the term microperoxisome is justified.

In guinea pig enterocytes carnitine acetyl- and palmitoyltransferases (similar to short-chain acyl CoA synthetase) are exclusively mitochondrial whereas long-chain acyl CoA synthetase is located on the endoplasmic reticulum (MARTIN *et al.* 1979); in rat liver carnitine acetyltransferase and long-chain acyl-CoA synthetase are bound to peroxisomes and microsomes (Table 6). On the other hand in the intestinal mucosa of the goldfish carnitine acetyltransferase was allocated to mitochondria *and* peroxisomes whereas NADP-dependent isocitrate dehydrogenase, glycerol-3-phosphate

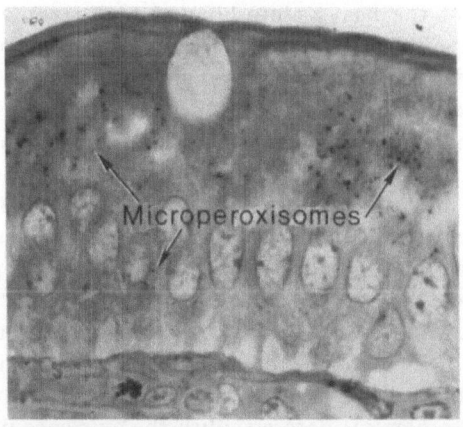

Fig. 51. Intestinal epithelial layer of mouse duodenum. Numerous microperoxisomes are mainly located in the supranuclear region of enterocytes. Glutaraldehyde fixed and incubated in alkaline DAB medium to demonstrate catalase. ×500.

dehydrogenase and xanthine dehydrogenase are soluble enzymes. Glycolate oxidase was not detectable in this tissue (TEMPLE *et al.* 1979).

Microperoxisomes have been prepared from isolated enterocytes of guinea pig small intestine by gradient centrifugation. The method is based on the relatively high density of these organelles (greater than 1.20 g/cm³) when compared with particles of equal size (CONNOCK *et al.* 1974). The relative specific activity of catalase in the microperoxisome-rich fraction is considerably high (about 25%). (The method is reviewed in detail in Section VI.B.1.) Electron microscopic examination of catalase-rich fractions subjected to histochemical DAB staining has shown that the positive particles are 0.15–0.25 μm in size; thus they resemble microperoxisomes as identified in histochemical studies (NOVIKOFF and NOVIKOFF 1972).

VI.B.5. Rat Preputial Gland and Bovine Adrenal Cortex

VI.B.5.a. Rat Preputial Gland

A large number of CPs has been identified in the epithelial cells of rat preputial glands by means of the histochemical catalase reaction (GOLDENBERG *et al.* 1975 b). Both the morphology and size of these particles are comparable to those observed in the Harderian gland (Böck *et al.* 1975). Most of them are worm- or rod-like in shape, about 0.1 μm in diameter, and up to 1 μm long. They are bordered by a unit membrane, are devoid of cores, and homogeneously filled with reaction product when incubated to demonstrate peroxidatic activity of catalase.

For biochemical studies preputial gland tissue from 20 rats was pooled (4–5 g tissue) and processed as described in Section II.A.3. Pellets from the individual fractions were controlled histochemically. The highest catalase activities sedimented with the microsomal fractions (relative specific activity 3.77). Considerable but lower activities were recorded in the light mitochondrial fractions (relative specific activity 2.98), whereas only small amounts occurred in the heavy mitochondrial fractions and postmicrosomal supernatants (relative specific activities 0.33 and 0.17, respectively). Thus the main catalase activity sediments with lighter particles, compared with its distribution in Harderian gland (Böck *et al.* 1975, Table 28).

Histochemical analysis of the pellets confirmed that catalase is located in smooth microsome-type vesicles. In larger vesicles a fringe of reaction product outlined the inner surface of the bordering membrane. It was assumed that the transformation of rod-like particles to spheres causes this particular distribution pattern (GOLDENBERG *et al.* 1975 b), as occurred in the case of the Harderian gland (see Section VI.B.3.).

D-Amino acid oxidase and L-α-hydroxy acid oxidase were also found in measurable amounts, mainly in the microsomal fraction (for the detailed distribution see Table 28). Urate oxidase activity was not detectable. Thus biochemical and histochemical data establish the existence of true microperoxisomes in the rat preputial gland.

VI.B.5.b. Bovine Adrenal Cortex

CPs have been found in adrenocortical cells of various species (BEARD 1972, BLACK and BOGART 1973, GOLDENBERG *et al.* 1975 b, HRUBAN *et al.* 1972 a, b, MAGALHAES and MAGALHAES 1971). These particles are predominantly spherical or ovoid and are 0.1–0.3 μm in diameter. Rod-like

---

Fig. 52 A–F. Purification of CPs from brown adipose tissue of cold-adapted rats by isopycnic density gradient centrifugation of a light mitochondrial fraction. Details of fractionated centrifugation are essentially as given in Fig. 44, except that M and L were pelleted at 4,500 g (10 minutes) and 20,000 g (20 minutes). First density gradient: 0.5 ml light mitochondrial fraction (about 7 mg protein) layered on top of a linear sucrose gradient (10 ml 1.2 M–1.6 M sucrose; additions: 0.1% ethanol, 2% dextran-10, 1 mM EDTA, and 3 mM imidazol-HCl, pH 7.2). MSE 6 × 14 swing out rotor or MSE SS 65 ultracentrifuge was used, 16 hours at 40,000 rpm (about 2×10^5 g). Second density gradient: 3 fractions containing the highest catalase activities layered on a linear sucrose gradient (10 ml 1.4–1.6 M sucrose); conditions as before. *A* Sucrose concentration in first gradient. *B* Carnitine acetyltransferase. *C* Cytochrome oxidase (———) and palmitoyl-CoA-dependent oxygen consumption (--------) in the mitochondrial and peroxisomal fractions. *Shaded area:* carbon monoxide-inhibitable portion of palmitoyl-CoA respiration. *D* Catalase. *E* Sucrose concentration in the second gradient (--------) and catalase distribution (———). *F* Cytochrome oxidase (--------) and reduction of APAD (acetylpyridine analogue of NAD by palmitoyl-CoA (———). From: KRAMAR *et al.* (1978).

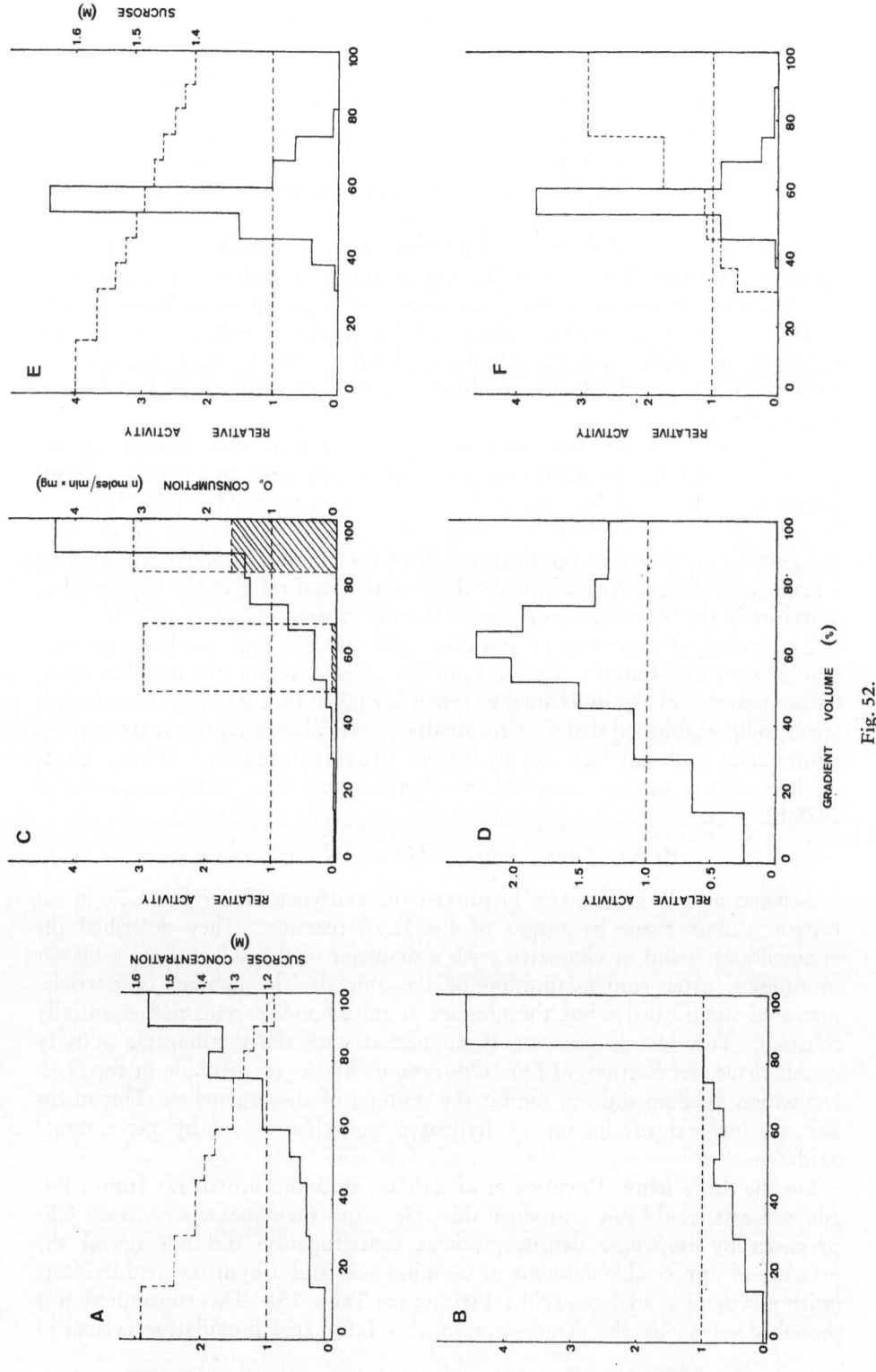

Fig. 52.

organelles are not observed. According to the morphological and histo-chemical criteria, these particles resemble microperoxisomes. In addition, bio-chemical studies of subcellular fractions prepared from bovine adrenal cortex (GOLDENBERG et al. 1975 b) indicate the concomitant occurrence of catalase, D-amino acid oxidase, and L-α-hydroxy acid oxidase in particulate form.

Subcellular fractions were prepared according to the differential centrifuga-tion scheme (see Section VI.B.1.). The greatest catalase activity was found in the lysosomal fraction (relative specific activity 1.69), while the activities in the mitochondrial and microsomal fractions were remarkably lower (relative specific activities 0.35 and 0.78, respectively). Considerable amounts of catalase were measured in the postmicrosomal supernatant (relative specific activity 1.08). These results indicate a) the greater size of these organelles, compared to those from the preputial gland of the rat (corresponding to electron microscopic findings), and b) the greater vulnerability of their border-ing membranes.

When pellets of the individual fractions were controlled histochemically, peroxidatic activity of catalase was seen to be restricted to smooth-surfaced, membrane-bound vesicles. Most of these vesicles contained a faint fringe of reaction product attached to the inner surface of the membrane. In this case the authors propose that the distribution of the reaction product is caused by a leakage of the majority of the catalase content and refer to the high catalase activities in the postmicrosomal supernatant (GOLDENBERG et al. 1975 b).

The principal activities of D-amino acid oxidase and L-α-hydroxy acid oxidase were also found in the light mitochondrial fraction (the detailed distri-bution patterns of the individual enzymes are given in Table 28). Therefore, it seems to be established that CPs from adrenocortical cells represent true micro-peroxisomes with catalase and hydrogen peroxide-producing oxidase. Urate oxidase activity was not detectable in adrenocortical tissue (GOLDENBERG et al. 1975 b).

VI.B.6. Brown Adipose Tissue and Lung Tissue

AHLABO and BARNARD (1971) proved the existence of coreless CPs in rat brown adipose tissue by means of the DAB reaction. They described the organelles as round or elongated with a diameter of 0.1–0.8 μm. As a matter of interest, after cold adaptation of the animals, the number of particles increased significantly, but the number of mitochondria remained essentially constant. This fact is conceivably connected with the thermogenic activity in this tissue (see Section VI.D.). Omission of hydrogen peroxide in the DAB incubation medium did not inhibit the staining of the organelles. This might indicate internal production of hydrogen peroxide caused by peroxisomal oxidases.

On the other hand, PAVELKA et al. (1976), studying brown fat from cold-adapted rats, could not reproduce this effect and their measurements on CPs prepared by isopycnic density-gradient centrifugation did not reveal the presence of appreciable amounts of D-amino acid and α-hydroxy acid oxidases (with glycolate as hydroxyacid substrate; see Table 28). This contradiction is probably solved by the demonstration of a fatty acid β-oxidation system in

brown fat tissue that is supposed to be capable of producing hydrogen peroxide during the acyl-CoA dehydrogenase step (see below).

With middle-chain α-hydroxy acids (α-hydroxycaproate, α-hydroxyisocaproate, and α-hydroxydecanoate), Pavelka et al. (1976) succeeded in

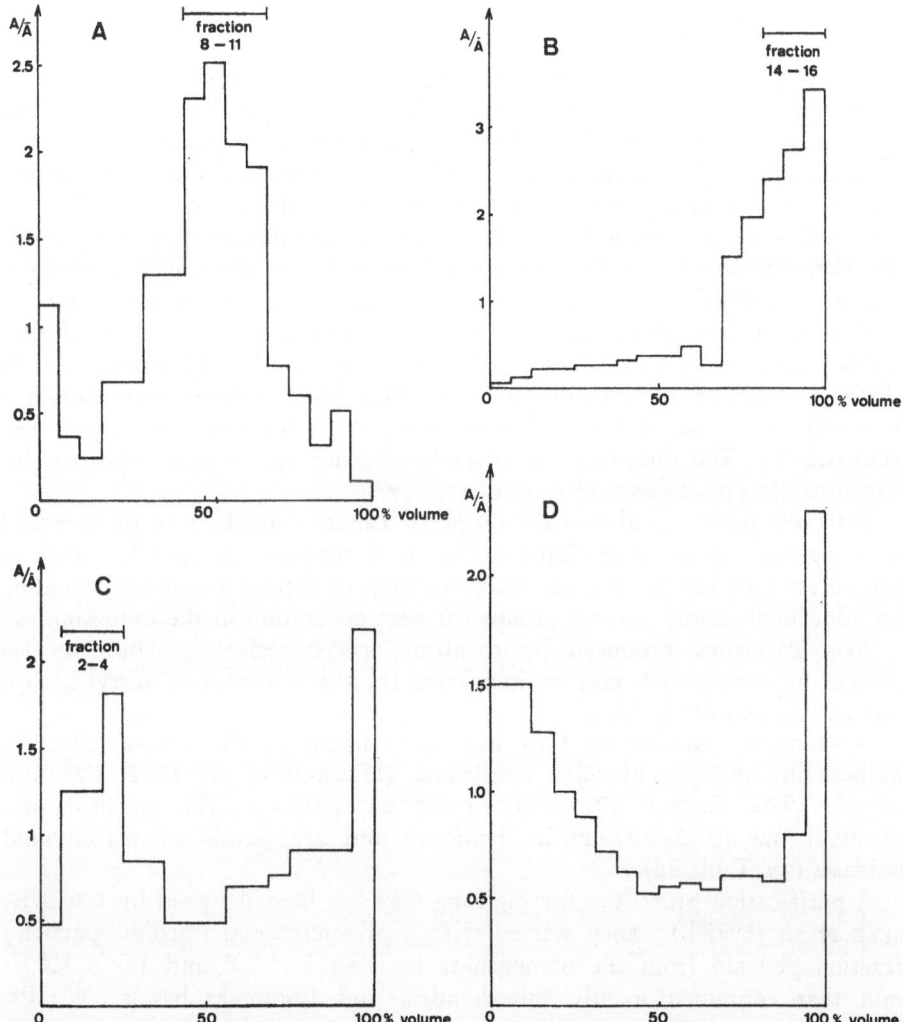

Fig. 53 A–D. Purification of CPs from pig lung by rate zonal centrifugation of a "medium"-
particle fraction (MSE zonal rotor; see text for conditions). A Catalase. B Cytochrome
oxidase. C Acid phosphatase. D Protein. Ordinate: Relative concentration or relative
activity (see Section II.A.1.a.). From: Goldenberg et al. (1978 b).

demonstrating a weak α-hydroxy acid oxidase activity. This finding draws
attention to the fact that rat kidney microbodies contain neither glycolate nor
lactate oxidase activity, whereas α-hydroxyisocaproate and, to some extent,
higher hydroxyacids are rapidly oxidized (Baudhuin et al. 1965 b,
McGroarty et al. 1974).

NAD-dependent glycerol-3-phosphate dehydrogenase—which in rat liver is located in the cytosol *and* in peroxisomes (see Section I.D.1.)—has been found only in the cytosol in brown fat (PAVELKA et al. 1976). This observation is of interest because glycerol-3-phosphate dehydrogenase can cut down triglyceride synthesis (by lowering the glycerol-3-phosphate concentration) in favor of lipolysis, which delivers the fuel for chemical thermogenesis in brown fat tissue.

The physiological significance of peroxisomes in brown adipose tissue seems to be indicated by the recent demonstration of fatty acid β-oxidation in these particles (KRAMAR et al. 1978). Isopycnic gradient centrifugation (1.2–1.6 M sucrose) of a light mitochondrial fraction from brown fat of cold-adapted rats resulted in the separation of CPs from mitochondria. Further purification was achieved by a second density-gradient centrifugation (Fig. 52). Either fraction was able to oxidize palmitoyl-CoA, but only the mitochondrial one exhibited inhibition of palmitoyl-CoA-dependent oxygen comsumption by carbon monoxide, which is known to be a powerful inhibitor of cytochrome oxidase (KEILIN and HARTREE 1939) but not of catalase (THEORELL 1951). Thus the CPs of brown adipose tissue (like liver peroxisomes) contain a β-oxidation system that is independent of the respiratory chain (see Section A.D.2.). This dual location of the β-oxidation system also holds true for carnitine acetyltransferase (KRAMAR et al. 1978).

Although peroxisomal β-oxidation *per se* cannot contribute to the chemical thermogenesis to an appreciable extent, it is tempting to speculate that in brown fat CPs fulfill some auxiliary function in thermogenesis by providing an additional supply of acyl groups for heat generation in the mitochondria.

Acetylcarnitine, produced by carnitine acetyltransferase, which is also present in peroxisomes, may be important for the transport of acetyl groups from CPs into mitochondria.

A moderate number of CPs has been shown in the mouse (Fig. 54), guinea pig and pig alveolar epithelium (HRUBAN et al. 1972, PAVELKA et al. 1976, PETRIK 1971, SCHNEEBERGER 1972 a). The particles are of usual size (0.13–0.27 µm in diameter) and are devoid of peroxisomal oxidases (see Table 28).

A purification procedure for pig lung CPs has been designed by GOLDEN-BERG et al. (1978 b). They started with a premicrosomal (medium-particle) fraction pelleted from the homogenate between 5×10^3 and $1.5 \times 10^6 \, g \cdot$ min that contained mainly mitochondria and lysosomes beside of CPs. Due to their small size, rate sedimentation in a B-14 zonal rotor (linear sucrose gradient 0.4–1.4 M sucrose, density range 1.053–1.170 g/cm³) could separate CPs from mitochondria and some of the lysosomes that are collected at the heavy end of the gradient. The major portion of lysosomal markers (acid phosphatase and β-glucuronidase), along with the microsomal marker cytochrome P-450, is found on top of the gradient; it seems (according to morphological controls) to correspond to lamellar bodies of type II pneumocytes, which might be considered to be of lysosomal nature (CORRIN et al. 1969, DIAUGUSTINE 1974, GOLDFISCHER et al. 1968). Fig. 53 shows the position of CPs in the middle of the separation path.

Whereas D-amino acid oxidase and α-hydroxy acid oxidase are missing in pig lung, a very low activity of fatty acid β-oxidation has been found. Generally, the demonstration of β-oxidation in CPs that do not possess "characteristic" peroxisomal oxidases (brown fat tissue and, possibly, particles from lung tissue) places the question as to their peroxisomal nature in a new light, since during the acyl-CoA dehydrogenase step of β-oxidation in rat liver peroxisomes, oxygen serves as hydrogen acceptor, forming hydrogen

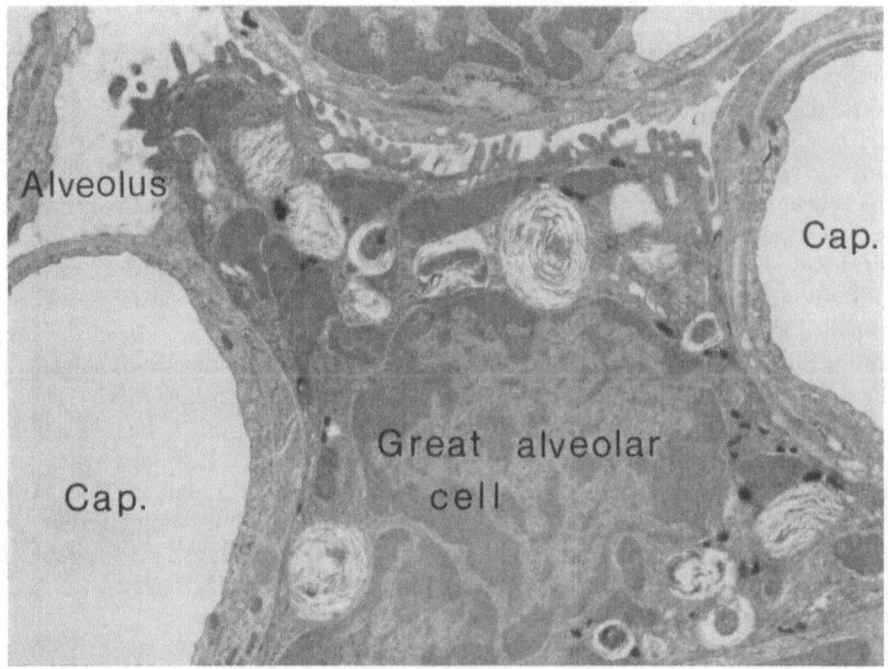

Fig. 54. Mouse pulmonary tissue fixed by perfusion with glutaraldehyde and incubated to demonstrate catalase. Numerous microperoxisomes are present in great alveolar cells (type II pneumocytes). *Cap* Capillary. ×10,000.

peroxide (LAZAROW and DE DUVE 1976). There seems to be good reason to assume that hydrogen peroxide is also produced during fatty acid β-oxidation in CPs from extrahepatic tissues, rendering these particles peroxisomal systems and justifying the term microperoxisome.

VI.B.7. Hepatocytes

VI.B.7.a. Human Hepatocytes and Terminology

NOVIKOFF et al. (1973 a) also used the term "microperoxisome" for the anucleated type of peroxisome that is found in human hepatocytes. These organelles range from 0.22 to 0.86 μm in diameter (average 0.43 μm) in the case of circular profiles, and elongated forms reach a length of 1.1 μm. It is not clear why these organelles are termed *micro*peroxisomes and not anucleated peroxisomes or at least anucleated microbodies, as a great majority of them are of the same size as regular peroxisomes. Discussing this problem, Novi-KOFF *et al.* (1973 a) stated that size alone would be insufficient to warrent a new term (originally the small CPs in tissues other than liver and kidney were termed microperoxisomes). They claim that microperoxisomes are char-acterized in the first place by their proximity and special relationship to the endoplasmic reticulum. Numerous slender connections between microperoxi-somes and the smooth endoplasmic reticulum are characteristic, whereas continuities between core-containing peroxisomes and smooth endoplasmic reticulum, are rare and are wider.

These essential slender continuities between microperoxisomes and the endo-plasmic reticulum are not convincingly demonstrated in their paper. In con-trast, the published micrographs show these slender connections as being more the result of the superimposition of membranes within the section (Novikoff and Novikoff 1973, Novikoff *et al.* 1973 a) (see Section I.B.3.). We regard the few connections of the wide type as representing the only true continuities between peroxisomes (or microperoxisomes) and the endoplasmic reticulum (*cf.*, Figs. 3 and 8–10). This is in agreement with observations of Reddy and Svoboda (1971 a, 1973 b), who did not refer to the existence of the so-called slender connections of Novikoff *et al.* (1973 a) but reported the rare connections of the wide type. Therefore, the absence of a core seems to remain the only difference between peroxisomes and large microperoxisomes.

Undoubtedly, a high number of small microperoxisomes (smaller than 0.3 μm in diameter) occurs in human hepatocytes, too. It is evident that any distinction between these small microperoxisomes and larger ones (the anucleated peroxisomes) would be arbitrary. Therefore, we suggest that the term microperoxisome should be extended to denote anucleated peroxisomes in general. This seems to comply with the intentions of Novikoff *et al.* (1973 a) but is based on different considerations.

Microperoxisomes (as well as regular peroxisomes) also occur in rat (Novi-KOFF *et al.* 1973 a) and mouse (Reddy *et al.* 1969 b) hepatocytes. Almost all peroxisomes in carp liver seem to be microperoxisomes; *i.e.*, they are devoid of cores (Kramar *et al.* 1974).

Biochemical data on large microperoxisomes can be obtained only in cases in which the tissue under investigation contains no peroxisomes. Peroxisomal enzymes sedimented in the microsomal fraction probably derive from small microperoxisomes. However, the effect of contamination by fragments of large microperoxisomes or peroxisomes cannot be excluded.

Sternlieb and Quintana (1977) recently observed marginal plates and amorphous densities in peroxisomes of human hepatocytes. The mean dia-meter of the microbodies has been estimated by these authors in four normal subjects as 0.618 ± 0.143 μm. They noted striking morphological alterations in number, size, and matrix density of peroxisomes in patients suffering from various deseases. A certain similarity is seen between peroxisome prolifera-tion during hepatocellular injury and peroxisome proliferation induced by hypolipidemic drugs in various laboratory animals. Careful morphological analysis of the peroxisome compartment in human liver biopsies may yield interesting results in the future. The observations of Sternlieb and Quintana

Table 29. *Morphological characteristics of hepatocellular peroxisomes in 20 subjects*

Subject no. [a]	Sex	Age (years)	Diagnosis	Size of peroxisomes [b] (μm)	Predominant profile	Appearance of matrix; remarks
1	M	12	Normal	0.62 ± 0.14	Round, ellipsoid, and angular	Granular matrix
2	M	20	Normal	0.64 ± 0.19	Angular	Granular matrix
3	M	20	Normal	0.62 ± 0.13	Round and irregular	Granular matrix
4	F	17	Normal	0.58 ± 0.12	Round	Granular matrix
5	M	1/3	Menkes' disease	0.49 ± 0.15	Round and ellipsoid	Condensed matrix
6	M	1	Menkes' disease	0.39 ± 0.12	Round	Granular matrix; increased numbers per cell
7	F	44	Analbuminemia	0.46 ± 0.13	Round	Granular matrix
8	F	2½	Familial cholestatic cirrhosis	0.78 ± 0.18	Round	Flocculent matrix
9	M	34	Chronic active hepatitis	0.60 ± 0.18	Irregular	Condensed matrix; increased numbers per cell
10	M	33	Chronic active hepatitis	0.77 ± 0.27	Irregular	Condensed matrix; increased numbers per cell
11	M	27	Chronic active hepatitis	0.83 ± 0.23	Irregular	Condensed matrix; increased numbers per cell
12	F	26	Cholestatic hepatitis	0.71 ± 0.19	Irregular	Flocculent matrix
13	F	38	Alcoholic hepatitis	0.73 ± 0.28	Irregular	Granular or flocculent matrix
14	F	36	Cryptogenic cirrhosis	0.61 ± 0.17	Irregular	Condensed or flocculent matrix
15	M	64	Isoniazid hepatitis	0.68 ± 0.14	Round and irregular	Condensed matrix
16 a	M	9	Wilson's disease	0.49 ± 0.14	Irregular	Variable electron density of matrix
17 a	M	8	Wilson's disease	1.10 ± 0.31	Irregular and ellipsoid	Variable electron density of matrix
18 a	F	9	Wilson's disease	0.86 ± 0.22	Round	Condensed or flocculent matrix
18 b		13		0.65 ± 0.16	Round	Condensed or flocculent matrix
19 a	M	12	Wilson's disease	0.44 ± 0.10	Irregular	Condensed matrix
19 b		16		0.69 ± 0.18	Irregular	Condensed matrix
20 a	M	7	Wilson's disease	0.89 ± 0.24	Irregular	Flocculent matrix
20 c		8		0.52 ± 0.13	Irregular	Granular matrix

From: Sternlieb and Quintana 1977.

[a] a Before treatment; b after 4 years of D-penicillamine therapy; c after 1 year of D-penicillamine therapy.

[b] Mean of 30 observations ± SD.

(1977) are summarized in Table 29. Crystalloid cores in human hepatic microbodies have been observed under pathological conditions (BIEMPICA 1966).

VI.B.7.b. Microperoxisomes from Carp Hepatocytes

Hepatocytes in *Cyprinus carpio* contain no crystalline cores and a density within their matrix is observed in less than 1% of the organelles (KRAMAR et al. 1974). They are 0.2–0.75 µm in diameter. The particles are bounded by a unit membrane, and the matrix material displays moderate electron density. At higher magnifications a distinct granular structure (30-nm granules) becomes prominent. Cisternae of the smooth endoplasmic reticulum

Table 30. *Peroxisomal enzymes in subcellular fractions from carp liver*

Enzyme	Units/g liver [a]	Recovery [b] (%)			Relative specific activity [c]		
		M	L	Mc	M	L	Mc
Catalase	1,570	15	26	11	1.6	6.4	1.1
Urate oxidase	130	28	35	15	3.0	8.6	1.5
D-Amino acid oxidase	134	31	22	10	3.3	5.4	1.0
L-α-Hydroxy acid oxidase	153	24	24	13	2.6	5.9	1.3

From: KRAMAR et al. 1974.

[a] Catalase in k-units (see II.A.2.a.). Oxidases in nmol O_2/min.

[b] Percentage of total enzyme activity of the liver.

[c] Percentage of enzyme recovered in the fraction/percentage of protein recovered in the fraction.

Abbreviations: M heavy mitochondrial fraction; L light mitochondrial fraction; Mc microsomal fraction.

are situated parallel to the surface of the microbodies and wide continuities between the smooth endoplasmic reticulum and the organelles are occasionally seen (Fig. 55). Peroxidatic activity of catalase is restricted to these organelles. According to morphological and histochemical criteria discussed above, the particles resemble microperoxisomes.

KRAMAR et al. (1974) prepared subcellular fractions from carp liver by differential centrifugation. The preparation scheme (Fig. 44) was varied only slightly from that given in Section II.A.1.b. (Fig. 26). The results are summarized in Table 30. Pellets of the individual fractions were controlled morphologically and after incubation for histochemical catalase staining. In every fraction, osmiophilic reaction product was observed in smooth-surfaced vesicles. In the microsomal fraction these vesicles measured 0.2–0.5 µm in diameter. In some of the larger vesicles in particular, the electron-dense precipitates were attached to the inner surface of the limiting membrane, most probably as a result of leakage of the enzyme. Slow diffusion of substrate can be eliminated as a cause of this staining pattern, since organelles of identical size homogeneously stain by means of the same technique when having been fixed in intact liver tissue.

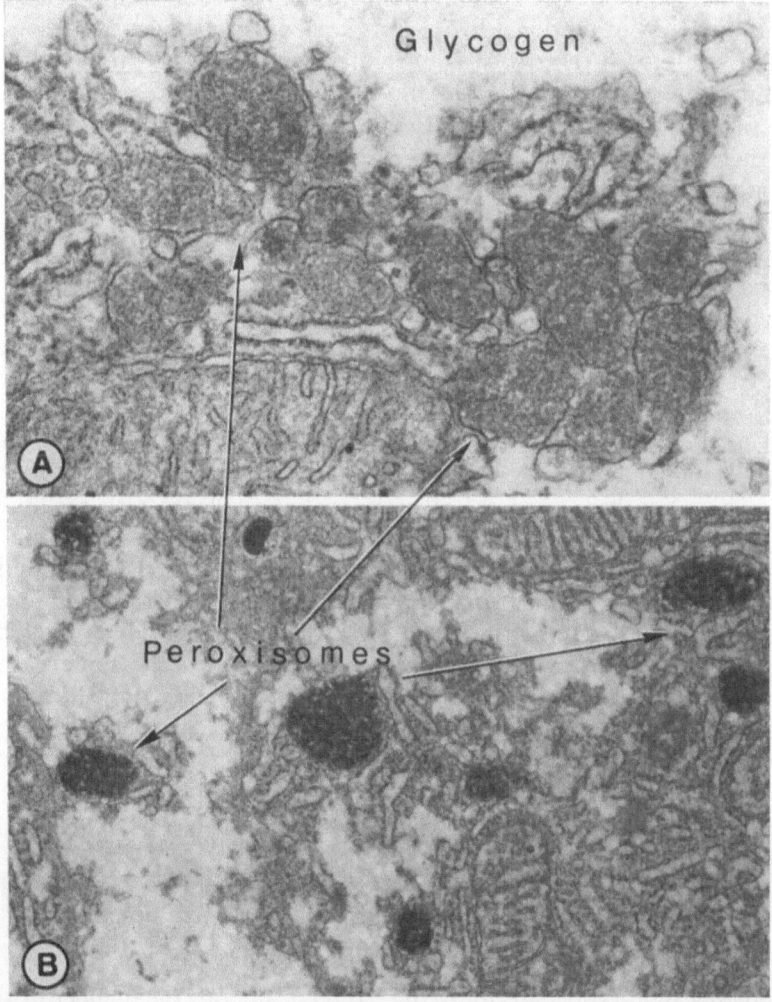

Fig. 55 A, B. Coreless liver peroxisomes from *Cyprinus carpio*. *A* In unreacted specimens. ×30,000. *B* After catalase staining. ×15,000. Although uricase is present in this type of organelle, nucleoids are absent. Adapted from: KRAMAR *et al.* (1974): Histochemistry 40, 141 and 144.

Since they are small (as compared to rat liver peroxisomes), microperoxisomes were found not only in the lysosomal fraction but also in the microsomal fraction. Biochemical data indicate that these small organelles (like the larger ones) are perfectly equipped with all the peroxisomal enzymes tested.

However, the ratios of enzyme activities measured in the individual fractions differ considerably. KRAMAR et al. (1974) suggest that the microperoxisome population might be heterogeneous (see Section VI.B.7.c.). The great stability of urate oxidase and its strong binding to the particles may account for the relatively high activities found. Microperoxisomes from carp liver contain urate oxidase although they are devoid of cores (the significance of crystalline cores is discussed in Section I.B.2.). This amazing fact became understandable when it was shown that in carp liver urate oxidase most probably is bound to the peroxisomal membrane (GOLDENBERG 1977 a, b).

Recently GOLDENBERG et al. (1978 a) purified peroxisomes from a carp liver light mitochondrial fraction (Fig. 56; cf., Fig. 26). The duvogram demonstrates the enrichment of peroxisomal enzymes in the light mitochondrial fraction.

After isopycnic density-gradient centrifugation urate oxidase, D-amino acid oxidase, and glycolate oxidase follow catalase in the gradient (Fig. 57). Some of the NADH-cytochrome c reductase, NADP-linked isocitrate dehydrogenase, and carnitine acetyltransferase is bound in the peroxisomal fraction, which in addition contains a β-oxidation system. Thus these "microperoxisomes" resemble rat liver peroxisomes in all biochemical characteristics known. On the other hand, it has been shown that apart from urate oxidase no purine-catabolizing enzyme is present in carp liver peroxisomes (GOLDENBERG 1977 a, b). Xanthine oxidase, allantoinase, and allantoicase are recovered in the soluble fraction. Purine degradation thus takes a detour from the soluble compartment into the peroxisome, resulting in the catalatic breakdown of hydrogen peroxide produced by urate oxidase (Figs. 58 and 59).

As a conflicting result NOGUCHI et al. (1979 a) reported recently that in liver (or hepatopancreas, respectively) of certain fish and crustaceans (mackerel, yellow mackerel, prawn and mantis club) not only urate oxidase but also allantoinase and allantoicase are located in the peroxisomes. Thus at least in certain cases a considerable part of purine degradation, namely the span between uric acid and urea, might take place inside the microbodies (cf., Fig. 58). In animal evolution more and more of the complete purine catabolism which has been ascribed to the "ancestral peroxisome" (DE DUVE 1969) was lost or transferred into the soluble compartment of the hepatocyte. The latter change might have happened in some fish and amphibians.

VI.B.7.c. Regenerating Rat Liver

Following subtotal hepatectomy (60–70%) in the rat, the number of smaller and coreless microbodies in liver tissue increases. The number reaches a

---

Fig. 56 A–I. Distribution of enzymes after differential centrifugation of a carp liver homogenate. Details of fractionation essentially as in Fig. 44, except that M and L were pelleted at 6,000 g for 10 minutes and 20,000 g for 15 minutes, respectively. Fractions: N nuclear; M heavy mitochondrial; L light mitochondrial; Mc microsomal; S soluble cytoplasmic. A Cytochrome oxidase. B Catalase. C Aryl esterase. D Acid phosphatase. E Urate oxidase. F D-Amino-acid oxidase. G Glycolate oxidase. H (NADP)-isocitrate dehydrogenase. I Carnitine acetyltransferase. From: GOLDENBERG et al. (1978 a): Histochem. J. 10, 106.

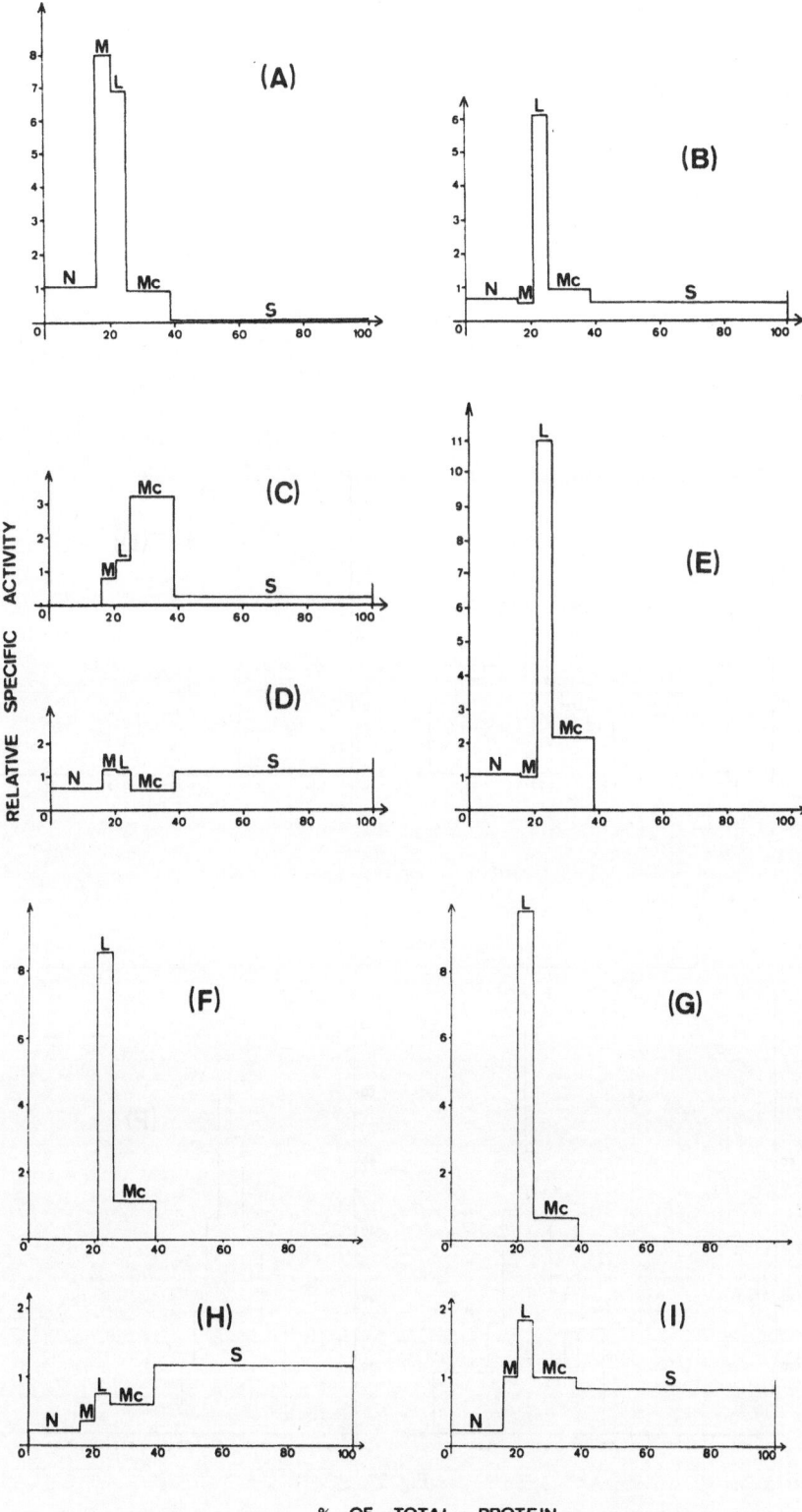

% OF TOTAL PROTEIN

Fig. 56.

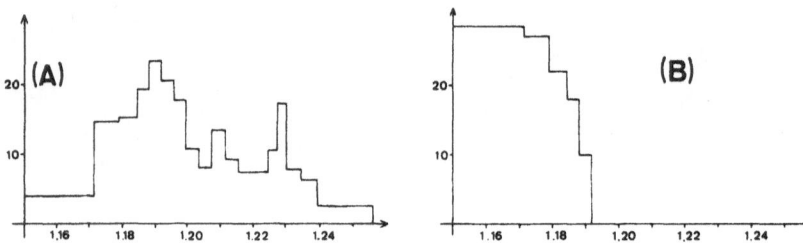

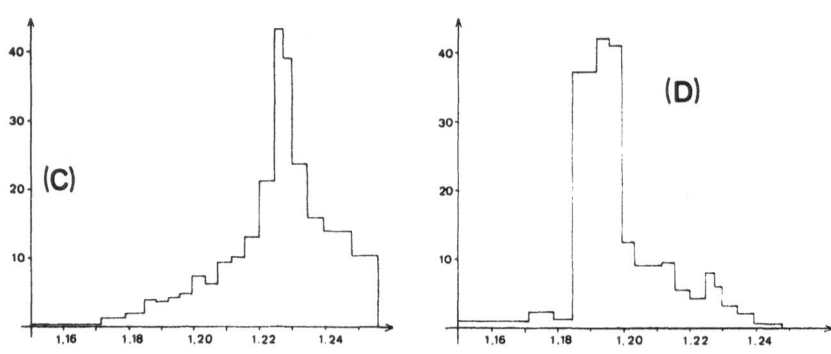

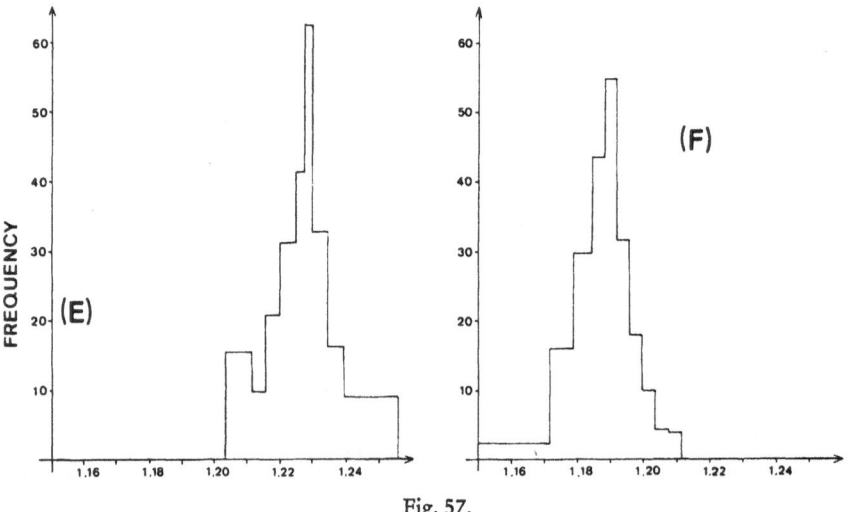

FREQUENCY

Fig. 57.

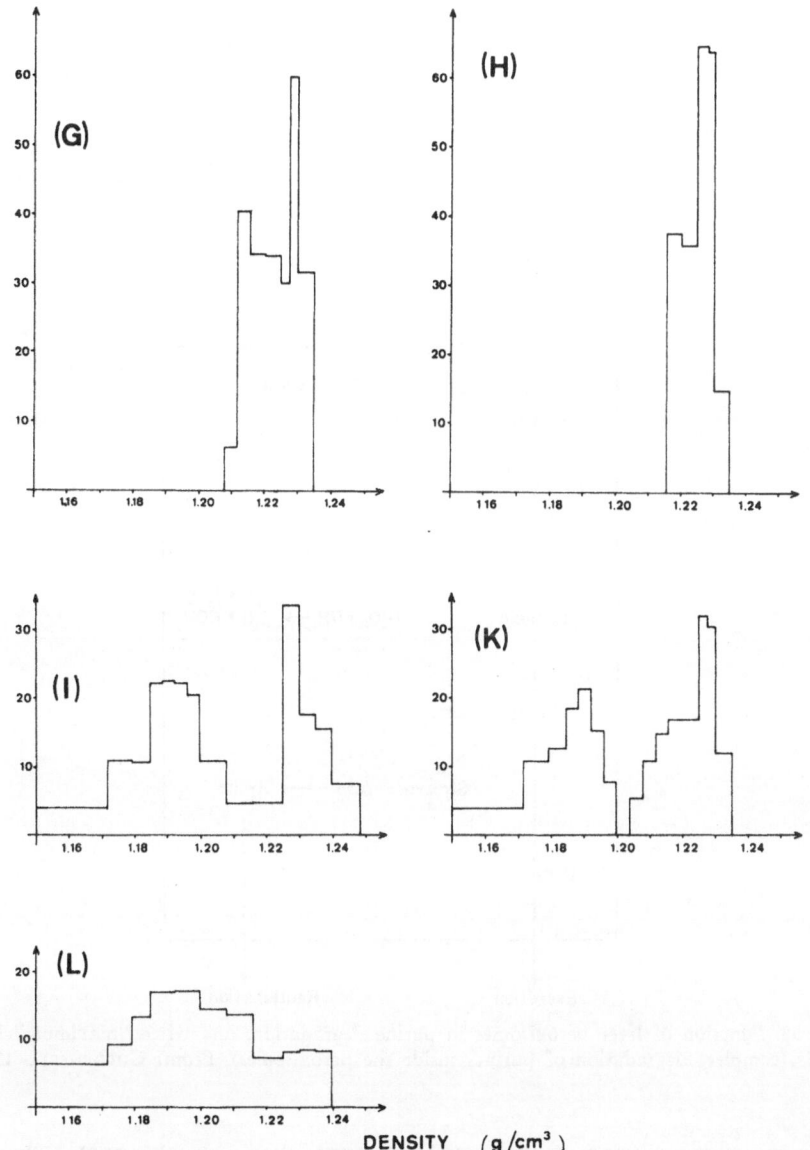

Fig. 57 A–L. Distribution of carp liver enzymes after density gradient centrifugation of the L-fraction (for preparation see Fig. 56) in a MSE B-14 aluminum zonal rotor. An amount of light mitochondrial fraction corresponding to 25 g of liver was layered onto 400 ml of a linear sucrose gradient (1.2–1.9 M sucrose; additions: 0.1% ethanol, 2% dextran-10, 1 mM EDTA, 3 mM imidazole-HCl, pH 7.2; density range 1.15–1.25 g/cm³). The gradient cushion 2.0 M sucrose. 4 hours at 25,000 rpm in an MSE SS 50 ultracentrifuge. 20 fractions of 20 ml collected from the rotor center by replacing the gradient with 2 M sucrose. *A* Protein (85% recovery). *B* Aryl esterase (110%). *C* Catalase (105%). *D* NADH-cytochrome C reductase (90%). *E* Urate oxidase (73%). *F* Cytochrome oxidase (92%). *G* D-Amino acid oxidase (94%). *H* Glycolate oxidase (74%). *I* Carnitine acetyltransferase (108%). *K* (NADP)-isocitrate dehydrogenase (89%). *L* Acid phosphatase (120%). From: GOLDENBERG *et al.* (1978 a): Histochem. J. 10, 108.

maximum about 24 hours after the operation (GOLDENBERG *et al.* 1975 a, RIGATUSO *et al.* 1970, SAITO *et al.* 1973, STENGER and CONFER 1966, VIRÁGH and BARTÓK 1966). These newly formed particles are relatively small, as observed under the electron microscope and indicated by their sedimentation behavior (GOLDENBERG *et al.* 1975 a). They seem to originate from large peroxisomes through a process of fragmentation or budding (FAHIMI and

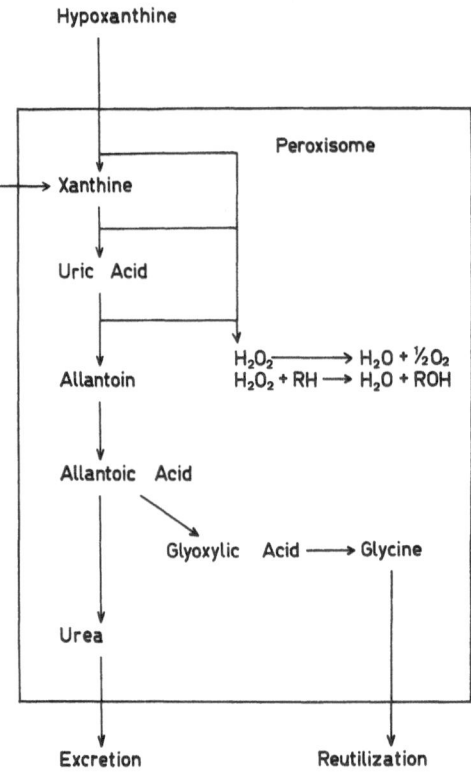

Fig. 58. Function of liver peroxisomes in purine degradation, unverified "maximum" hypothesis (complete degradation of purines inside the peroxisomes). From: GOLDENBERG 1977 b.

VENKATACHALAM 1970, RIGATUSO *et al.* 1970, SAITO *et al.* 1973). The vast majority of these newly formed organelles are devoid of cores and therefore they resemble microperoxisomes. It has been established that regeneration of peroxisomes is initiated by the formation of microperoxisomes. Microperoxisomes have been suggested as progenitors of peroxisomes under normal circumstances, *i.e.*, during the normal turnover of these organelles (NOVIKOFF *et al.* 1973 a).

The enzyme activities in individual fractions, prepared at various intervals from regenerating rat liver, do not behave in a comparable manner. This can be explained by assuming either different regeneration (or turnover) rates for catalase and the oxidases, or the existence of different populations of peroxisomes and/or microperoxisomes. The latter interpretation would mean that organelles with different levels of enzyme activities exist. These would probably be microperoxisomes because they sediment with the light mitochondrial or microsomal fraction (GOLDENBERG *et al.* 1975 a). Results obtained from studies on carp liver (KRAMAR *et al.* 1974) also seem to indicate the heterogeneity of microperoxisomes. It is possible to solve this problem only with histochemical methods for the fine structural localization of peroxisomal oxidases. Biochemical data obtained from regenerating rat liver 0, 1, 2, and

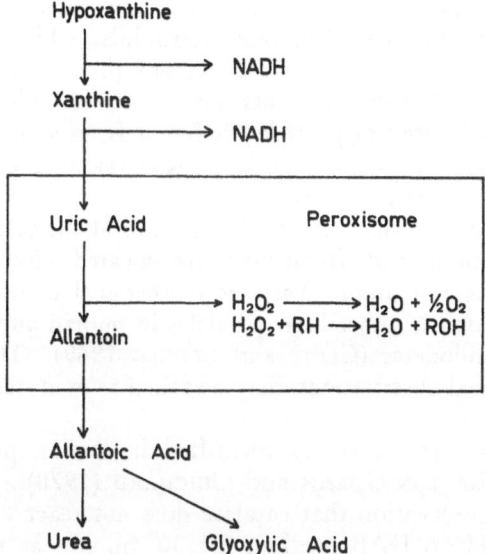

Fig. 59. Function of peroxisomes in purine degradation in carp liver (according to results of cell fractionation studies). From: GOLDENBERG 1977 b.

4 days after subtotal hepatectomy in the heavy and light mitochondrial, and microsomal fractions are compiled in Fig. 38.

VI.B.8. Blood Cells

VI.B.8.a. Erythropoietic Series

Small CPs (0.09–0.2 µm in diameter) have been identified in all cells of the erythropoietic line except mature erythrocytes (BRETON-GORIUS and GUICHARD 1975 a). Addition of 10^{-2} M potassium cyanide to the incubation medium for cytochemical catalase staining sufficiently reduces peroxidatic activities of hemoglobin and peroxisomes can be discerned. The cytochemical reaction is inhibited by the addition of 2×10^{-2} M aminotriazole. The number and distribution of CPs is similar in normal and in myeloperoxidase-deficient human subjects. The organelles are numerous in proerythroblasts and gradually become diminished in polychromatophilic erythroblasts. CPs of elongated shape have been observed frequently; these particles are up to 0.45 µm long. In material obtained from myeloperoxidase-deficient patients, the presence of CPs has also been confirmed in neutrophils (see below).

CPs are formed in most immature erythropoietic cells. The mechanism of their disappearance during cell maturation has not been elucidated as yet (BRETON-GORIUS and GUICHARD 1975 a).

VI.B.8.b. Neutrophilic Granulocytes

BRETON-GORIUS et al. (1978) succeeded in identifying CPs in immature and mature neutrophilic granulocytes from the bone marrow and blood of patients with hereditary myeloperoxidase deficiency by means of histochemical DAB methods. Part of the azurophilic granule population was found to stain for catalase (only small, ellipsoid azurophils), thus suggesting a heterogeneity of the azurophilic granules. Chicken heterophils, which normally do not possess peroxidase, also lack CPs. It is evident that studies on granulocytes of myeloperoxidase-deficient patients are most suitable for determining whether catalase is located in particles different from azurophils (MORIKAWA and HARADA 1969, SZMIGIELSKY 1972, VERCAUTEREN 1962) or identical to azurophils (NISHIMURA et al. 1976).

CPs in immature and mature granulocytes are about 0.15 μm in diameter, are membrane bound, and frequently are located close to segments of smooth endoplasmic reticulum. There is biochemical evidence that D-amino acid oxidase is located in subcellular particles in human and guinea pig polymorphnuclear granulocytes (CLINE and LEHRER 1969). This finding, along with the cytochemical observations, may be taken as evidence for the existence of microperoxisomes.

Microperoxisomes were recently identified in human promyelocytes and promonocytes by BRETON-GORIUS and GUICHARD (1978). The authors took advantage of the observation that catalase does not react when incubation is performed in a pH 7.6 DAB medium for 30 minutes at room temperature (original Graham and Karnovsky medium), whereas myeloperoxidase readily stains under these conditions. These observations confirm earlier ones obtained in cases of myeloperoxidase deficiency. Moreover, the findings of BRETON-GORIUS and GUICHARD (1978) show that catalase activity is associated with young myeloic cells, thereby explaining the observation of high catalase activity in leukocytes of patients with myelogenous leukemia, and low catalase activity in leukocytes obtained during inflammatory processes. However, the presence of Φ bodies (HANKER et al. 1978) may also contribute to the high catalase level of leukemic leukocytes.

VI.B.8.c. Monocytes and Related Cells

CPs have been demonstrated in tissue macrophages (DAEMS et al. 1975, 1976, FAHIMI et al. 1976, NOVIKOFF et al. 1973 b), in mouse, rat, and guinea pig peritoneal exudate cells (DAEMS et al. 1975, 1976), and in monocytes (BODEL et al. 1977, BRETON-GORIUS and GUICHARD 1975 b, BRETON-GORIUS et al. 1978). The presence of CPs in circulating rat blood monocytes was refuted by VAN DER RHEE et al. (1977), although these authors observed membrane-bound organelles that positively react with alkaline DAB media, but not at neutral pH values. However, this type of staining was not inhibited by aminotriazole ("secondary granules"). Another group of granules ("primary granules") also displayed unusual staining behavior, being reactive in both neutral and alkaline incubation media. VAN DER RHEE et al. (1977) concluded that primary granules contain peroxidase but secondary granules do not, whereas catalase is present in secondary granules but may be absent in primary granules.

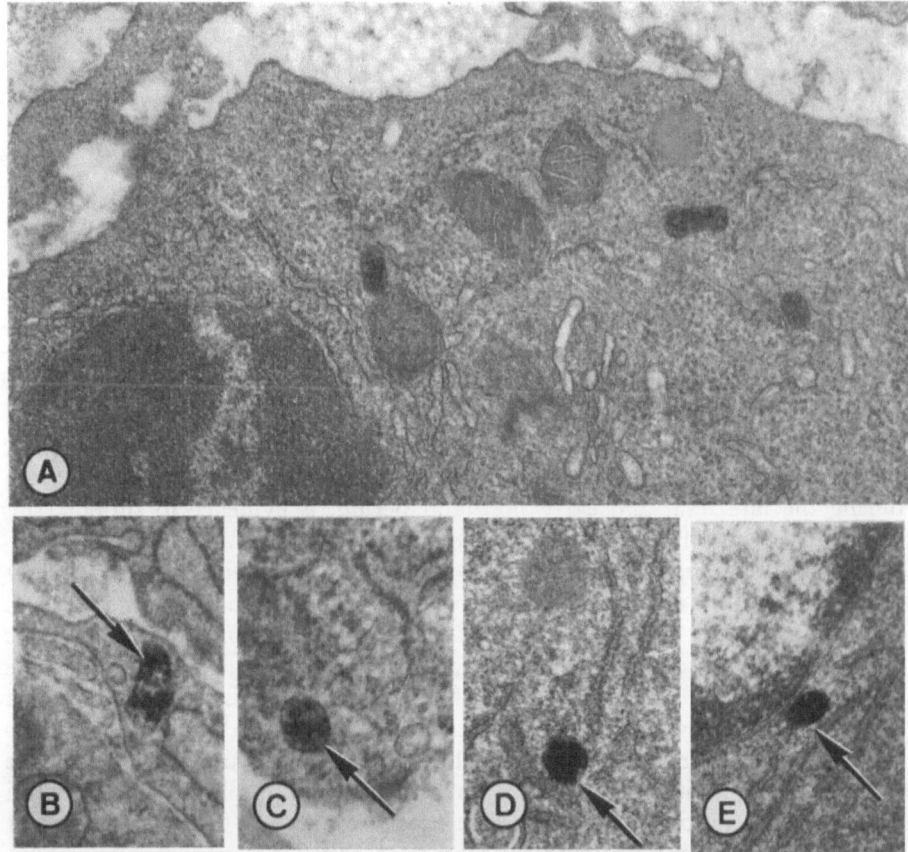

Fig. 60 A–E. Microperoxisomes in the cytoplasm of mouse fibroblasts. *A* Tissue incubated for peroxidatic activity of catalase. *B, C* Catalase-positive particles (*arrows*) in fibroblasts from the mouse tongue, unstained ultrathin sections. *D, E* Positively reacting particles (*arrows*) in fibrocytes of the stomach mucosa, ultrathin sections stained with uranyl acetate and lead citrate. The electron density of the reaction product is enhanced by the staining procedure and the granular appearance of the reaction product becomes less prominent. Endoplasmic reticulum and nuclear envelope are more easily discerned in stained sections.
From: BÖCK (1973 c): Z. Zellforsch. 144, 541 and 543 (labeling altered).

VI.B.8.d. Platelets and Megakaryocytes

Peroxidatic activity has been reported to be present in the tubular system of megakaryocytes and platelets (BRETON-GORIUS and GUICHARD 1972).

Catalase, on the other hand, has been identified in small, membrane-bound organelles in human platelets and megakaryocytes (BRETON-GORIUS and GUICHARD 1975 b) and in chicken thrombocytes (BRETON-GORIUS et al. 1978). These organelles have been interpreted as microperoxisomes.

VI.C. Catalase Positive Particles as Ubiquitous Cells Organelles

CPs are present in widely distributed cells (e.g., in fibroblasts, Fig. 60) and have been identified in almost all types of cells studied thus far. Their detection has become possible by means of improved cytochemical methods for the demonstration of catalase via its peroxidatic activity. A summary is given in Table 27 and can also be found in extensive earlier studies (HRUBAN et al. 1972 a, b, NOVIKOFF et al. 1973 b). It is important to realize that the absence of CPs in some cells of cytochemical preparations (sectioned material) is no proof that these cells are indeed devoid of the organelles. The observation of rarely occurring, small organelles is a matter of favorably orientated sections and depends on the number of cells and sections investigated. Furthermore, the frequency of CPs may be related to the degree of differentiation of a given cells, as been reported for cells of the erythropoietic series (BRETON-GORIUS and GUICHARD 1975 a). If CPs (possibly as microperoxisomes) are assumed to be significantly involved in cell function and metabolism, their development should be considered to be related to cellular activity and specific function. The frequency of CPs in hormonally controlled cells, or in cells during the late fetal and early postnatal period, is evidence to support this view (BERCHTOLD 1975 b, GOECKERMANN and VIGIL 1975).

CPs (and microperoxisomes) often have been reported to be continuous with, or at least intimately related to the smooth endoplasmic reticulum (NOVIKOFF et al. 1973 b). There is no doubt that smooth endoplasmic reticulum is an essential and generally occurring cell organelle, although its development significantly varies in different cells. There is also no doubt that cells that are nearly devoid of endoplasmic reticulum, such as small lymphocytes, instantly form smooth and rough endoplasmic reticulum when they are stimulated (e.g., with PHA). Since physiological and unphysiological stimuli for the proliferation of CPs and microperoxisomes are unknown, it is not possible to trigger the development of these organelles as yet. Assuming that CPs and microperoxisomes are a direct outgrowth of the endoplasmic reticulum, all cells should be capable of forming CPs and microperoxisomes and, additionally, CPs and microperoxisomes should be accepted as constituent cell organelles.

There are, however, examples of certain cells that are extremely rich in CPs or microperoxisomes—e.g., those involved in the synthesis and metabolism of steroids, cholesterol, and lipids. These observations can be taken as evidence for the significance of CPs and microperoxisomes in cell metabolism. Obviously, every cell metabolizes lipids and cholesterol. If CPs and microperoxisomes are assumed to participate in these metabolic pathways (as one example), they may be considered to be ubiquitous and common cell organelles.

VI.D. Hypotheses on the Metabolic Role of Microperoxisomes

It is clear from the preceding sections of this chapter that very little is known about microperoxisomal enzymes. Likewise, there have been very few metabolic studies on isolated organelles. A successful approach to this field depends on progress in purification procedures. Thus there is no substantiated knowledge about the role of microperoxisomes in cellular metabolism. Nevertheless, several authors have been working on theories on this subject that are mainly based on histochemical and pharmacological observations.

HRUBAN and VIGIL (1972) and HRUBAN et al. (1972 a, b) pointed out that tissues that are concerned, in particular, with cholesterol metabolism or absorption are generally rich in microperoxisomes (e.g., sebaceous glands, adrenal cortex, corpus luteum, Leydig cells, epithelial cells of small intestine, placenta, adipose tissue). In this respect the observations on microperoxisomes in brown adipose tissue seem to be significant. AHLABO and BARNARD (1971) showed that the number of microperoxisomes is correlated with the thermogenic phase in this tissue. They speculated that microperoxisomes might be involved in chemical thermogenesis. This has been confirmed by the demonstration of fatty acid β-oxidation in purified microperoxisomes of this tissue (KRAMAR et al. 1978). Although it is generally accepted that fatty acids are the fuel for chemical thermogenesis in brown fat tissue (SMITH and HORWITZ 1969), peroxisomal degradation of fatty acids per se can contribute little to heat production by catalatic decomposition of hydrogen peroxide generated in the acyl-CoA dehydrogenase reaction and, eventually, by peroxisomal reoxidation of NADH (see below).

Most of the heat is produced during the terminal degradation of acetate in the tricarboxylic acid cycle and respiratory chain after the acyl groups have been transported into the inner mitochondrial compartment. Considering that the acetyl shuttle via citrate is only able to transport acetyl groups out of the mitochondria (SRERE and BHADURI 1962), we suppose that acetylcarnitine formed by the peroxisomal carnitine acetyltransferase might be the proper intermediate to transport acyl groups (which arise from the peroxisomal β-oxidation) into the mitochondria.

The question remains as to whether there is a real demand for acetyl groups produced in microperoxisomes in addition to acetyl-CoA formed during mitochondrial β-oxidation. On the one hand, additional acetyl-CoA might be necessary during active thermogenic phases in order to take advantage of the full capacity of the tricarboxylic acid cycle; furthermore one is tempted to speculate that acetyl-CoA (besides its action as a positive allosteric effector of pyruvate carboxylase) might also play a role in the regulation of mitochondrial respiratory activity, e.g., by uncoupling phosphorylation from respiration. A comparable coupling effect is known to be induced by purine nucleotides, which, when bound to the inner membrane, lower its proton conductance (for a review see NICHOLLS 1977).

In contrast, in glandular tissues synthesizing steroids, the peroxisomal formation of acetyl-CoA via β-oxidation may lead to increased formation of cholesterol. In other words, triglycerides stored in these tissues are rebuilt into steroids. The proximity of microperoxisomes and endoplasmic membranes should facilitate this transformation:

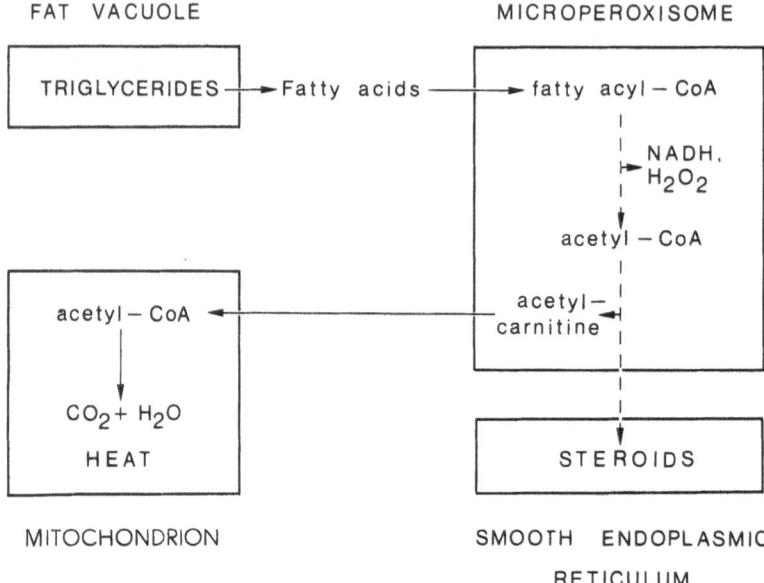

It might well be that in brown fat tissue microperoxisomes are involved in the catabolic part of this scheme, whereas in other tissues (*e.g.*, in the Harderian gland) cholesterol formation is triggered. KÜHNEL (1971) has stressed a morphological relationship between brown fat and the Harderian gland, and both tissues are rich in CPs. However, such speculations are idle as long as β-oxidation has not been proved to be a general capacity of CPs in extrahepatic tissues.

It is generally accepted that most of the hydroxylases responsible for the formation of steroid hormones from cholesterol are located in mitochondria and microsomes (ROSENTHAL and NARASIMHULU 1969). Nevertheless, nobody has disproved the possibility that one or the other of these enzymes might be located partially in microperoxisomes, or that microperoxisomes have auxiliary functions in these reactions. Such speculations could only be discouraged by experiments with microsomes and mitochondria, which are virtually free of peroxisomal contamination. The proximity of endoplasmic reticulum and microperoxisomes also seems to indicate a functional relationship.

In addition, it seems to be possible that microperoxisomes participate in lipid metabolism by indirectly oxidizing pyridine nucleotides (via the redox couple keto-hydroxy acids), thus acting as a sink for abundant cytoplasmic hydrogen and taking part in the regulation of fat and steroid metabolism. It may be presumed that this function would be particularly important in tissues that lack shuttle mechanisms for the transport of reducing equivalents from cytoplasm into mitochondria. It would be interesting to study the relative efficiency of peroxisomal and mitochondrial pyridine nucleotide reoxidation in tissues that are rich in microperoxisomes.

Hypothesizing on the role of microperoxisomes in lipid metabolism, HRUBAN et al. (1972 a) suggested indirect participation via prostaglandins. In fact, the *in vitro* systems that have been used for studying prostaglandin synthesis may well be rich in microperoxisomes in view of their preparation.

In conclusion, it is probable that microperoxisomes are not independent organelles housing complete metabolic chains. Instead, they perform auxiliary and complementary functions, being part of a higher functional and structural unit that also comprises the endoplasmic membranes.

VII. Concluding Remarks on Metabolic Aspects of Animal Peroxisomes

In the preceding chapters a considerable number of metabolic capacities have been reported. On the other hand, the functions of animal peroxisomes are far from being classifiable under a common principle in spite of the peroxisome concept (Section I.C.). Nevertheless, in the light of new developments (*e.g.*, the detection of a peroxisomal fatty acid β-oxidation), some unifying views of the metabolic functions of peroxisomes are presented in this chapter.

VII.A. Generation of Hydrogen Peroxide

In the nonprimate liver production of hydrogen peroxide connected with urate oxidation plays an important role and justifies the common compartmentation of urate oxidase and catalase. In contrast, the function of D-amino acid oxidase is somewhat puzzling. D-Amino acids, which do not originate physiologically in the animal cell, have been found in blood, conceivably from the intestinal flora (HOEPRICH 1965). On the other hand, it is known that glycine, as an optically inactive amino acid, is deaminated slowly by the kidney enzyme (Section II.A.2.c.); furthermore, DE MARCHI and JOHNSTON (1968), stressing the effect of glycine as an inhibitory neurotransmitter, proved its degradation by D-amino acid oxidase from spinal cord and cerebellum.

Another hydrogen peroxide-producing oxidase, glycolate oxidase, offers an alternative pathway for reoxidation of cytoplasmic NADH (see Section VII.B.4.). Glycolate oxidase is also able to oxidize glyoxylate to give oxalate, an activity that deserves consideration in connection with primary hyperoxaluria (GIBBS et al. 1977, RICHARDSON and TOLBERT 1961). Fatty acid β-oxidation must also be mentioned as a peroxisomal source of hydrogen peroxide that is produced during the β-hydroxyacyl-CoA dehydrogenase step (Section II.A.2.l.).

Is should be stressed that in the aerobic cell peroxisomes are not the only site of hydrogen peroxide production. Hydrogen peroxide is also generated in the endoplasmic reticulum at the expense of NADPH (THURMAN et al. 1972) as well as in the mitochondria (BOVERIS and CHANCE 1973). Mitochondrial hydrogen peroxide is thought to originate to a considerable extent from superoxide radical anions (O_2^-) by the superoxide dismutase reaction:

$$2 O_2^- + 2 H^+ \rightarrow O_2 + H_2O_2$$

Superoxide is produced in the oxidation of various mitochondrial redox carriers by molecular oxygen (see BOVERIS and CADENAS 1975).

In the cytoplasm, superoxide (formed, *e.g.*, by xanthine oxidase; FRIDOVICH 1970) seems to be the main source of hydrogen peroxide. At least in polymorphonuclear leukocytes, the plasma membrane has also been recognized as a center forming superoxide radicals with NADPH as electron donor (DEWALD *et al.* 1979). Therefore superoxide from the plasma membrane is another source of hydrogen peroxide and probably the main cause of the respiratory burst during phagocytosis (see below).

The overall cellular production of hydrogen peroxide has been assessed to be about 50 nmol/(min · g) in the perfused liver; this is on the order of about 1% of the total oxygen consumption. However, an appropriate peroxisomal substrate (urate or glycolate) increases the hydrogen peroxide formation to 750 or 450 nmol/(min · g), respectively (for a review see SIES 1974). Obviously, under certain conditions the peroxisomal hydrogen peroxide formation may amount to an important share of the total oxygen metabolism of the cell.

VII.B. Disposal of Hydrogen Peroxide

Catalase is preferentially responsible for the degradation of hydrogen peroxide produced in the peroxisomes. Due to the low affinity of catalase for hydrogen peroxide (high Michaelis constant, see Section II.A.2.a.) some hydrogen peroxide escapes decomposition but its concentration is kept low enough (see Section I.C.) to avoid harmful effects on the cell; catalase "buffers" the cell with respect to hydrogen peroxide. At higher hydrogen peroxide concentrations free radicals, *e.g.*, the very destructive hydroxyl-radical, are produced (see Section I.C.). Under certain conditions (*e.g.*, after infusion of hypoxanthine) a considerable portion of the hydrogen peroxide formed in the cytoplasm may be detoxicated by the peroxisomes, whereas intraperoxisomally produced hydrogen peroxide hardly escapes into the cytoplasm (OSHINO and CHANCE 1977, POOLE 1975). Cytoplasmic hydrogen peroxide is decomposed mainly through the cooperation of glutathione peroxidase and glutathione reductase, a system that is located mainly in the cytoplasm and up to about 30% in the mitochondrial matrix (FLOHÉ and SCHLEGEL 1971):

$$2 GSH + H_2O_2 \rightarrow GSSG + 2 H_2O$$

$$GSSG + NADPH + H^+ \rightarrow 2 GSH + NADP^+$$

where GSH and GSSG are reduced and oxidized glutathione, respectively. Instead of hydrogen peroxide, an organic peroxide may react as well.

Obviously, this sequence of reactions represents a means of reoxidation of NADPH by hydrogen peroxide. Therefore, reactions reducing NADP should be promoted by hydrogen peroxide formation. The hexose monophosphate shunt is stimulated, as has been well described for phagocytizing polymorphonuclear leukocytes, for example. Catalase competes with the glutathione peroxidase system for hydrogen peroxide and is able to suppress the hexose monophosphate pathway, which consequently is stimulated during the inhibition of catalase by azide or aminotriazole (REED 1969). As the cellular production of hydrogen peroxide controls the cytoplasmic NADPH, it may be important in the regulation of the biosynthesis of steroids and fatty acids.

It is known that phagocytosis is accompanied by an explosion of oxygen consumption and by an increased release of superoxide radicals and hydrogen peroxide (BABIOR et al. 1973, PAUL et al. 1970). The role of plasma membrane in these events has been mentioned above. Oxygen may exert a direct bactericidal effect when microorganisms are sequestered by the plasma membrane into a phagocytic vacuole. On the other hand in rat peritoneal macrophages peroxisomes have been observed in apparent fusion with phagosomes [EGUCHI et al. 1979; cf., the detection of CPs in pulmonary alveolar macrophages (DAVIES et al. 1979) and in polymorphonuclear neutrophils (Section VI.B.b.)]. Thus catalase, like peroxidase, may fulfill bactericidal functions in oxidizing halide ions (e.g., iodide to free iodine; KLEBANOFF et al. 1969). The peroxidatic reaction of catalase has been studied with alcohols, nitrite, formaldehyde, and formate as substrates. On the other hand, in phagocytic cells catalase may act mainly to protect the cell against excessive hydrogen peroxide formation; this interpretation helps explain the induction of catalase and superoxide dismutase by oxygen in pulmonary macrophages of neonatal rats (STEVENS and AUTOR 1977).

In general, an assessment of the physiological significance of the peroxidatic reaction of catalase is difficult: known natural hydrogen donors do not seem to reach a concentration necessary for their peroxidatic degradation to an appreciable extent (see Section I.C.).

VII.C. Peroxisomal Electron Transport

Peroxisomes can oxidize NADH indirectly via the L-α-hydroxy acid oxidase-glyoxylate reductase (or lactate dehydrogenase) couple. The question arises as to whether the reduction of the corresponding α-keto acid by NADH is mainly due to extraperoxisomal dehydrogenases, or whether it may proceed in the peroxisomes themselves in order that intraperoxisomal NADH that arises (e.g., from fatty acid β-oxidation) may be oxidized (see Section I.D.1.). Peroxisomal NADH reoxidation may also be useful when the mitochondria are in the controlled state or when the mitochondrial respiratory chain has reached its full capacity.

The detection of NADH-cytochrome c reductase in animal peroxisomes (see Section I.D.) and of cytochrome b_5 (FOWLER et al. 1976), which is also incorporated into peroxisomes in vitro (REMACLE 1978), indicates that there is an alternative pathway for the oxidation of intraperoxisomal NADH. In our laboratory isolated rat liver peroxisomes were used to demonstrate the palmitoyl-CoA-dependent reduction of cytochrome b_5; the NADH required for this step apparently is produced during the β-hydroxyacyl-CoA dehydrogenase reaction.

The second reductant of this electron transport chain is unknown, but it is plausible that the system resembles the microsomal chain in this respect too (for review see Jansson and Schenkman 1977): Fatty acid desaturation might be coupled to NADH oxidation:

$$\text{stearoyl-CoA} + \text{NADH} + \text{H}^+ + \text{O}_2 \rightarrow \text{oleyl-CoA} + \text{NAD}^+ + 2\,\text{H}_2\text{O}$$

or, in a flow scheme:

$$\text{NADH} \rightarrow \text{NADH-cytochrome b}_5 \text{ reductase} \rightarrow$$
$$\rightarrow \text{cytochrome b}_5 \rightarrow \text{CN}^- \text{-sensitive factor} \rightarrow \text{O}_2$$

$$\text{stearoyl-CoA} \qquad \text{oleyl-CoA}$$

It is remarkable that monounsaturated long-chain fatty acids are a preferential substrate of peroxisomal β-oxidation (see Section I.D.1.).

In contrast to the mitochondrial respiratory chain, all oxidative reactions in peroxisomes are energy wasting. This fact gives rise to the speculation that peroxisomes are involved in chemical heat production; in this respect, brown fat has been discussed elsewhere (Sections VI.B.6. and VI.D.). Furthermore, in the liver of cold-acclimated rats and hibernating hamsters an increase in the number of peroxisomes has been observed (Horwitz 1976). Similarly, Murphy et al. (1979) suggested that the dissipation of energy in peroxisomes might be important for the maintenance of body weight; however, a higher activity of β-oxidation was found in isolated liver peroxisomes of obese mice than in peroxisomes from normal animals.

VII.D. Amino Acid Metabolism

The peroxisomal aminotransferase reactions that have been studied in some detail (see Section I.D.1., Table 6) are distinguished on the one hand by specificity for a few ketoacids as acceptors (glyoxylate, pyruvate), and on the other hand by relative interness with certain amino acids (see Section II.A.2.g., Table 11). Pyruvate as amino group acceptor in peroxisomes is easily available from lactate, connecting the aminotransferase reaction with carbohydrate catabolism. Glyoxylate, which is formed abundantly in plant peroxisomes (Section I.E.1.), in animal peroxisomes may arise from glycine and sarcosine (via amino acid oxidase reactions) or from hydroxyproline. An oxidation-amination cycle may take place between glycine and glyoxylate; glycine can also be used for certain biosyntheses. In any case, the metabolism of some amino acids for use as amino group donors seems to be connected in some way with peroxisomes. Thus hydroxypyruvate formed from serine (serine: alanine aminotransferase reaction) may serve as a precursor in gluconeogenesis (Noguchi and Takada 1978 a).

The removal of glyoxylate by glyoxylate aminotransferase reaction conceivable prevents excessive formation of oxalate. As alanine: glyoxylate aminotransferase was found in human liver peroxisomes, Noguchi and Takada (1979) suspect that primary hyperoxaluria might arise from defective peroxisomes.

VII.E. Lipid Metabolism

Whereas the glycerol-3-phosphate pathway of phosphatide biosynthesis is located in the endoplasmic reticulum, the alternative DHAP pathway, by which ether phosphatides are also synthesized, seems to occur in the peroxisomes (see Section I.D.1.). In recent studies on guinea pig liver, HAJRA *et al.*

Table 31. *Distribution of different enzymes [nmol/(min · mg protein)] in the guinea pig liver subcellular fractions prepared by differential centrifugation*

Fraction	DHAP acyltransferase	Acyl-DHAP–NADPH oxidoreductase	Alkyl-DHAP synthase	G-3-P acyltransferase
Nuclear	0.04	1.3	0.03	0.17
Mitochondrial	0.25	18.6	0.15	0.92
Lysosomal and peroxisomal	2.36	76.5	1.80	1.55
Microsomal	0.11	28.5	0.80	2.03
Cytosol	0.00	0.0	0.02	0.00

From: HAJRA *et al.* 1978.

(1978, 1979, Table 31) were able to allocate (with some probability) the following enzymes of the DHAP pathway to the peroxisomes:

DHAP acyltransferase:

$$DHAP \xrightarrow{\text{acyl-CoA} \quad \text{CoA}} 1\text{-acyl-DHAP}$$

alkyl-DHAP synthase:

$$1\text{-acyl-DHAP} \xrightarrow{\text{alcohol} \quad \text{fatty acid}} 1\text{-}O\text{-alkyl-DHAP}$$

acyl/alkyl-DHAP-NADPH oxidoreductase:

$$1\text{-}O\text{-alkyl-DHAP (or 1-acyl-DHAP)} \xrightarrow{\text{NADPH} \quad \text{NADP}^+}$$
$$1\text{-}O\text{-alkyl-glycerol-3-phosphate (or 1-acyl-glycerol-3-phosphate)}$$

where the alcohol is a long-chain alcohol such as *n*-hexadecanol.

This segregation of the two pathways of phospholipid synthesis into the smooth endoplasmic reticulum and peroxisomes conceivably causes the incorporation of distinctly composed phospholipids into different membranes or into a particular side of a membrane (HAJRA *et al.* 1978). The long-chain alcohols necessary for the formation of ether lipids (see above) might also be generated in (micro-)peroxisomes. The Harderian gland is known for its high content of lipids, especially of ether lipids (KASAMA *et al.* 1973, RADOMINSKA-PYREK *et al.* 1979). ROCK *et al.* (1978) reported that microsomes from the Harderian gland of the rabbit contain an NADPH-dependent acylreductase that delivers higher alcohols. The location of this enzyme in microperoxisomes

13*

seems possible, as it has been shown that a large number of the CPs of the Harderian gland sediment in the microsomal fraction (Section VI.B.3.).

In conclusion, there is good reason to believe that the synthesis of certain phosphatides is an important domain of peroxisomes, especially in lipid-secreting cells. The fact that NADPH is necessary for the reduction of acyl/alkyl-DHAP and acyl groups suggests a function for NADP-dependent iso-citrate dehydrogenase found in liver peroxisomes (Section I.D.1., Table 6). Whether the peroxisomal pool of NADPH is important for other lipid bio-syntheses, and whether peroxisomes in general play a part in lipid biosyntheses (as has been claimed by several authors; see Section VI.D.) remain open questions.

The capacity of peroxisomes from different tissues to perform β-oxidation of fatty acids and the carnitine acyl-(and acetyl-) transferase activities (see Section I.D., Table 6) probably are the best described contributions of peroxi-somes to lipid metabolism—contributions that they share with mitochondria. In peroxisomes β-oxidation never seems to proceed to completion (see Sec-tion II.A.2.l.); long-chain fatty acids, especially the monounsaturated ones (see Section I.D.1.), react preferentially. Thus, peroxisomes acting only as chain shorteners presumably require the aid of mitochondria for full degradation of fatty acids to acetyl-CoA. On the other hand, peroxisomes might be in-dispensable when the cell is presented with high concentrations of long-chain fatty acids.

Such an auxiliary function of peroxisomes could be of particular importance in some metabolic situations or in specialized tissues (e.g., in brown fat using fatty acids as a fuel for thermogeneses; see Sections VI.B.6. and VI.D.). The marked increase in peroxisomal β-oxidation caused by hypolipidemic drugs strengthens this hypothesis (see Section III.G.1.), although the physiological correlate of this action is not yet known. At least during starvation and in diabetes no changes in peroxisomal β-oxidation have been detected (MANNAERTS et al. 1979).

The peroxisomal fatty acid-degrading system is able to deal with the free fatty acids themselves: A rather active acyl-CoA synthetase reacting prefer-entially with C_{16}–C_{18} acids is present in the particles from liver (Sec-tion I.D.1., Table 6). Therefore, long-chain fatty acids seem to be activated in the peroxisomes (and in the smooth endoplasmic reticulum) in particular; the long-chain acyl-CoA's can be shortened in the particles or can serve for the formation of cholesterol esters and phosphatides (peroxisomes acting as a shunting station for long-chain acyls). This action points once more to the significance of peroxisomes in lipid synthesis, especially in tissues that store and secrete lipids (see Section VI.A.1.).

Acyl-CoA's produced in the peroxisomes by chain shortening are trans-ferred into the inner mitochondrial compartment for further β-oxidation. Presumably, carnitine octanoyltransferase (medium-chain transferase; see Sec-tion I.D.1., Table 6) delivers the corresponding acylcarnitine, which can be taken up by the mitochondria and thereby mediates this acyl transport. Similarly, the acetyl groups produced by (incomplete) peroxisomal β-oxidation are transferred due to the action of carnitine acetyltransferase : ac(et)yl-carnitine shuttle (as it has also been described in yeast; KAWAMOTO et al. 1978).

References

AEBI, H., 1970: Katalase. In: Methoden der enzymatischen Analyse (BERGMEYER, H. U., ed.), 2nd ed., Vol. 1, pp. 836—847. Weinheim/Bergstr.: Verlag Chemie.

— FREI, E., KNAB, R., SIEGENTHALER, P., 1957: Untersuchungen über die Formiatoxydation in der Leber. Helv. Physiol. Acta 15, 150—167.

— JEUNET, F., RICHTERICH, R., SUTER, H., BÜTLER, R., FREI, J., MARTI, H. R., 1962: Observations in two Swiss families with acatalasemia. Enzymol. Biol. Clin. 2, 1—22.

— BAGGIOLINI, M., DEWALD, B., LAUBER, E., SUTER, H., MICHELI, A., FREI, J., 1964: Observations in two Swiss families with acatalasemia. II. Enzymol. Biol. Clin. 4, 121—151.

— SUTER, H., FEINSTEIN, R. N., 1968: Activity and stability of catalase in head and tissues of normal and acatalasemic mice. Biochem. Genet. 2, 245—251.

AFZELIUS, B., 1965: The occurrence and structure of microbodies. A comparative study. J. Cell Biol. 26, 835—843.

— SCHOENTAL, R., 1967: The ultrastructure of enlarged hepatocytes induced in rat with a single oral dose of Retrorsine, a pyrrolizidine (Senecio) Alkaloid. J. Ultrastruct. Res. 20, 328—345.

AGNER, K., 1938: The preparation and properties of a highly active catalase from horse liver. Biochem. J. 32, 1702—1706.

AHLABO, I., 1972: Observations on the peroxisomes in the sustentocytus (Sertoli cells) of the cat. J. Submicrosc. Cytol. 4, 83—88.

— BARNARD, T., 1971: Observations on peroxisomes in brown adipose tissue of the rat. J. Histochem. Cytochem. 19, 670—675.

ALLEN, J. M., BEARD, M. E., 1965: α-Hydroxyacid oxidase: Localization in renal microbodies. Science 149, 1507—1509.

ALLMANN, D. W., GALZIGNA, L., McCAMAN, R. E., GREEN, D. E., 1966: The membrane systems of the mitochondrion. IV. The localization of the fatty acid oxidizing system. Arch. Biochem. Biophys. 117, 413—422.

AMBROSE, A. M., 1964: Toxicologic studies on pyrethrin-type esters of chrysanthemumic acid. II. Chrysanthemumic acid 2,4-dimethyl-benzyl ester. Toxicol. Appl. Pharmacol. 6, 112—120.

ANDERSON, N. G., HARRIS, W. W., BARBER, A. A., RANKIN, C. T., JR., CANDLER, E. L., 1966: Separation of subcellular components and viruses by combined rate- and isopycnic-zonal centrifugation. Nat. Cancer Inst. Monogr. 21, 253—283.

ANTHONY, L. E., SCHMUCKER, D. L., MOONEY, J. S., JONES, A. L., 1978: Quantitative analysis of fine structure and drug metabolism in livers of clofibrate-treated young adult and retired breeder rats. J. Lipid Res. 19, 154—166.

ANTONINI, E., BRUNORI, M., BRUZZESI, M. R., CHIANCORE, E., MASSEY, V., 1966: Association-dissociation phenomena of D-amino acid oxidase. J. Biol. Chem. 241, 2358—2366.

APPELMANS, F., WATTIAUX, R., DE DUVE, C., 1955: Tissue fractionation studies. 5. The association of acid phosphatase with a special class of cytoplasmic granules in rat liver. Biochem. J. 59, 438—445.

ARAKAWA, M., MIYAJIMA, H., MATSUMURA, H., IZUKAWA, M., IMAI, Y., 1978: Morphological and enzymatic alterations in the rat liver caused by administration of a hypochol-esterolemic agent AT-308 and its related compounds. Biochem. Pharmacol. 27, 167—171.

ARNOLD, G., HOLTZMAN, E., 1975: Peroxisomes in rat sympathetic ganglia and adrenal medulla. Brain Res. **83**, 509—515.

— — 1978: Microperoxisomes in the central nervous system of the postnatal rat. Brain Res. **155**, 1—17.

— LISCUM, L., HOLTZMAN, E., 1977: Cytochemistry of D-aminoacid oxidase in rat cerebellum and kidney. J. Cell Biol. **75**, 202 A.

ASADA, K., URANO, M., TAKAHASHI, M., 1973: Subcellular location of superoxide dismutase in spinach leaves and preparation and properties of crystalline spinach superoxide dismutase. Eur. J. Biochem. **36**, 257—266.

AUDETTE, A., 1975: Quoted from MÜLLER 1975.

AUTIAN, J., 1973: Toxicity and health threats of phthalate esters: Review of the literature. Environ. Health Perspect. 3—26.

AVERON, F., ROTHSTEIN, M., 1974: Nematode biochemistry—XIII. Peroxisomes in the free-living nematode, *Turbatrix aceti.* Comp. Biochem. Physiol. **49 B**, 261—271.

AVERS, C. J., 1971: Peroxisomes of yeast and other fungi. Subcell. Biochem. **1**, 25—37.

AVOY, D. R., SWYRYD, E. A., GOULD, R. G., 1965: Effect of α-*p*-chlorophenoxyisobutyryl ethylester (CPIB) with and without androsterone on cholesterol biosynthesis in rat liver. J. Lipid Res. **6**, 369—376.

AZARNOFF, D. L., SVOBODA, D., 1966: Changes in microbodies in the rat induced by ethyl-*p*-chlorophenoxyisobutyrate. J. Lab. Clin. Med. **68**, 854.

— — 1969: Microbodies in experimentally altered cells. VI. Thyroxine displacement from plasma problems and clofibrate effect. Arch. Int. Pharmacodyn. Rev. **181**, 386—393.

— TUCKER, D. R., 1965: The effect of clofibrate on liver enzymes and substrates. Fed. Proc. **25**, 388.

— — BARR, G., 1965: Studies with ethyl chlorophenoxyisobutyrate (clofibrate). Metabolism **14**, 959—965.

BABIOR, B. M., KIPNES, R. S., CURNUTTE, J. T., 1973: Biological defense mechanisms. The production by leukocytes of superoxide, a potential bactericidal agent. J. Clin. Invest. **52**, 741—744.

BARRETT, M., HEIDGER, P. M., JR., 1975: Microbodies of the rat renal proximal tubule: Ultrastructural and cytochemical investigations. Cell Tissue Res. **157**, 283—305.

BAUDHUIN, P., 1969: Liver peroxisomes, cytology and function. Ann. N.Y. Acad. Sci. **168**, 214—228.

— 1974: Isolation of liver peroxisomes. In: Methods in enzymology (COLOWICK, S. P., KAPLAN, N. O., eds.), Vol. 31 A, pp. 356—368. New York: Academic Press.

— BEAUFAY, H., 1963: Examen au microscope électronique de fractions purifiées d'organites cytoplasmiques de foie de rat. Arch. Intern. Physiol. Biochim. **71**, 119—120.

— — RAHMAN-LI, Y., SELLINGER, O. Z., WATTIAUX, R., JAQUES, P., DE DUVE, C., 1964: Tissue fractionation studies. 17. Intracellular distribution of monoamine oxidase, aspartate aminotransferase, alanine aminotransferase, D-amino acid oxidase and catalase in rat-liver tissue. Biochem. J. **92**, 179—184.

— — DE DUVE, C., 1965 a: Combined biochemical and morphological study of particulate fractions from rat liver. J. Cell Biol. **26**, 219—243.

— MÜLLER, M., POOLE, B., DE DUVE, C., 1965 b: Non-mitochondrial oxidizing particles (microbodies) in rat liver and kidney and in *Tetrahymena pyriformis.* Biochem. Biophys. Res. Commun. **20**, 53—59.

BAYNE, R. A., MUSE, K. E., ROBERTS, J. F., 1969: Isolation of bodies containing the cyanide-insensitive glycerophosphate oxidase of *Trypanosoma equiperdum.* Comp. Biochem. Physiol. **30**, 1049—1054.

BEARD, M. E., 1972: Identification of peroxisomes in the rat adrenal cortex. J. Histochem. Cytochem. **20**, 173—179.

— NOVIKOFF, A. B., 1969: Distribution of peroxisomes (microbodies) in the nephron of the rat. A cytochemical study. J. Cell Biol. **42**, 501—518.

BEATTIE, D. S., 1968: The submitochondrial distribution of the fatty acid oxidizing system in rat liver mitochondria. Biochem. Biophys. Res. Commun. 30, 57—62.

BEAUFAY, H., 1966: La centrifugation en gradient de densité. Thesis, Université Catholique de Louvain.

— HERS, H. G., BERTHET, J., DE DUVE, C., 1954: Le système hexose-phosphatasique. IV. Specificité de la glucose-6-phosphatase. Bull. Soc. Chim. Biol. (Paris) 36, 1525—1537.

— JACQUES, P., BAUDHUIN, P., SELLINGER, O. Z., BERTHET, J., DE DUVE, C., 1964: Tissue fractionation studies. 18. Resolution of mitochondrial fractions from rat liver into three distinct populations of cytoplasmic particles by means of density equilibration in various gradients. Biochem. J. 92, 184—205.

BECKER, F. F., 1970: The normal hepatocyte in division: Regeneration of the mammalian liver. In: Progress in liver disease (POPPER, H., SCHAFFNER, F., eds.), Vol. III, pp. 60—76. New York-London: Grune & Stratton.

BECKETT, R. B., WEISS, R., STITZEL, R. E., CENERELLA, R. J., 1972: Studies on the hepatomegaly caused by the hypolipidemic drugs nafenopin and clofibrate. Toxicol. Appl. Pharmacol. 23, 42—53.

BEERS, R. F., SIZER, I. W., 1952: A spectrophometric method for measuring the breakdown of hydrogen peroxide by catalase. Biochem. J. 195, 133—140.

— — 1953: Catalase assay with special reference to manometric methods. Science 117, 710—712.

BEEVERS, H., 1961: Metabolic production of sucrose from fat. Nature 191, 433—436.

— 1969: Glyoxysomes of castor beans endosperm and their relation to gluconeogenesis. Ann. N.Y. Acad. Sci. 168, 313—324.

BERCHTOLD, J.-P., 1975 a: Observations on peroxisomes in interrenal (adrenocortical) cells of *Triturus cristatus* and *Salamandra salamandra* (Urodele Amphibians). Cell Tissue Res. 162, 349—356.

— 1975 b: Cytophysiologie — Les peroxysomes des cellules interrénales de *Triturus cristatus* (Amphibien, Urodele): variation de leur nombre en fonction de l'activité steroidogenique. C. R. Acad. Sci. (Paris) D 281, 543—545.

BERGMEYER, H. U., GAWEHN, K., GRASSL, M., 1970: Enzyme. In: Methoden der enzymatischen Analyse (BERGMEYER, H. U., ed.), 2nd ed., Vol. 1, pp. 388—483. Weinheim/Bergstr.: Verlag Chemie.

BERKOWITZ, D., 1965: The effect of chlorophenoxyisobutyrate with and without androsterone on the serum lipids, fat tolerance and uric acid. Metabolism 14, 966—975.

— 1969: Clinical experiences with two new lipid lowering agents. Circulation 40, 44.

BERTHET, J., DE DUVE, C., 1951: Tissue fractionation studies. 1. The existence of a mitochondria-linked, enzymically inactive form of acid phosphatase in rat-liver tissue. Biochem. J. 50, 174—181.

BEST, M., DUNCAN, C., 1964: Hypolipidemia and hepatomegaly from ethyl chlorophenoxyisobutyrate (CPIB) in the rat. J. Lab. Clin. Med. 64, 634—642.

— — 1968: Lipid effects of a tetralinphenoxy isobutyric acid. Circulation 38, Suppl. 6, 2.

— — 1969: Preliminary evaluation of hypolipidemic effects in man of a new phenolic ether (SH-1343 CBA). J. Atheroscler. Res. 10, 103—106.

— — 1970: Lipid effects of a phenolic ether (Su-13437) in the rat. Comparison with CPIB. Atherosclerosis 12, 185—192.

BHAWAN, J., FRIEDELL, G. H., JACOBS, J. B., 1975: Ultrastructural changes in livers of tumour bearing rats. Br. J. Exp. Pathol. 56, 561—569.

BIEGLMAYER, C., RUIS, H., 1974: Protein composition of the glyoxysomal membrane. FEBS Lett. 47, 53—55.

— GRAF, J., RUIS, H., 1973: Membranes of glyoxysomes from castor-bean endosperm. Enzymes bound to purified membrane preparations. Eur. J. Biochem. 37, 553—562.

BIEMPICA, L., 1966: Human hepatic microbodies with crystalloid cores. J. Cell Biol. **29**, 383—386.

— KOSOWER, N., NOVIKOFF, A., 1967: Cytochemical and ultrastructural changes in rat liver in experimental porphyria. I. Effects of a single injection of allylisopropyl acetamide. Lab. Invest. **17**, 171—189.

— — ROHEIM, P., 1971: Cytochemical and ultrastructural changes in rat liver in experimental porphyria. II. Effects of repeated injections of allylisopropyl acetamide. Lab. Invest. **24**, 110—117.

BING, R. J., 1965: Cardiac metabolism. Physiol. Rev. **45**, 171—213.

BLACK, V. H., 1974: Peroxisomes in fetal testicular interstitial cells of guinea pigs. Anat. Rec. **178**, 507.

— BOGART, B. I., 1973: Peroxisomes in inner adrenocortical cells of fetal and adult guinea pigs. J. Cell Biol. **57**, 345—358.

BLAIR, O. M., STENGER, R. J., HOPKINS, R. W., SIMEONE, F. A., 1968: Hepatocellular ultra-structure in dogs with hypolipidemic shock. Lab. Invest. **18**, 172—178.

BLANCHARD, M., GREEN, D. E., NOCITO, V., RATNER, S., 1944: L-Amino acid oxidase of animal tissue. J. Biol. Chem. **155**, 421—440.

BLOBEL, G., DOBBERSTEIN, B., 1975 a: Transfer of proteins across membranes. I. Presence of proteolytically processed and unprocessed nascent immunoglobulin light chains on membrane-bound ribosomes of murine myeloma. J. Cell Biol. **67**, 835—851.

— — 1975 b: Transfer of proteins across membranes. II. Reconstitution of functional rough microsomes from heterologous components. J. Cell Biol. **67**, 852—862.

BÖCK, P., 1972 a: DAB-Färbung von Ribosomen. Acta Histochem. (Jena) **43**, 92—97.

— 1972 b: Peroxysomen im Ovar der Maus. Z. Zellforsch. **133**, 131—140.

— 1973 a: Cytochemischer Nachweis von Katalaseaktivität in Microbodies (Peroxysomen?) von Flimmerzellen. Z. Anat. Entwicklungsgesch. **141**, 103—114.

— 1973 b: Cytochemischer Nachweis von Katalaseaktivität in der Magenschleimhaut. Z. Anat. Entwicklungsgesch. **141**, 265—273.

— 1973 c: Cytochemical demonstration of catalase positive particles (peroxisomes?) in fibro-blasts. Z. Zellforsch. **144**, 539—547.

— GOLDENBERG, H., HÜTTINGER, M., KOLAR, M., KRAMAR, R., 1975: Preparation and characterization of catalase-positive particles ("microperoxisomes") from Harder's gland of the rat. Exp. Cell Res. **90**, 15—19.

BODEL, P. T., NICHOLS, B. A., BAINTON, D. F., 1977: Appearance of peroxidase reactivity within rough endoplasmic reticulum of blood monocytes after surface adherence. J. Exp. Med. **145**, 264.

BONNICHSEN, R. K., CHANCE, B., THEORELL, H., 1947: Catalase activity. Acta Chem. Scand. **1**, 685—708.

BOURNE, G. H., 1953: Histochemistry of xanthine oxidase. Nature **172**, 193—195.

BOVERIS, A., CADENAS, E., 1975: Production of superoxide anions and its relationship to the antimycin insensitive respiration. FEBS Lett. **54**, 311—314.

— CHANCE, B., 1973: The mitochondrial generation of hydrogen peroxide. General properties and effect of hyperbaric oxygen. Biochem. J. **134**, 707—716.

BOWDEN, L., LORD, J. M., 1976: Similarities in the polypeptide composition of glyoxysomal and endoplasmic reticulum membranes from castor-bean endosperm. Biochem. J. **154**, 491—499.

BOWEN, P., LEE, C. S. N., ZELLWEGER, H., 1964: A familial syndrome of multiple congenital defects. Bull. Johns Hopkins Hosp. **114**, 402—414.

BRDICZKA, D., PETTE, D., BRUNNER, G., MILLER, F., 1968: Kompartimentierte Verteilung von Enzymen in Rattenlebermitochondrien. Eur. J. Biochem. **5**, 294—304.

BREIDENBACH, W. W., BEEVERS, H., 1967: Association of the glyoxylate cycle enzymes in a novel subcellular paticle from castor bean endosperm. Biochem. Biophys. Res. Commun. **27**, 462—469.

BRESSLER, R., KATZ, R. I., 1965: The effect of carnitine on the rate of incorporation of precursors into fatty acids. J. Biol. Chem. **240**, 622—627.

BRETON-GORIUS, J., GUICHARD, J., 1972: Ultrastructural localization of peroxidase activity in human platelets and megakaryocytes. Am. J. Pathol. 66, 277—294.

— — 1975 a: Fine structural and cytochemical identification of microperoxisomes in developing human erythrocytic cells. Am. J. Pathol. 79, 523—536.

— — 1975 b: The different types of granules in megakaryocytes and platelets as revealed by the diaminobenzidine method. J. Microsc. Biol. Cell. 23, 197—202.

— — 1978: Cytochemical distinction between azurophils and catalase-containing granules in leukocytes: Distribution in human promyelocytes and promonocytes. J. Reticulo-endothel. Soc. 24, 637—646.

— COQUIN, Y., GUICHARD, J., 1978: Cytochemical distinction between azurophils and catalase-containing granules in leukocytes. I. Studies in developing neutrophils and monocytes from patients with myeloperoxidase deficiency: Comparison with peroxidase deficient chicken heterophils. Lab. Invest. 38, 21—31.

BRIGGS, R. T., DRATH, D. B., KARNOVSKY, M. L., KARNOVSKY, M. J., 1975 a: Localization of NADH oxidase on the surface of human polymorphonuclear leukocytes by a new cytochemical method. J. Cell Biol. 67, 566—586.

— KARNOVSKY, M. L., KARNOVSKY, M. J., 1975 b: Cytochemical demonstration of hydrogen peroxide in polymorphonuclear leukocyte phagosomes. J. Cell Biol. 64, 254—260.

BRONFMAN, M., BEAUFAY, H., 1973: Alteration of subcellular organelles induced by com-pression. FEBS Lett. 36, 163—168.

— INESTROSA, N. C., LEIGHTON, F., 1979: Fatty acid oxidation by human liver peroxisomes. Biochem. Biophys. Res. Commun. 88, 1030—1036.

BROWN, G., 1964 a: Allantoinase assay procedure. Am. Zool. 4, 310.

BROWN, G. W., JR., 1964 b: The metabolism of amphibia. In: Physiology of the amphibia (MOORE, J. A., ed.), pp. 1—98. New York-London: Academic Press.

BROWN, G. W., JAMES, J., HENDERSON, R. J., THOMAS, W. N., ROBINSON, R. O., THOMPSON, A. L., BROWN, E., BROWN, S. G., 1966: Uricolytic enzymes in liver of the dipnoan Protopterus aethiopicus. Science 153, 1653—1654.

BROWN, R. H., LORD, J. M., MERRETT, M., 1974: Fractionation of the proteins of plant microbodies. Biochem. J. 144, 559—566.

BRUNI, C., PORTER, K. R., 1965: The fine structure of the parenchymal cell of the normal rat liver. I. General observations. Am. J. Pathol. 46, 691—756.

BURGER, P. C., HERDSON, P. B., 1966: Phenobarbital-induced fine-structural changes in rat liver. Am. J. Pathol. 48, 793—810.

BURKE, J. J., TRELEASE, R. N., 1975: Cytochemical demonstration of urate synthase and glycolate oxidase in microbodies of cucumber cotyledons. Plant Physiol. 56, 716—717.

BURTON, K., 1955: D-Amino acid oxidase from kidney. In: Methods in enzymology (COLOWICK, S. P., KAPLAN, N. O., eds.), Vol. 2, pp. 199—204. New York: Academic Press.

CALVERT, R., MENARD, D., 1978: Cytochemical and biochemical studies on the differentiation of microperoxisomes in the small intestine of the fetal mouse. Dev. Biol. 65, 342—351.

— MALKA, D., MÉNARD, D., 1979: Effect of clofibrate on the small intestine of fetal mice. Histochemistry 63, 7—14.

CANONICO, P. G., WHITE, J. D., POWANDA, M. C., 1975: Peroxidase depletion in rat during pneumococcal sepsis. Lab. Invest. 33, 147—150.

— RILL, W., AYALA, E., 1977: Effects of inflammation on peroxisomal enzyme activities, catalase synthesis and lipid metabolism. Lab. Invest. 37, 479—487.

CARAVACA, J., MAY, M. D., 1964: The isolation and properties of an active peroxidase from hepatocatalase. Biochem. Biophys. Res. Commun. 16, 528—534.

— DIMOND, E. G., SEMMERS, S. C., WENK, R., 1967: Prevention of induced atherosclerosis by peroxidase. Science 155, 1284—1287.

CARPENTER, C. P., WEIT, C. S., SMYTH, H. F., 1953: Chronic oral toxicity of di(2-ethyl-hexyl)phthalate for rats, guinea pigs and dogs. Arch. Ind. Hyg. 8, 219—226.

CASSIER, P., FAIN-MAUREL, M.-A., 1972: Caractères infrastructureaux et cytochimiques des Oenocytes de Locusta migratoria migratorioides en rapport avec les mues et les cycles ovariens. Arch. Anat. Microsc. Morphol. Exp. 61, 357—380.

CHANCE, B., 1947: An intermediate compound in the catalase hydrogen peroxide reaction. Acta Chem. Scand. 1, 236—267.
— 1950: The reactions of catalase in the presence of the notatin system. Biochem. J. 46, 387—402.
— HERBERT, D., 1950: The enzyme–substrate compounds of bacterial catalase and peroxides. Biochem. J. 46, 402—414.
— MAEHLEY, A. C., 1955: Assay of catalases and peroxidases. In: Methods in enzymology (COLOWICK, S. P., and KAPLAN, N. O., eds.), Vol. 2, pp. 764—775. New York: Academic Press.
— GREENSTEIN, D. S., ROUGHTON, F. J. W., 1952: The mechanism of catalase action. I. Steady state analysis. Arch. Biochem. Biophys. 37, 301—321.
CHANG, C. H., SCHILLER, B., GOLDFISCHER, S., 1971: Small cytoplasmic bodies in the loop of Henle and distal convoluted tubule that resemble peroxisomes. J. Histochem. Cytochem. 19, 56—62.
CHANG, C. N., BLOBEL, G., MODEL, P., 1978: Detection of procaryotic signal peptidase in an Escherichia coli membrane fraction: Endoproteolytic cleavage of nascent f1 pre-coat protein. Proc. nat. Acad. Sci. (U.S.A.) 75, 361—365.
CHANTRENNE, H., 1955: Effects d'un inhibiteur de la catalase sur la formation induite de cet enzyme chez la levure. Biochim. Biophys. Acta 16, 410—417.
CHIGA, M., REDDY, J., SVOBODA, D., 1971: Degradation kinetics of liver catalase in rats treated with ethyl-α-p-chlorophenoxyisobutyrate. Lab. Invest. 25, 49—52.
CHILDS, G. E., 1973 a: Diaminobenzidine reactivity of peroxisomes and mitochondria in a parasitic amoeba, Hartmanella culbertsoni. J. Histochem. Cytochem. 21, 26—33.
— 1973 b: Hartmanella culbertsoni: Enzymatic, ultrastructural, and cytochemical characteristics of peroxisomes in a density gradient. Exp. Parasitol. 34, 44—45.
CHRISTIANSEN, R. Z., 1978: The effect of clofibrate feeding on hepatic fatty acid metabolism. Biochim. Biophys. Acta 530, 314—324.
CITKOWITZ, E., HOLTZMAN, E., 1973: Peroxisomes in dorsal root ganglia. J. Histochem. Cytochem. 21, 34—41.
CLARKSON, A. B., 1975: Microbodies of Trypanosoma equiperdum. Ph.D. Thesis, Univ. of Georgia, Athens, Georgia, U.S.A. Ann Arbor: Microfilms Ltd.
CLAUDE, A., 1946 a: Fractionation of mammalian liver cells by differential centrifugation. I. Problems, methods, and preparation of extract. J. Exp. Med. 84, 51—59.
— 1946 b: Fractionation of mammalian liver cells by differential centrifugation. II. Experimental procedures and results. J. Exp. Med. 84, 61—89.
CLINE, M. J., LEHRER, R. I., 1969: D-Aminoacid oxidase in leukocytes: A possible D-aminoacid linked antimicrobial system. Proc. nat. Acad. Sci. (U.S.A.) 62, 756—763.
COLLINS, C., MERRETT, M. J., 1975: Microbody-marker enzymes during transition from phototrophic to organotrophic growth in Euglena. Plant Physiol. 55, 1018—1022.
COLLINS, N., BROWN, R. H., MERRETT, M. J., 1975: Oxidative phosphorylation during glycolate metabolism in mitochondria from phototrophic Euglena gracilis. Biochem. J. 150, 373—377.
CONNOCK, M. J., 1973: Intestinal peroxisomes in the goldfish (Carassius auratus). Comp. Biochem. Physiol. 45 A, 945—951.
— KIRK, P. R., 1973: Identification of peroxisomes in the epithelial cells of the small intestine of the guinea pig. J. Histochem. Cytochem. 21, 502—503.
— POVER, W., 1970: Catalase particles in the epithelial cells of the guinea pig small intestine. Histochem. J. 2, 371—380.
— KIRK, P. R., STURDEE, A. P., 1974: A zonal rotor method for the preparation of microperoxisomes from epithelial cells of guinea pig small intestine. J. Cell Biol. 61, 123—133.
COOPER, T. G., BEEVERS, H., 1969: Oxidation in glyoxisomes from castor bean endosperm. J. Biol. Chem. 244, 3514—3520.
CORNFORD, M. E., OUTKA, D. E., 1970: Peroxisomes as organelles of terminal oxidation in trichomonads. J. Cell Biol. 47, 41 A, Abstract 102.
CORRIN, B., CLARK, A. E., SPENCER, H., 1969: Ultrastructural localization of acid phosphatase in rat lung. J. Anat. 104, 65—70.
COSSEL, L., 1964: Die menschliche Leber im Elektronenmikroskop. Untersuchungen an Leberpunktaten. Jena: Fischer.

CRAIG, G. M., 1972: A comparison of clofibrate and its derivative methyl clofenapate. Atherosclerosis 15, 265—271.

CRANE, F. L., MII, S., HAUGE, J. G., GREEN, D. E., BEINERT, H., 1956: On the mechanism of dehydrogenation of fatty acyl derivatives of coenzyme A. I. The general fatty acyl coenzyme A dehydrogenase. J. Biol. Chem. 218, 701—716.

CUADRADO, R. L., BRICKER, L. A., 1973: An abnormality of hepatic lipogenesis in a mutant strain of acatalasemic mice. Biochem. Biophys. Acta 306, 168—172.

DAEMS, W. TH., 1975: The fine structure of mouse liver microbodies. J. Microsc. 5, 295—304.

— WISSE, E., BREDEROO, P., EMEIS, J. J., 1975: Peroxidatic activity in monocytes and macrophages. In: Mononuclear phagocytes in immunity, infection and pathology (FURTH, R., ed.), pp. 57—77. London: Blackwell.

— KOERTEN, H. K., SORANZO, M. R., 1976: Differences between monocyte-derived and tissue macrophages. In: The reticulo-endothelial system in health and disease: functions and characteristics (REICHARD, S. M., ESCOBAR, M. R., FRIEDMANN, H., eds.), pp. 27—40. New York: Plenum Press.

DALTON, A. J., 1964: An electronmicroscopic study of a series of chemically induced hepatomas. In: Cellular control mechanisms and cancer (EMMELOT, P., MÜHLBOCK, O., eds.), pp. 211—225. Amsterdam: Elsevier.

DANKS, D. M., TIPPETT, P., ADAMS, C., CAMPBELL, P., 1975: Cerebro-hepato-renal syndrome of Zellweger. J. Pediatr. 86, 382—387.

DANNEN, E., BEARD, M. E., 1977: Peroxisomes in pulmonate gastropodes. J. Histochem. Cytochem. 25, 319—328.

DAVIES, P., DRATH, P. B., ENGEL, E. E., HUBER, G. L., 1979: The localization of catalase in the pulmonar alveolar macrophage. Lab. Invest. 40, 221—226.

DAY, E. D., GABRIELSON, F. C., LIPKIND, J. B., 1954: Depressions in the activity of liver catalase in mice injected with homogenates of normal mouse spleen. J. nat. Cancer Inst. 15, 239—252.

DECKER, K., 1970: Acetyl-Coenzyme A. UV-spektrometrische Bestimmung. In: Methoden der enzymatischen Analyse (BERGMEYER, H. U., ed.), 2nd ed., Vol. 1, pp. 1922—1927. Weinheim/Bergstr.: Verlag Chemie.

DE DUVE, C., 1960: Intracellular localization of enzymes. Nature 187, 836—853.

— 1965: Function of microbodies (peroxisomes). J. Cell Biol. 27, 25 A—26 A.

— 1969: Evolution of the peroxisome. Ann. N.Y. Acad. Sci. 168, 369—381.

— 1973: Biochemical studies on the occurrence, biogenesis and life history of mammalian peroxisomes. J. Histochem. Cytochem. 21, 941—948.

— BAUDHUIN, P., 1966: Peroxisomes (microbodies and related particles). Physiol. Rev. 46, 323—357.

— BERTHET, J., 1953: Reproducibility of differential centrifugation experiments in tissue fractionation. Nature 172, 1142.

— BERTHET, J., BEAUFAY, H., 1959: Gradient centrifugation of cell particles. Theory and application. Prog. Biophys. Chem. 9, 325—369.

— BEAUFAY, H., JACQUES, P., RAHMAN-LI, Y., SELLINGER, O. Z., WATTIAUX, R., DE CONINCK, S., 1960: Intracellular localization of catalase and of some oxidases in rat liver. Biochim. Biophys. Acta 40, 186—187.

— PRESSMAN, B. C., GIANETTO, R., WATTIAUX, R., APPELMANS, F., 1955: Tissue fractionation studies. 6. Intracellular distribution patterns of enzymes in rat liver tissue. Biochem. J. 60, 604—617.

DE LA IGLESIA, F. A., 1969: Comparative analysis of hepatic microbodies (a review). Acta Hepatosplenol. (Stuttg.) 16, 141—160.

— PORTA, E. A., HARTROFT, W. S., 1966: Histochemical urate oxidase activity and microbodies in non human primate liver. J. Histochem. Cytochem. 14, 685—687.

DELLA CORTE, E., STIRPE, F., 1970: The regulation of xanthine oxidase. Inhibition by reduced nicotinamide-adenine dinucleotide of rat liver xanthine oxidase type D and of chick liver xanthine dehydrogenase. Biochem. J. 117, 97—100.

DELMON, G., BABIN, R., BIRABEN, J., 1958: A propos des inhibiteurs de la catalase extraite des tumeurs malignes. Bull. Soc. Pharm. Bordeaux 97, 7—12.

DE MARCHI, W. J., JOHNSTON, G. A. R., 1969: The oxidation of glycine by D-amino acid oxidase in extracts of mammalian central nervous tissue. J. Neurochem. 16, 355—361.

DE PALMA, R. G., COIL, J., DAVID, J. H., HOLDEN, W. D., 1967: Cellular and ultrastructural changes in endotoxemia: A light and electron microscopic study. Surgery 62, 505—515.

DEWALD, B., BAGGIOLINI, M., CURNUTTE, J. T., BABIOR, B. M., 1979: Subcellular localization of the superoxide-forming enzyme in human neutrophils. J. Clin. Invest. 63, 21—29.

DIAUGUSTINE, R. P., 1974: Lung concentric lamellar organelle. Hydrolase activity and composition analysis. J. Biol. Chem. 294, 584—593.

DIXON, M., KLEPPE, K., 1965: D-Amino acid oxidase. II. Specificity, competitive inhibition and reaction sequence. Biochim. Biophys. Acta 96, 368—382.

DONALDSON, R. P., TOLBERT, N. E., SCHNARRENBERGER, C., 1972: A comparison of microbody membranes with microsomes and mitochondria from plant and animal tissue. Arch. Biochem. Biophys. 152, 199—215.

DOYLE, J., TWETO, J., 1975: Measurement of protein turnover in animal cells. Methods Cell Biol. 10, 235—260.

DULEY, J., HOLMES, R. S., 1974: α-Hydroxyacid oxidase genetics in the mouse: Evidence for two genetic loci and a tetrameric subunit structure for the liver isoenzyme. Genetics 76, 93—97.

DVOŘÁK, M., KONEČNA, H., ŠTASTNÁ, J., 1967 a: Differentiation of microbodies of hepatic cell during ontogenesis. Arch. Anat. Gist. Embriol. 52, 86—93.

— — — 1967 b: Differentiation of microbodies and lysosomes of the liver cell during ontogenesis. Scr. Med. Fac. Med. Brun. 40, 49—50.

EGUCHI, M., SANNES, P. L., SPICER, S. S., 1979: Peroxisomes of rat peritoneal macrophage during phagocytosis. Am. J. Pathol. 95, 281—294.

ERICSSON, J. L. E., TRUMP, B. F., 1966: Electron microscopic studies of the proximal tubule of the rat kidney. III. Microbodies, multivesicular bodies and the Golgi apparatus. Lab. Invest. 15, 1610—1623.

ESSNER, E., 1967: Endoplasmic reticulum and the origin of microbodies in fetal mouse liver. Lab. Invest. 17, 71—87.

— 1969: Localization of peroxidase activity in microbodies of fetal mouse liver. J. Histochem. Cytochem. 17, 454—466.

— 1970: Observations on hepatic and renal peroxisomes (microbodies) in the developing chicken. J. Histochem. Cytochem. 18, 80—92.

FAHIMI, H. D., 1968: Cytochemical localization of peroxidase activity in the rat hepatic microbodies (peroxisomes). J. Histochem. Cytochem. 16, 547—550.

— 1969: Cytochemical localization of catalase in rat hepatic microbodies (peroxisomes). J. Cell Biol. 43, 275—288.

— 1973: Diffusion artifacts in cytochemistry of catalase. J. Histochem. Cytochem. 21, 999—1009.

— 1974: Effect of buffer storage on the fine structure and catalase cytochemistry of peroxisomes. J. Cell Biol. 63, 675—683.

— VENKATACHALAM, M. A., 1970: Microbody regeneration and catalase synthesis in rat liver. J. Cell Biol. 47, 58 A.

— GRAY, B. A., HERZOG, V. K., 1976: Cytochemical localization of catalase and peroxidase in sinusoidal cells of rat liver. Lab. Invest. 34, 192—201.

— KINO, M., HICKS, L., THORP, K. A., ABELMAN, W. H., 1979: Increased myocardial catalase in rats fed ethanol. Am. J. Pathol. 96, 373—390.

FARBER, E., STERNBERG, W. H., PEARCE, N. A. M., 1958: Histochemical localization of choline oxidase and D-aminoacid oxidase with tetrazolium salts and phenazine methosulfate. J. Histochem. Cytochem. 6, 389.

FEINSTEIN, R. N., BERLINER, S., GREEN, F. O., 1958: Mechanism of inhibition of catalase by 3-amino,1,2,4 triazole. Arch. Biochem. Biophys. 76, 32—44.
— HOWARD, J. B., BRAUN, J. T., SEAHOLM, J. E., 1966: Acatalasemic and hypocatalasemic mouse mutants. Genetics 53, 923—933.
— BRAUN, J. T., HOWARD, J. B., 1967: Acatalasemic and hypotalasemic mouse mutants. II. Mutational variations in blood and solid tissue catalases. Arch. Biochem. 120, 165—169.
— SUTER, H., JAROSLAW, B. N., 1968: Blood catalase polymorphism: Some immunological aspects. Science 159, 638—639.
FISCHER, E., HORVÁTH, I., 1978: Evidence of the presence of extraperoxisomal catalase in chloragogen cells of the earthworm, Lumbricus terrestris L. Histochemistry 56, 165—171.
FLAKS, B., 1968: Formation of membrane-glycogen array in rat hepatoma cells. J. Cell Biol. 36, 410—414.
FLOHÉ, L., SCHLEGEL, W., 1971: Glutathion-Peroxidase. IV. Intrazelluläre Verteilung des Glutathion-Peroxidase-Systems in der Rattenleber. Hoppe-Seylers Z. Physiol. Chem. 352, 1401—1410.
FLORKIN, M., DUCHÂTEAU-BOSSON, G., 1943: Forms of uricolytic enzyme systems and the evolution of purine metabolism in the animal kingdom. Arch. Intern. Physiol. 53, 267—307.
FOK, A. K., ALLEN, R. D., 1975: Cytochemical localization of peroxisomes in Tetrahymena pyriformis. J. Histochem. Cytochem. 23, 599—606.
FOMINA, W. A., ROGOVINE, V. V., PIRUZYAN, A., MURAVIEFF, R. A., PODOVINNIKOVA, E. A., 1975: Electron-microscopic investigation on peroxisomes in the epithelia of mice gall-bladder. Experientia 31, 1030—1031.
FOOTE, C. S., WEXLER, S., ANDO, W., HIGGINS, R., 1968: Chemistry of singlet oxygen. IV. Oxygenations with hypochlorite-hydrogen peroxide. J. Am. Chem. Soc. 90, 975—981.
FOWLER, S., REMACLE, J., TROUET, A., BEAUFAY, H., BERTHET, J., WIBO, M., HAUSER, P., 1976: Analytical study of microsomes and isolated subcellular membranes from rat liver. V. Immunological localization of cytochrome b$_5$ by electron microscopy: Methodology and application to various subcellular fractions. J. Cell Biol. 71, 535—550.
FRANKE, W. W., MORRÉ, D. J., DEUMLING, B., CHEETHAM, R. D., KARTENBECK, J., JARASCH, E.-D., ZENTGRAF, H.-W., 1971: Synthesis and turnover of membrane proteins in rat liver: An examination of the membrane flow hypothesis. Z. Naturforsch. 26 b, 1031—1039.
FREDERICK, S. E., GRUBER, P. J., NEWCOMB, E. H., 1975: Plant microbodies. Protoplasma 84, 1—29.
FRIDOVICH, I., 1965: A new class of xanthine oxidase inhibitors isolated from guanidinium salts. Biochemistry 4, 1098—1101.
— 1970: Quantitative aspects of the production of superoxide anion radical by milk xanthine oxidase. J. Biol. Chem. 245, 4053—4057.
FRITZ, I. B., SCHULTZ, S. K., SRERE, P. A., 1963: Properties of partially purified carnitine acetyltransferase. J. Biol. Chem. 238, 2509—2517.
FUJIWARA, F., 1964: Influences of fractionation media on the release of enzymes in light mitochondrial fraction in rat and mouse liver cells, with special reference to uricase-containing particles. Sapporo Med. J. 26, 102—113.
FUKUDA, K., SHINDO, H., YAMASHINA, S., MIZUHIRA, V., 1978: The fine structural changes in the hepatic microbodies of rats treated with hypolipidemic agents, gemfibrozil and clofibrate. Acta Histochem. Cytochem. 11, 432—443.
FUKUI, S., KAWAMOTO, S., YASUHARA, S., TANAKA, A., OSUMI, M., IMAIZUMI, F., 1975 a: Microbody of methanol-grown yeast. Localization of catalase and flavin-dependent alcohol oxidase in the isolated microbody. Eur. J. Biochem. 59, 561—566.
— TANAKA, A., KAWAMOTO, S., YASUHARA, S., TERANISHI, Y., OSUMI, M., 1975 b: Ultra-structure of methanol-utilizing yeast cells: Appearance of microbodies in relation to high catalase activity. J. Bacteriol. 123, 317—328.

GÄNSLER, H., ROUILLER, C., 1956: Modifications physiologiques et pathologiques du chondriome. Schweiz. Z. Allg. Pathol. Bacteriol. 19, 217—243.
GEE, R., MCGROARTY, E., HSIEH, B., WIED, D. M., TOLBERT, N. E., 1974: Glycerol phosphate dehydrogenase in mammalian peroxisomes. Arch. Biochem. Biophys. 161, 187—193.

GERHARDT, B., 1978: Microbodies/Peroxisomen pflanzlicher Zellen. (Cell Biology Monographs, Vol. 5.) Wien-New York: Springer.

— BEEVERS, H., 1970: Developmental studies on glyoxysomes in *Ricinus* endosperm. J. Cell Biol. **44**, 94—102.

GHADIALLY, F. N., PARRY, E. W., 1966: Ultrastructure of human hepatocellular carcinoma and surrounding non-neoplastic liver. Cancer **19**, 1989—2004.

GIBBS, D. A., HAUSCHILDT, S., WATTS, R. W. E., 1977: Glyoxylate oxidation in rat liver and kidney. J. Biochem. **82**, 221—230.

GIBBS, M., 1969: Photorespiration, Warburg effect and glycolate. Ann. N.Y. Acad. Sci. **168**, 356—368.

GILCHRIST, K. W., GILBERT, E. F., GOLDFARB, S., GOLL, U., SPRANGER, J. W., OPITZ, J. M., 1976: Studies of malformation syndromes of man XIB: The cerebro-hepato-renal syndrome of Zellweger: Comparative pathology. Eur. J. Pediatr. **121**, 99—118.

GIORGI, F., DERI, P., 1976: Cytochemistry of late ovarian chambers of *Drosophila melanogaster*. Histochemistry **48**, 325—334.

GIRAUD, G., CZANINSKI, Y., 1971: Ultrastructural localization of oxidase activities in *Chlamydomonas reinhardtii*. C. R. Acad. Sci. (Paris) D **273**, 2500—2503.

GLUECK, C. S., 1971: Effects of oxandrolone on plasma triglycerides and postheparin lipolytic activity in patients with types III, IV, and V familial hyperlipoproteinemia. Metabolism **20**, 691—702.

GNATZKY, W., 1970: Struktur und Entwicklung des Integuments und der Oenocyten von *Culex pipiens* L. (Dipt.). Z. Zellforsch. **110**, 401—443.

GOECKERMANN, J. A., VIGIL, E. L., 1975: Peroxisome development in the metanephric kidney of the mouse. J. Histochem. Cytochem. **23**, 957—973.

GOLDBERG, A., RIMINGTON, C., 1955: Experimentally produced porphyria in animals. Proc. R. Soc. (London) B **143**, 257—280.

GOLDENBERG, H., 1977 a: Organization of purine degradation in the liver of a teleost (carp; *Cyprinus carpio* L.). A study of its subcellular distribution. Mol. Cell. Biochem. **16**, 17—21.

— 1977 b: Subcellular distribution of purine degrading enzymes in the liver of the carp (*Cyprinus carpio* L.). In: Purine metabolism in man (MÜLLER, M. M., KAISER, E., eds.), Vol. 2, pp. 254—264. New York: Plenum Press.

— HÜTTINGER, M., BÖCK, P., KRAMAR, R., 1975 a: Influence of subtotal hepatectomy on peroxisomes and peroxisomal enzymes of rat liver and isolated liver cell fractions. Histochemistry **44**, 47—56.

— — LUDWIG, R., KRAMAR, R., BÖCK, P., 1975 b: Catalase-positive particles ("microperoxisomes") from rat preputial gland and bovine adrenal cortex. Exp. Cell Res. **93**, 438—442.

— — KAMPFER, P., KRAMAR, R., PAVELKA, M., 1976: Effect of clofibrate on morphology and enzyme content of liver peroxisomes. Histochemistry **46**, 189—196.

— — — 1978 a: Preparation of peroxisomes from carp liver by zonal rotor density gradient centrifugation. Histochem. J. **10**, 103—113.

— — KOLLNER, U., KRAMAR, R., PAVELKA, M., 1978 b: Catalase-positive particles from pig lung. Biochemical preparation and morphological studies. Histochemistry **56**, 253—264.

GOLDFISCHER, S., ESSNER, E., 1969: Further observations on the peroxidatic activities of microbodies (peroxisomes). J. Histochem. Cytochem. **17**, 681—685.

— — 1970: Peroxidase activity in peroxisomes (microbodies) of acatalasemic mice. J. Histochem. Cytochem. **18**, 482—489.

— KIKKAWA, Y., HOFFMAN, L., 1968: The demonstration of acid hydrolase activities in the inclusion bodies of type II alveolar cells and other lysosomes in the rabbit lung. J. Histochem. Cytochem. **16**, 102—109.

— ROHEIM, P. S., EDELSTEIN, D., ESSNER, E., 1971: Hypolipidemia in a mutant strain of "acatalasemic" mice. Science **173**, 65—66.

— MOORE, C. L., JOHNSON, A. B., SPIRO, A. J., VALSAMIS, M. P., WISNIEWSKI, H. K., RITCH, R. H., NORTON, W. T., RAPIN, I., GARTNER, L. M., 1973: Peroxisomal and mitochondrial defects in the cerebro-hepatorenal syndrome. Science **182**, 62—64.

GOLDMAN, B. M., BLOBEL, G., 1978: Biogenesis of peroxisomes: Intracellular site of synthesis of catalase and uricase. Proc. nat. Acad. Sci. (U.S.A.) 75, 5066—5070.

GOTOH, M., GRIFFIN, C., HRUBAN, Z., 1975: Effects of citrate and aminotriazole on matrical plates induced in hepatic microbodies. Virchows Arch. (Zellpathol.) 17, 279—294.

GRAHAM, R. C., KARNOVSKY, M. J., 1965: The histochemical demonstration of uricase activity. J. Histochem. Cytochem. 13, 448—453.

— — 1966: The early stages of absorption of injected horseradish peroxidase in the proximal tubules of mouse kidney: Ultrastructural cytochemistry by a new technique. J. Histochem. Cytochem. 14, 291—302.

GRANICK, S., 1965: Hepatic porphyria and drug-induced or chemical porphyria. Ann. N.Y. Acad. Sci. 123, 188—197.

— 1966: The induction in vitro of the synthesis of δ-aminolevulinic acid synthetase in chemical porphyria: A response to certain drugs, sex hormones, and foreign chemicals. J. Biol. Chem. 241, 1359—1375.

GRANT, P. T., SARGENT, J. R., 1960: Properties of L-α-glycerophosphate oxidase and its role in the respiration of Trypanosoma rhodesiense. Biochem. J. 76, 229—237.

— — 1961: L-α-Glycerophosphate dehydrogenase, a component of an oxidase system in Trypanosoma rhodesiense. Biochem. J. 81, 206—214.

— — RYLEY, J. F., 1961: Respiratory systems in the Trypanosomidae. Biochem. J. 81, 200—206.

GRAVES, L. B., BECKER, W. M., 1974: Beta-oxidation in glyoxysomes from Euglena. J. Protozool. 21, 771—773.

— HANZELY, L., TRELEASE, R. N., 1971: The occurrence and fine structural characterization of microbodies in Euglena gracilis. Protoplasma 72, 141—152.

GREEN, D. E., MII, S., MAHLER, H. R., BOCK, R. M., 1954: Studies on the fatty acid oxidizing system of animal tissues. III. Butyryl coenzyme A dehydrogenase. J. Biol. Chem. 206, 1—12.

GREENFIELD, R. E., PRICE, V. E., 1956: Liver catalase. III. Isolation of catalase from mitochondrial fractions of polyvinylpyrrolidine–sucrose homogenates. J. Biol. Chem. 220, 607—618.

GREENSTEIN, J. P., 1954: Biochemistry of cancer, 2nd ed. New York: Academic Press.

— 1955: The in-vivo effect on liver catalase by a tumor. J. nat. Cancer Inst. 15, 1603—1605.

GRUBER, P. J., FREDERICK, S. E., 1977: Cytochemical localization of glycolate oxidase in microbodies of Klebsormidium. Planta 135, 45—49.

GULYAS, B. J., YUAN, L. C., 1975: Microperoxisomes in the late pregnancy corpus luteum of rhesus monkeys (Macaca mulatta). J. Histochem. Cytochem. 23, 359—368.

— — 1977: Association of microperoxisomes with the endoplasmic reticulum in the granulosa lutein cells of the rhesus monkey (Macaca mulatta). Cell Tissue Res. 179, 357—366.

HAJRA, A. K., 1968 a: Biosynthesis of acyl dihydroxyacetone phosphate in guinea pig mitochondria. J. Biol. Chem. 243, 3458—3465.

— 1968 b: Biosynthesis of phosphatidic acid from dihydroxyacetone phosphate. Biochem. Biophys. Res. Commun. 33, 929—935.

— AGRANOFF, B. W., 1968 a: Reduction of palmitoyl dihydroxyacetone phosphate by mitochondria. J. Biol. Chem. 243, 3542—3543.

— — 1968 b: Acyl dihydroxyacetone phosphate. Characterization of a ^{32}P-labeled lipid from guinea pig liver mitochondria. J. Biol. Chem. 243, 1617—1622.

— JONES, C. L., DAVIS, P. A., 1978: Studies on the biosynthesis of the O-alkyl bound in glycerol ether lipids. In: Advances in experimental medicine and biology (GATT, S., FREYSZ, L., MANDEL, P., eds.), Vol. 101, pp. 369—378. New York: Plenum Press.

— BURKE, C. L., JONES, C. L., 1979: Subcellular localization of acyl coenzyme A: dihydroxyacetone phosphate acyltransferase in rat liver peroxisomes (microbodies). J. Biol. Chem. 254, 10896—10900.

HALBACH, S., 1977: Catalase activity measured with a micro oxygen electrode in a pressurized reaction vessel. Anal. Biochem. 80, 383—391.

HALLIWELL, B., 1974: Superoxide dismutase, catalase and glutathione peroxidase: Solutions to the problems of living with oxygen. New Phytol. 73, 1075—1086.

HAMOR, T., GARSIDE, E. T., 1973: Peroxisome-like vesicles and oxidative activity in the zona radiata and yolk of the ovum of the Atlantic salmon (*Salmo salar* L.). Comp. Biochem. Physiol. 45 B, 147—151.

HAND, A. R., 1973: Morphological and cytochemical identification of peroxisomes in the rat parotid and other exocrine glands. J. Histochem. Cytochem. 21, 131—141.

— 1974: Peroxisomes (microbodies) in striated muscle cells. J. Histochem. Cytochem. 22, 207—209.

— 1975: Ultrastructural localization of L-α-hydroxyacid oxidase in rat liver peroxisomes. Histochemistry 41, 195—206.

— 1976: Ultrastructural localization of catalase and L-α-hydroxy acid oxidase in microperoxisomes of *Hydra*. J. Histochem. Cytochem. 24, 915—925.

HANKER, J. S., ROMANOVICZ, D. K., 1977: Phi-bodies: Peroxidatic particles that produce crystalloidal cellular inclusions. Science 197, 895—898.

— ANDERSON, W. A., BLOOM, F. E., 1972 a: Osmiophilic polymer generation: Catalysis by transition metal compounds in ultrastructural cytochemistry. Science 175, 991—993.

— YATES, P. E., CLAPP, D. H., ANDERSON, W. A., 1972 b: New methods for the demonstration of lysosomal hydrolases by the formation of osmium blacks. Histochemie 30, 201—214.

— PREECE, J. W., BURKES, E. J., JR., ROMANOVICZ, D. K., 1977 a: Catalase in salivary gland striated and excretory duct cells. I. The distribution of cytoplasmic and particulate catalase and the presence of catalase-positive rods. Histochem. J. 9, 711—728.

— SILVERMAN, M. S., ROMANOVICZ, D. K., 1977 b: Catalase in salivary gland striated and excretory duct cells. II. Φ-body: An elliptoidal peroxisomal organelle with crystalloid axial projections. Histochem. J. 9, 729—744.

— YATES, P. E., METZ, C. B., RUSTIONI, A., 1977 c: A new specific, sensitive and noncarcinogenic reagent for the demonstration of horseradish peroxidase. Histochem. J. 9, 789—792.

— LASZLO, J., MOORE, J. O., 1978: The light microscopic demonstration of hydroperoxidase-positive phi-bodies and rods in leukocytes in acute myeloid leukemia. Histochemistry 58, 241—252.

HANNA, CH. H., HOPKINS, T. A., BUCK, J., 1976: Peroxisomes of the firefly lantern. J. Ultrastruct. Res. 57, 150—162.

HANNIG, K., 1964: Eine Neuentwicklung der trägerfreien kontinuierlichen Elektrophorese. Zur Trennung hochmolekularer und grobdisperser Teilchen. Hoppe-Seylers Z. Physiol. Chem. 338, 211—227.

— HEIDRICH, H. G., 1974: The use of continous preparative free-flow electrophoresis for dissociating cell fractions and isolation of membranous components. In: Methods in enzymology (COLOWICK, S. P., KAPLAN, N. O., eds.), Vol. 31 A, pp. 746—761. New York: Academic Press.

HAUGE, J. G., CRANE, F. L., BEINERT, H., 1956: On the mechanism of dehydrogenation of fatty acyl derivatives of coenzyme A. III. Palmityl CoA dehydrogenase. J. Biol. Chem. 219, 727—733.

HAVEL, R. J., KANE, J. P., 1973: Drugs and lipid metabolism. Ann. Rev. Pharmacol. 13, 287—308.

HAYASHI, H., SUGA, T., 1978: Some characteristics of peroxisomes in the slime mould, *Dictyostelium discoideum*. J. Biochem. 84, 513—520.

— NIINOBE, S., 1971: Studies on peroxisomes. I. Intraparticulate localization of peroxisomal enzymes in rat liver. Biochim. Biophys. Acta 252, 58—68.

— — 1973: Studies on peroxisomes. III. Further studies on the intraparticular localization of peroxisomal components in the liver of the rat. Biochim. Biophys. Acta 297, 110—119.

— — 1975: Studies on peroxisomes. V. Effect of ethyl *p*-chlorophenoxy isobutyrate on the centrifugal behaviour of rat liver peroxisomes. J. Biochem. 77, 1199—1204.

HAYASHI, H., SUGA, T., NIINOBE, S., 1976 a: Studies on peroxisomes. VII. Effect of the light component on the formation of the core DAAO complex of rat liver peroxisomes. Cell Struct. Function 1, 367—376.

— TAYA, K., SUGA, T., NIINOBE, I., 1976 b: Studies on peroxisomes. TI. Relationship between the peroxisomal core and urate oxidase. J. Biochem. 79, 1029—1034.

HAZEU, W., BATENBURG-VAN DER VEGTE, W. H., NIEUWDORP, P. J., 1975: The fine structure of microbodies in the yeast Pichia pastoris. Experientia 31, 926—927.

HAZLETT, L. D., HAZLETT, J. C., IRELAND, M., BRADLEY, R. H., 1978: Microperoxisomes in retinal epithelium and tapetum lucidum of the American opossum. Exp. Eye Res. 27, 343—348.

HEIM, W. G., APPLEMAN, D., 1956: Effects of 3-amino-1,2,4-triazole (AT) on catalase and other components. Am. J. Physiol. 186, 19—23.

— — PYTRON, H. T., 1955: Production of catalase changes in animals with 3-amino-1,2,4-triazole. Science 122, 693—694.

HELENIUS, A., SIMONS, K., 1975: Solubilization of membranes by detergents. Biochim. Biophys. Acta 415, 29—79.

HELLMAN, L., ZUMOFF, B., KESSLER, G., KARA, E., RUBIN, I. L., ROSENFELD, R. S., 1963: Reduction of cholesterol and lipids in man by ethyl p-chlorophenoxyisobutyrate. Amer. Int. Med. 59, 477—494.

HEPPEL, L. A., PORTERFIELD, U. T., 1949: Metabolism of inorganic nitrite and nitrate esters. I. The coupled oxidation of nitrite by peroxide-forming systems and catalase. J. Biol. Chem. 178, 549—556.

HERZOG, V., FAHIMI, H. D., 1973: An improved cytochemical method for demonstration of the peroxidatic activity of beef liver catalase (BLC). J. Histochem. Cytochem. 21, 412.

— — 1974 a: Microbodies (peroxisomes) containing catalase in myocardium: Morphological and biochemical evidence. Science 185, 271—273.

— — 1974 b: The effect of glutaraldehyde on catalase. Biochemical and cytochemical studies with beef liver catalase and rat liver peroxisomes. J. Cell Biol. 60, 303—311.

— — 1974 c: Peroxisomes in mouse myocardium, biochemical identification. J. Cell Biol. 63, 136 A.

— — 1974 d: Peroxisomes (microbodies) containing catalase in myocardium. Cytochemical, biochemical and morphological evidence. Eighth International Congress on Electron Microscopy, Canberra, Vol. II, pp. 390—391.

— — 1975: Identification of peroxisomes (microbodies) in mouse myocardium. J. Mol. Cell Cardiol. 8, 271—281.

— and MILLER, F., 1972: Endogenous peroxidase in the lacrimal gland of the rat and its differentiation against injected catalase and horseradish-peroxidase. Histochemie 30, 235—246.

HESS, R., BENCZE, W. L., 1968: Hypolipidemic properties of a new tetralin derivative (Ciba 13437-Su). Experientia (Basel) 24, 418—419.

— STÄUBLI, W., RIESS, W., 1965: Nature of the hepatomegalic effect produced by ethyl-chlorophenoxyisobutyrate in the rat. Nature 208, 856—858.

HEUSEQUIN, E., 1973: Sur les éléments mis en évidence par la 3,3'-diamino-benzidine dans la foie de Lebistes reticulatus. Arch. Biol. (Liége) 84, 243—279.

HICKS, L., FAHIMI, H. D., 1977: Peroxisomes (microbodies) in the myocardium of rodents and primates. A comparative ultrastructural cytochemical study. Cell Tissue Res. 175, 457—481.

HIGASHI, T., PETERS, T., 1963 a: Studies on rat liver catalase. I. Combined immunochemical and enzymatic determination of catalase in liver cell fractions. J. Biol. Chem. 238, 3945—3951.

— — 1963 b: Studies on rat liver catalase. II. Incorporation of ^{14}C-leucine into catalase of liver cell fractions in vivo. J. Biol. Chem. 238, 3952—3954.

HIGGINS, J. A., BARRNETT, R. J., 1970: Cytochemical localization of transferase activities: Carnitine acetyltransferase. J. Cell Sci. 6, 29—51.

HINTON, R., DOBROTA, M., 1976: Density gradient centrifugation. In: Laboratory techniques in biochemistry and molecular biology (WORK, T. S., WORK, E., eds.), Vol. 6, Part I, p. 69. Amsterdam: North Holland.

HIRAI, K. I., 1968: Specific affinity of oxidizied amine dye (radical intermediates) for heme enzymes. Study in microscopy and spectrophotometry. Acta Histochem. Cytochem. 1, 43—55.

— 1969: Light microscopic study of the peroxidatic activity of catalase in formaldehyde fixed rat liver. J. Histochem. Cytochem. 17, 585—590.

— 1971: Comparison between 3,3'-diamino benzidine and auto-oxidized 3,3'-diamino benzidine in the cytochemical demonstration of oxidative enzymes. J. Histochem. Cytochem. 19, 434—442.

— 1974: Distribution of peroxidase activity in *Tetrahymena pyriformis* mitochondria. J. Histochem. Cytochem. 22, 189—202.

— OGAWA, K., 1973: Appearance of giant peroxisomes in mouse hepatocytes treated with a hypolipidemic drug. Acta Histochem. Cytochem. 6, 195.

— — 1975: Ultrastructural studies on the morphogenesis of peroxisomes in mouse hepatocytes treated with simfibrate. Acta Histochem. Cytochem. 8, 18—29.

HOEPRICH, P. D., 1965: Alanine: Cycloserine antagonism. VI. Demonstration of D-alanine in the serum of guinea pigs and mice. J. Biol. Chem. 240, 1654—1660.

HOFFMAN, H. A., GRIESHABER, K., 1976: Genetic studies of murine catalase: Regulation of multiple molecular forms of kidney catalase. Biochem. Genet. 14, 59—66.

HOGEBOOM, G. H., 1955: Fractionation of cell components of animal tissues. In: Methods in enzymology (COLOWICK, S. P., KAPLAN, N. O., eds.), Vol. 1, pp. 16—22. New York: Academic Press.

— SCHNEIDER, W. C., PALADE, G. E., 1948: Cytochemical studies of mammalian tissues. I. Isolation of intact mitochondria from rat liver; some biochemical properties of mitochondria and submicroscopic particulate material. J. Biol. Chem. 172, 619—636.

HOLMES, R. S., 1971: Ontogeny of mouse liver peroxisomes and catalase isoenzymes. Nature 232, 218—220.

— MASTERS, C. J., 1970: On the latency, multiplicity and subcellular distribution of catalase activity in mammalian tissues. Int. J. Biochem. 1, 474—482.

— — 1972: Species specific features of the distribution and multiplicity of mammalian liver catalase. Arch. Biochem. Biophys. 148, 217—223.

HOLT, S. J., HICKS, R. M., 1961: The localization of acid phosphatase in rat liver cells as revealed by combined cytochemical staining and electron microscopy. J. Biophys. Biochem. Cytol. 11, 47—66.

HOLTZMAN, E., TEICHBERG, S., ABRAHAM, S. J., CITKOWITZ, E., CRAIN, ST. M., KAWAI, N., PETERSON, E. R., 1973: Notes on synaptic vesicles and related structures, endoplasmic reticulum, lysosomes and peroxisomes in nervous tissue and the adrenal medulla. J. Histochem. Cytochem. 21, 349—385.

HORWITZ, B. A., 1976: The effect of cold exposure on liver mitochondrial and peroxisomal distribution in the rat, hamster and bat. Comp. Biochem. Physiol. 54 A, 45—48.

HOWARD, K. P., ALAVPOVIC, P., BRUSCO, O. J., FURMAN, R. H., 1963: Effects of ethyl chlorophenoxyisobutyrate, alone or with androsterone (Atromid) on serum lipids, lipoproteins, and related metabolic parameters in normal and hyperlipidaemic subjects. J. Atheroscler. Res. 3, 482—499.

HRUBAN, Z., RECHCIGL, M., JR., 1967: Comparative ultrastructure of microbodies. Fed. Proc. 26, 513.

— — 1969: Microbodies and related particles: Morphology, biochemistry and physiology. Int. Rev. Cytol., Suppl. 1, pp. 1—296.

— SWIFT, H., 1964: Uricase: Localization in hepatic microbodies. Science 146, 1316—1318.

— VIGIL, E. L., 1972: Microbodies, cholesterol and steroids. Gastroenterology 62, 194.

— SWIFT, H., WISSLER, R. W., 1963: Alterations in the fine structure of hepatocytes produced by β-3-thienylalanine. J. Ultrastruct. Res. 8, 236—250.

— — RECHCIGL, M., JR., 1965 a: Fine structure of transplantable hepatomas of the rat. J. nat. Cancer Inst. 35, 459—495.

— — SLESERS, A., 1965 b: Effect of azaserine on the fine structure of the liver and pancreatic acinar cells. Cancer Res. 25, 708—724.

— — — 1966: Ultrastructural alterations of hepatic microbodies. Lab. Invest. 15, 1884—1901.

HRUBAN, Z., VIGIL, E. L., SLESERS, A., 1972 a: Widespread distribution of microbodies (peroxisomes) in animal tissue. Fed. Proc. 31, 641.
— — HOPKINS, E., 1972 b: Microbodies. Constituent organelles of animal cells. Lab. Invest. 27, 184—191.
— GOTOH, M., SLESERS, A., CHOU, S. F., 1974 a: Structure of hepatic microbodies in rats treated with acetylsalcylic acid, clofibrate and dimethrin. Lab. Invest. 30, 64—75.
— MOCHIZUKI, Y., GOTOH, M., SLESERS, A., CHOU, S. H., 1974 b: Effects of some hypocholesteremic agents on hepatic ultrastructure and microbody enzymes. Lab. Invest. 30, 474—485.
HRYB, D. J., HOGG, J. F., 1979: Chain length specifities of peroxisomal and mitochondrial β-oxidation in rat liver. Biochem. Biophys. Res. Commun. 87, 1200—1206.
HSIEH, B., TOLBERT, N. E., 1976: Glyoxylate aminotransferase in peroxisomes from rat liver and kidney. J. Biol. Chem. 251, 4408—4415.
HUANG, A. H. C., BEEVERS, H., 1971: Isolation of microbodies from plant tissues. Plant Physiol. 48, 637—641.
— — 1973: Localization of enzymes within microbodies. J. Cell Biol. 58, 379—389.
HÜTTINGER, M., GOLDENBERG, H., KRAMAR, R., 1979: A characteristic membrane protein of liver peroxisomes inducible by clofibrate. Biochim. Biophys. Acta 558, 251—254.

IMAI, Y., SHIMAMOTO, K., 1973: AT-308 or a related compound produced an enlargement of the liver like that of clofibrate. Atherosclerosis 17, 121—129.
INESTROSA, N. C., BRONFMAN, M., LEIGHTON, F., 1979: Detection of peroxisomal fatty acyl-coenzyme A oxidase activity. Biochem. J. 182, 779—788.
ITABASHI, M., MOCHIZUKI, K., TSUKADA, H., 1975: Peroxisomes in liver tumors of rats induced by 3'-methyl-4-(dimethylamino)azobenzene. Gann 66, 463—472.
— — — 1977: Changes in peroxisomes in preneoplastic liver of rats induced by 3'methyl-4-dimethylaminoazobenzene. Cancer Res. 37, 1035—1043.

JACOB, H. E., 1964: Polarographic determination of catalase activity of microorganisms. Z. Chem. 4, 189—190.
JANSSON, I., SCHENKMAN, J. B., 1977: Studies on three microsomal electron transfer enzyme systems. Specificity of electron flow pathways. Arch. Biochem. Biophys. 178, 89—107.
JONES, C. L., HAJRA, A. K., 1977: The subcellular distribution of acyl CoA : dihydroxyacetone phosphate acyl transferase in guinea pig liver. Biochem. Biophys. Res. Commun. 76, 1138—1143.
JONES, G. L., MASTERS, C. J., 1974: On the synthesis and degradation of the multiple forms of catalase in mouse liver. Arch. Biochem. Biophys. 161, 601—609.
— — 1975: On the nature and characteristics of the multiple forms of catalase in mouse liver. Arch. Biochem. Biophys. 169, 7—21.
— — 1976: On the turnover and proteolysis of catalase in tissues of the guinea pig and acatalasemic mice. Arch. Biochem. Biophys. 173, 463—489.

KÄHÖNEN, M. T., 1976: Effect of clofibrate treatment on carnitine acyltransferases in different subcellular fractions of rat liver. Biochim. Biophys. Acta 28, 690—701.
— YLIKAHRI, R. H., 1974: Effect of clofibrate treatment on the activity of carnitine acetyltransferase in rat tissues. FEBS Lett. 43, 297—299.
— — 1979: Effect of clofibrate and gemfibrozil on the activities of mitochondrial carnitine acyltransferases in liver. Dose-response relations. Atherosclerosis 32, 47—56.
KALCKAR, H. M., 1947: Differential spectrophotometry of purine compounds by means of specific enzymes. III. Studies of the enzymes of purine metabolism. J. Biol. Chem. 167, 461—475.
KALMBACH, P., FAHIMI, H. D., 1978: Peroxisomes: Identification in freeze-etch preparations of rat kidney. Cell Biol. Int. Rep. 2, 389—396.
KAMPSCHMIDT, K., 1965: Mechanism of liver catalase degression in tumor-bearing animals. A review. Cancer Res. 25, 34—45.
KANEKO, A., SAKAMOTO, S., MORITA, M., ONOE, T., 1969: Morphological and biochemical changes in rat liver during early stages of ethyl chlorophenoxyisobutyrate administration. Tohoku J. Exp. Med. 99, 81—101.

212   References

KARIYA, T., PARKER, R., GRISAR, J., MARTIN, J., WILLE, L., 1975: Laboratory studies with RMI 14,514, a new hypolipidemic agent. Fed. Proc. **34**, 789.

KARNOVSKY, M. J., 1965: A formaldehyde-glutaraldehyde fixative of high osmolality for use in electron microscopy. J. Cell Biol. **27**, 137 A.

— 1971: Use of ferrocyanide-reduced osmium tetroxide in electron microscopy. J. Cell Biol. **51**, 146—284 A.

KARTENBECK, J., FRANKE, W. W., 1974: Membrane relationship between endoplasmic reticulum and peroxisomes in rat hepatocytes and Morris hepatoma cells. Cytobiologie **10**, 152—156.

KASAMA, K., RAINEY, W. T., SNYDER, F., 1973: Chemical identification and enzymic synthesis of a newly discovered lipid class. Hydroxyalkylglycerols. Arch. Biochem. Biophys. **154**, 648—658.

KATZ, F. N., ROTHMAN, J. E., LINGAPPA, V. R., BLOBEL, G., LODISH, H. F., 1977: Membrane assembly *in vitro:* Synthesis, glycosylation, and asymmetric insertion of a transmembrane protein. Proc. nat. Acad. Sci. (U.S.A.) **74**, 3278—3282.

KAUFMANN, F. C., NAKANISHI, S., SCHOLAR, E. M., SCHULMAN, M. P., 1966: Stimulation of hepatic protein synthesis by the porphyrinogenic drug, allylisopropylacetamide (AIA). Fed. Proc. **25**, 195.

KAWAMATA, F., SAKURA, T., HIGASHI, T., 1975 a: Biosynthesis of liver catalase in rats treated with allylisopropyl-acetylcarbamide. I. Immunochemical assay of catalase in liver cell fractions. J. Biochem. **78**, 969—974.

— — — 1975 b: Biosynthesis of liver catalase in rats treated with allylisopropylacetyl-carbamide. II. Double-labeling of catalase with ^{14}C-leucine and delta-^{3}H-aminoleucinic acid. J. Biochem. **78**, 975—980.

KAWAMOTO, S., UEDA, M., NOZAKI, C., YAMAMURA, M., TANAKA, A., FUKUI, S., 1978: Localization of carnitine acetyltransferase in peroxisomes and in mitochondria of *n*-alkane-grown *Candida tropicalis.* FEBS Lett. **96**, 37—40.

KEILIN, D., HARTREE, E. F., 1936 a: Uricase, amino acid oxidase and xanthine oxidase. Proc. R. Soc. (London) **B 119**, 114—140.

— — 1936 b: Coupled oxidation of alcohol. Proc. R. Soc. (London) **B 119**, 141.

— — 1939: Cytochrome and cytochrome oxidase. Proc. R. Soc. (London) **B 127**, 167—191.

— — 1945: Properties of catalase. Catalysis of coupled oxidation of alcohols. Biochem. J. **39**, 293—301.

KIRK, J. E., 1963: A rapid procedure for catalase determination in blood and tissue samples with the van Slyke manometric apparatus. Clin. Chem. **9**, 763—775.

KISAKI, T., TOLBERT, N. E., 1969: Glycolate and glyoxylate metabolism by isolated peroxisomes or chloroplasts. Plant Physiol. **44**, 242—250.

KITANO, R., MORIMOTO, S., 1975: Isolation of peroxisomes from the dog kidney cortex. Biochem. Biophys. Acta **411**, 113—120.

KLEBANOFF, S. J., 1969: Antimicrobial activity of catalase at acid pH. Proc. Soc. Exp. Biol. Med. **132**, 571—574.

KOBAYASHI, K., YAMAMOTO, T., 1978: Evidence for glycogen particles in the peroxisome of riboflavin deficient mouse kidney. J. Histochem. Cytochem. **26**, 434—440.

KOLDE, G., ROESSNER, A., THEMANN, H., 1976: Effects of clofibrate (alpha-*p*-chlorophenoxy-isobutyrylethyl-ester) on male rat liver. Virchows Arch. (Zellpathol.) **B 22**, 73—87.

KOPPENOL, W. H., BUTLER, J., 1977: Mechanism of reactions involving singlet oxygen and the superoxide anion. FEBS Lett. **83**, 1—6.

KORNBERG, A., 1955: Isocitric dehydrogenase of yeast (TPN). In: Methods in enzymology (COLOWICK, S. P., KAPLAN, N. O., eds.), Vol. 1, pp. 705—707. New York: Academic Press.

KRAHLING, J. B., GEE, R., MURPHY, P. A., KIRK, J. R., TOLBERT, N. E., 1978: Comparison of fatty acid oxidation in mitochondria and peroxisomes from rat liver. Biochem. Biophys. Res. Commun. **82**, 136—141.

KRAMAR, R., GOLDENBERG, H., BÖCK, P., KLOBUČAR, N., 1974: Peroxisomes in the liver of the carp (*Cyprinus carpio* L.). Electron microscopic, cytochemical and biochemical studies. Histochemistry **40**, 137—154.

KRAMAR, R., HÜTTINGER, M., GMEINER, B., GOLDENBERG, H., 1978: β-Oxidation in peroxisomes of brown adipose tissue. Biochim. Biophys. Acta **531**, 353—356.

KREBS, H. A., 1935: Metabolism of aminoacids. III. Deamination of aminoacids. Biochem. J. 29, 1620—1643.
— 1951: Oxidation of amino acids. In: The enzymes (SUMNER, J. B., MYRBÄCK, K., eds.), Vol. 2, pp. 499—535. New York: Academic Press.
— PERKINS, J. R., 1970: The physiological role of liver alcohol dehydrogenase. Biochem. J. 118, 635—644.
KRISHNAIAH, K. V., INANDER, A. R., RAMSARMA, T., 1967: Regulation of steroidgenesis by ubiquinon. Biochem. Biophys. Res. Commun. 27, 474—478.
KRISHNAKANTHA, T. P., RAMAKRISHNA-KURUP, C. K., 1972: Increase in hepatic catalase and glycerophosphate dehydrogenase activities on administration of clofibrate and clofenapate to the rat. Biochem. J. 130, 167—175.
KRITCHEVSKY, D., 1971: Newer hypolipidemic agents. Fed. Proc. 30, 835—840.
— 1974: New drugs effecting lipid metabolism. Lipids 9, 97—102.
KUHN, C., 1968: Particles resembling microbodies in normal and neoplastic perianal glands of dogs. Z. Zellforsch. 90, 554—562.
KÜHNEL, W., 1971: Struktur und Cytochemie der Harderschen Drüse vom Kaninchen. Z. Zellforsch. 119, 384—404.
KUN, E., DECHARY, J. M., PITOT, H. C., 1954: The oxidation of glycolic acid by a liver enzyme. J. Biol. Chem. 210, 269—289.
KURUP, C. K. R., AITHAL, H. N., KAMASARMA, T., 1970: Increase in hepatic mitochondria on administration of ethyl-α-p-chlorophenoxyisobutyrate to the rat. Biochem. J. 116, 773—779.

LABBE, R. F., HANAWA, Y., LOTTSFELD, F. I., 1961: Heme and fatty acid biosynthesis in experimental porphyria. Arch. Biol. 92, 373—374.
LA BELLE, E. F., JR., HAJRA, A. K., 1972: Biosynthesis of acyl dihydroxyacetone phosphate in subcellular fractions of rat liver. J. Biol. Chem. 247, 5835—5841.
LAKE, B. G., GANGOLLI, S. D., WRIGHT, M. G., GRASSO, P., LLOYD, A. G., 1974: Studies on the effects of oral administration of di(2-ethylhexyl)phthalate on some hepatic enzymes in the rat. Biochem. Soc. Trans. 2, 322—325.
— — GRASSO, P., LLOYD, A. G., 1975: Studies on the hepatic effects of orally administered di(2-ethylhexyl)phthalate in the rat. Toxicol. Appl. Pharmacol. 32, 355—367.
LAKSHMANAN, M. R., PHILLIPS, W. E., BRIAN, R. L., 1968: Effect of p-chlorophenoxyisobutyrate (CPIB) fed to rats on hepatic biosynthesis and catabolism of ubiquinone. J. Lipid Res. 9, 353—356.
LAMY, J. N., LAMY, J., SCHMITT, M., WEILL, J., 1970: Sur la diminution de la vitesse de biosynthèse de la catalase hepatique chez le rat après hepatectomie minimale. Bull. Soc. Chim. Biol. 52, 299—232.
— — — — 1973: Effet d'une hepatectomie minimale sur l'activité de la catalase et des oxydases peroxysomales du foie de rat. Biochimie 55, 1491—1494.
LANDRISCINA, C., PAPA, S., CORATELLI, P., MAZZARELLA, L., QUAGLIARIELLO, E., 1970: Enzymatic activities of the matrix and inner membrane of pigeon-liver mitochondria. Biochim. Biophys. Acta 205, 136—147.
LANGER, K. H., 1968: Feinstrukturen der Mikrokörper (microbodies) des proximalen Nierentubulus. Z. Zellforsch. 90, 432—446.
LASKOWSKI, M., 1951: Allantoinase and allantoicase. In: The enzymes (SUMNER, J. B., MYRBÄCK, K., eds.), Vol. I, pp. 946—950. New York: Academic Press.
LATA, G. F., MAMRAK, F., BLOCH, PH., BAKER, B., 1977: An electron microscopic and enzymic study of rat liver peroxisomal nucleoid core and its association with urate oxidase. J. Supramol. Struct. 7, 419—434.
LAZAROW, P. B., 1977: Three hypolipidemic drugs increase hepatic palmitoyl-coenzyme A oxidation in the rat. Science 197, 580—581.
— 1978: Rat liver peroxisomes catalyze the β-oxidation of fatty acids. J. Biol. Chem. 253, 1522—1528.
— DE DUVE, C., 1973 a: The synthesis and turnover of rat liver peroxisomes. IV. Biochemical pathway of catalase synthesis. J. Cell Biol. 59, 491—506.

LAZAROW, P. B., DE DUVE, G., 1973 b: The synthesis and turnover of rat liver peroxisomes. V. Intracellular pathway of catalase synthesis. J. Cell Biol. **59**, 507—524.

— — 1976: A fatty acyl-CoA oxidizing system in rat liver peroxisomes; enhancement by clofibrate, a hypolipidemic drug. Proc. nat. Acad. Sci. (U.S.A.) **73**, 2043—2046.

LEE, K. J., KIM, D. N., LEE, K. T., 1974: Ultrastructural study of hepatic mitochondrial abnormality in swine treated with clofibrate. Exp. Mol. Pathol. **20**, 387—396.

LEE, P. C., 1973: Developmental changes of adenosine deaminase, xanthine oxidase and uricase in mouse tissue. Dev. Biol. **31**, 227—233.

LEE, S. H., TORACK, R. M., 1968: Electron microscopic studies of glutamic oxaloacetic transaminase in rat liver cell. J. Cell Biol. **39**, 716—724.

LEGG, P. G., WOOD, R. L., 1970 a: New observations on microbodies. A cytochemical study on CPIB-treated rat liver. J. Cell Biol. **45**, 118—129.

— — 1970 b: Effects of catalase inhibitors on the ultrastructure and peroxidase activity of proliferating microbodies. Histochemie **22**, 262—276.

— — 1972: Effects of allylisopropylacetamide (AIA) on the fine structure and peroxidase activity of microbodies in rat hepatic cells. Z. Zellforsch. **128**, 19—30.

LEIGHTON, F., POOLE, B., BEAUFAY, H., BAUDHUIN, P., COFFEY, J. W., FOWLER, S., DE DUVE, C., 1968: The large-scale separation of peroxisomes, mitochondria, and lysosomes from the livers of rats injected with Triton WR-1339. J. Cell Biol. **37**, 482—513.

— — LAZAROW, P. B., DE DUVE, C., 1969: The synthesis and turnover of rat liver peroxisomes. I. Fractionation of peroxisome proteins. J. Cell Biol. **41**, 521—537.

— COLOMA, L., KOENIG, C., 1975: Structure, composition, physical properties, and turnover of proliferated peroxisomes. A study of the trophic effects of Su-13437 on rat liver. J. Cell Biol. **67**, 281—309.

LEUENBERGER, P. M., NOVIKOFF, A. B., 1973: Microperoxisomes in the retina of pigmented and albino rodents. Exp. Eye Res. **17**, 399—409.

— — 1975: Studies on microperoxisomes. VII. Pigment epithelial cells and other cell types in the retina of rodents. J. Cell Biol. **65**, 324—334.

LEVY, E., SLUSSER, J. R., RUEDNER, B. H., 1968: Hepatic changes produced by a single dose of endotoxin in the mouse. Am. J. Pathol. **52**, 477—502.

LEWIS, N. J., WITIAK, D. T., FELLER, D. R., 1974: Influence of clofibrate (ethyl-4-chloro-phenoxy-isobutyrate) on hepatic drug metabolism in male rats. Proc. Soc. Exp. Biol. Med. **145**, 281—285.

LINDMARK, D. G., MÜLLER, M., 1973: Hydrogenosome, a cytoplasmic organelle of the anaerobic flagellate *Tritrichomonas foetus*, and its role in pyruvate metabolism. J. Biol. Chem. **248**, 7724—7728.

— — 1974: Superoxide dismutase in the anaerobic flagellates, *Tritrichomonas foetus* and *Monocercomonas sp.* J. Biol. Chem. **249**, 4634—4637.

LINDQUIST, R. R., 1969: Studies on the pathogenesis of hepatolenticular degeneration. II. Cytochemical methods for the localization of copper. Arch. Pathol. **87**, 370—379.

LITWIN, J. A., 1979: Histochemistry and cytochemistry of 3,3′-diaminobenzidine. A review. Fol. Histochem. Cytochem. **17**, 3—28.

LOCKE, M., 1969: The ultrastructure of oenocytes in molt/intermolt cycle of an insect. Tissue Cell **1**, 103—154.

— McMAHON, J. T., 1971: The origin and fate of microbodies in the fat body of an insect. J. Cell Biol. **48**, 61—78.

LOTTSFELD, F. I., LABBE, R. F., 1965: Some cytologic changes of rat liver in experimental porphyria. Proc. Soc. Exp. Biol. **119**, 226—229.

LOUD, A. V., 1968: A quantitative sterological description of the ultrastructure of normal rat liver parenchymal cells. J. Cell Biol. **37**, 27—46.

LOWRY, O. H., ROSEBROUGH, N. J., FARR, A. L., RANDALL, R. J., 1951: Protein measurement with the Folin phenol reagent. J. Biol. Chem. **193**, 265—275.

LUDWIG, B., KINDL, H., 1976: Plant microbody proteins. II. Purification and characterization of the major protein component (SP-63) of peroxisome membranes. Hoppe-Seylers Z. Physiol. Chem. **357**, 177—186.

LYNEN, F., OCHOA, S., 1953: Enzymes of fatty acid metabolism. Biochim. Biophys. Acta **12**, 299—314.

MAGALHÃES, M. M., MAGALHÃES, M. C., 1971: Microbodies (peroxisomes) in rat adrenal cortex. J. Ultrastruct. Res. **37**, 563—573.

MAHLER, H. R., HÜBSCHER, G., BAUM, H., 1955: Studies on uricase. I. Preparation, purification and properties of a cuproprotein. J. Biol. Chem. **216**, 625—641.

— BAUM, H. M., HÜBSCHER, G., 1956: Enzymatic oxidation of urate in non-borate buffers. Science **124**, 705—708.

MALICK, L. E., 1972: Ultrastructure of transplantable mouse hepatomas with different growth rates. J. nat. Cancer Inst. **49**, 1039—1055.

MANNAERTS, G. P., THOMAS, J., DEBEER, L. J., McGARVY, J. D., FOSTER, D. W., 1978: Hepatic fatty acid oxidation and betagenesis after clofibrate treatment. Biochim. Biophys. Acta **529**, 201—211.

— DEBEER, L. J., THOMAS, J., DE SCHEPPER, P. J., 1979: Mitochondrial and peroxisomal fatty acid oxidation in liver homogenates and isolated hepatocytes from control and clofibrate-treated rats. J. Biol. Chem. **254**, 4585—4595.

MARGOLIASH, E., NOVOGRODSKY, A., 1958: A study of the inhibition of catalase by 3-amino-1 : 2 : 4-triazole. Biochem. J. **68**, 468—475.

— — SCHEIJTER, A., 1960: Irreversible reaction of 3-amino-1 : 2 : 4-triazole and related inhibitors with the protein of catalase. Biochem. J. **74**, 339—348.

MARKWELL, M. A. K., BIEBER, L. L., 1976: Localization and solubilization of a rat liver microsomal carnitine acetyltransferase. Arch. Biochem. Biophys. **172**, 502—509.

— McGROARTY, E. J., BIEBER, L. L., TOLBERT, N. E., 1973: The subcellular distribution of carnitine acyltransferases in mammalian liver and kidney. A new peroxisomal enzyme. J. Biol. Chem. **248**, 3426—3432.

— TOLBERT, N. E., BIEBER, L. L., 1976: Comparison of the carnitine acyltransferase activities from rat liver peroxisomes and microsomes. Arch. Biochem. Biophys. **176**, 479—488.

— BIEBER, L. L., TOLBERT, E., 1977: Differential increase of hepatic peroxisomal, mitochondrial and microsomal carnitine acyltransferases in clofibrate-fed rats. Biochem. Pharmacol. **26**, 1697—1703.

MARSI, M. S., HENDRICKSON, A. P., COX, A. J., JR., DAEDS, F., 1964: Subacute toxicity of two chrysanthemumic acid esters: Bathrin and dimethrin. Toxicol. Appl. Pharmacol. **6**, 716—725.

MARTIN, P. A., TEMPLE, N. J., CONNOCK, M. J., 1979: Zonal rotor study of the subcellular distribution of acyl-CoA synthetases, carnitine acyl transferases and phosphatidate phosphatase in the guinea pig small intestine. Eur. J. Cell Biol. **19**, 3—10.

MARVER, H. S., COLLINS, A., TSCHUDY, D. P., RECHCIGL, M., JR., 1966: γ-Aminolevulinic acid synthetase. II. Induction in rat liver. J. Biol. Chem. **241**, 4323—4329.

MASTERS, C., HOLMES, R., 1977: Peroxisomes: New aspects of cell physiology and biochemistry. Physiol. Rev. **57**, 816—882.

MAUNSBACH, A. B., 1966: Observations on the ultrastructure and acid phosphatase activity of the cytoplasmic bodies in rat kidney proximal tubule cells. J. Ultrastruct. Res. **16**, 197—238.

— WIRSEN, C., 1966: Ultrastructural changes in kidney, myocardium and skeletal muscle of dog during excessive mobilization of free fatty acids. J. Ultrastruct. Res. **16**, 35—54.

McGROARTY, E. J., TOLBERT, N. E., 1973: Enzymes in peroxisomes. J. Histochem. Cytochem. **21**, 949—954.

— HSIEH, B., WIED, D. M., GEE, R., TOLBERT, N. E., 1974: Alpha-hydroxyacid oxidation by peroxisomes. Arch. Biochem. Biophys. **161**, 194—210.

McKENNA, O., ARNOLD, G., HOLTZMAN, E., 1976: Microperoxisome distribution in the central nervous system of the rat. Brain Res. **117**, 181—194.

MII, S., GREEN, D. E., 1954: Studies on the fatty acid oxidizing system of animal tissues. VIII. Reconstruction of fatty acid oxidizing system with triphenyltetrazolium as electron acceptor. Biochim. Biophys. Acta **13**, 425—432.

MILES, J. L., HOLMES, R. S., 1975: The ontogeny of L-α-hydroxyacid oxidase isozymes in the mouse. J. Exp. Zool. **192**, 119—125.

MIZUTANI, A., BARRNETT, R. J., 1965: Fine structural demonstration of phosphatase activity at pH 9. Nature **206**, 1001—1003.

MOCHIZUKI, Y., HRUBAN, Z., MORRIS, H. P., SLESERS, A., VIGIL, E. L., 1971: Microbodies of Morris hepatomas. Cancer Res. 31, 763—773.

MONTALI, R. J., REHG, J. E., MARGOLIS, S., GILMORE, M. J., 1974: Cytoplasmic microbodies associated with renal tubular lipids in cats. Lab. Invest. 30, 16.

MOODY, D. E., REDDY, J. K., 1974: Increase in hepatic carnitine acetyltransferase activity associated with peroxisomal (microbody) proliferation induced by the hypolipidemic drugs clofibrate, nafenopin, and methyl clofenapate. Res. Commun. Chem. Pathol. Pharmacol. 9, 501—510.

— — 1975: Morphometric analysis of ultrastructural changes in rat liver induced by peroxisome proliferators. J. Histochem. Cytochem. 23, 317—318.

— — 1976 a: Morphometric analysis of the ultrastructural changes in rat liver induced by the peroxisome proliferator SaH 42-348. J. Cell Biol. 71, 768—780.

— — 1976 b: Comparative effects of hypolipidemic drugs on hepatic peroxisomal enzymes. Fed. Proc. 35, 381.

— — 1978 a: The hepatic effects of hypolipidemic drugs (clofibrate, nafenopin, tibric acid, Wy-14,643) on hepatic peroxisomes and peroxisome associated enzymes. Am. J. Pathol. 90, 435—448.

— — 1978 b: Hepatic peroxisome (microbody) proliferation in rats fed plasticizers and related compounds. Toxicol. and Appl. Pharmacol. 45, 497—505.

— AZARNOFF, D. L., REDDY, J. K., 1976: Induction of hepatic peroxisomes, peroxisome-associated enzymes and hypolipidemia in rats treated with 2-ethyl hexanoic acid and 2-ethyl hexanol. J. Cell Biol. 70, 363 A.

— SAMBASIVA RAO, M., REDDY, J. K., 1977: Mitogenic effect in mouse liver induced by a hypolipidemic drug, nafenopin. Virchows Arch. (Zellpathol.) 23 B, 291—296.

MOORIDAN, B. A., CUTLER, L. S., 1978: Developmental distribution of microperoxisomes in the rat submandibular gland. J. Histochem. Cytochem. 26, 989—999.

MORIKAWA, S., HARADA, T., 1969: Immunohistochemical localization of catalase in mammalian tissues. J. Histochem. Cytochem. 17, 30—35.

MORIMOTO, S., MATSUMURA, Y., ABE, R., FUKUHARA, A., 1978: Effects of 3-amino-1,2,4-triazole on plasma renin activity and renal peroxisomal enzymes in rats. Jpn. J. Pharmacol. 28, 793—796.

MORRIS, H., DYER, H. M., WAGNER, B. P., NIYAJI, H., RECHCIGL, M., JR., 1964: Some aspects of the development, biology and biochemistry of rat hepatomas of different growth rate. Adv. Enzyme Regul. 2, 321—333.

MOSES, H. L., STEIN, J. A., TSCHUDY, O. P., 1970: Hepatocellular changes associated with allylisopropylacetamide-induced hepatic porphyria in rats. Lab. Invest. 22, 432—442.

MÜLLER, M., 1973 a: Peroxisomes and hydrogenosomes in protozoa. J. Histochem. Cytochem. 21, 955—957.

— 1973 b: Biochemical cytology of trichomonad flagellates. I. Subcellular localization of hydrolases, dehydrogenases, and catalase in Tritrichomonas foetus. J. Cell Biol. 57, 453—474.

— 1975: Biochemistry of protozoan microbodies: Peroxisomes, α-glycerophosphate oxidase bodies, hydrogenosomes. Ann. Rev. Microbiol. 29, 467—483.

— MØLLER, K. M., 1969: Urate oxidase and its association with peroxisomes in Acanthamoeba sp. Eur. J. Biochem. 9, 424—430.

— HOGG, J. F., DE DUVE, C., 1968: Distribution of tricarboxylic acid cycle enzymes between mitochondria and peroxisomes in Tetrahymena pyriformis. J. Biol. Chem. 243, 5385—5395.

MURPHY, P. M., KRAHLING, J. B., GEE, R., KIRK, J. R., TOLBERT, N. E., 1979: Enzyme activities of isolated hepatic peroxisomes from genetically lean and obese mice. Arch. Biochem. Biophys. 193, 179—185.

MUSE, K. E., ROBERTS, J. F., 1973: Microbodies in Crithidia fasciculata. Protoplasma 78, 343—348.

NAKADA, H. I., WEINHOUSE, S., 1953: Studies on glycine oxidation in rat tissues. Arch. Biochem. Biophys. 42, 257—270.

NAKAHARA, W., FUKUOKA, F. A., 1948: A toxic cancer tissue constituent as evidenced by its effect on liver catalase activity. Jpn. Med. J. 1, 271—277.

— — 1961: Chemistry of cancer toxin: Toxohormone. Springfield, Ill.: Charles C Thomas.

NAKAJIMA, Y., BOURNE, G. H., 1976: Histochemical studies on urate oxidase in several mammals with special reference to uricolytic ability of primates. Histochemie 22, 20—24.

NAKAMURA, H., YOSIHIYA, M., KAZIRO, K., KIKUCH, G., 1952: On "anenzymia catalasea", a new type of constitutional abnormality. Proc. Jpn. Acad. 28, 59—64.

NAKANO, M., DANOWSKI, O., 1966: Crystalline mammalian L-amino acid oxidase from rat kidney mitochondria. J. Biol. Chem. 241, 2075—2083.

— TARUTANI, O., DANOWSKI, T. S., 1968 a: Molecular weight of mammalian L-aminoacid oxidase from rat kidney. Biochim. Biophys. Acta 168, 156—157.

— USHIJIMA, Y., SAGA, M., TSUTSUMI, Y., ASAMI, H., 1968 b: Aliphatic L-α-hydroxyacid oxidase from rat livers. Purification and properties. Biochim. Biophys. Acta 167, 9—22.

NICHOLLS, D. G., 1977: Cellular mechanisms in brown fat thermogenesis: Mitochondria. Experientia 33, 1130—1131.

NIKORONOW, M., MAZUR, H., PIEKECZ, H., 1973: Effect of orally administered plasticizers and polyvinylchloride stabilizers in the rat. Toxicol. Appl. Pharmacol. 26, 253—259.

NISHIMURA, E. T., ROSENHEIM, S., KLEIN, L., 1963: Depression of hepatic catalase in mice after subcutaneous injury. Lab. Invest. 12, 415—421.

— KOBARA, T. Y., DRUTY, J. J., 1964: Localization of catalase in tissues by fluorescent antibody technique. Lab. Invest. 13, 69—76.

— WHEST, G. M., YANG, H. Y., 1976: Ultrastructural localization of peroxidatic catalase in human peripheral blood leukocytes. Lab. Invest. 34, 60—68.

NOGUCHI, T., TAKADA, Y., 1978 a: Peroxisomal localization of serine : pyruvate amino-transferase in human liver. J. Biol. Chem. 253, 7598—7600.

— — 1978 b: Purification and properties of peroxisomal pyruvate (glyoxylate) amino-transferase from rat liver. Biochem. J. 175, 765—786.

— — 1979: Peroxisomal localization of alanine : glyoxylate aminotransferase in human liver. Arch. Biochem. Biophys. 196, 645—647.

— — FUJIWARA, S., 1979 a: Degradation of uric acid to urea and glyoxylate in animal peroxisomes. J. Biol. Chem. 254, 5272—5275.

— — OOTA, Y., 1979 b: Intraperoxisomal and intramitochondrial localization, and assay of pyruvate (glyoxylate) aminotransferase from rat liver. Hoppe-Seylers Z. Physiol. Chem. 360, 919—927.

NOHL, H., HEGNER, D., 1978: Evidence for the existence of catalase in the matrix space of rat heart mitochondria. FEBS Lett. 89, 126—130.

NORMAN, R. O. C., WEST, P. R., 1969: Electron spin resonance studies. XIV. Oxidation of organic radicals and the occurrence of chain processes during the reactions of some organic compounds with the hydroxyl radical derived from hydrogen peroxide and metal ions. J. Chem. Soc. (B), 389—399.

NORUM, K. E., BREMER, J., 1967: The localization of acyl coenzyme A-carnitine acyltrans-ferases in rat liver cells. J. Biol. Chem. 242, 407—411.

NOVIKOFF, A. B., GOLDFISCHER, S., 1968: Visualization of microbodies for light and electron microscopy. J. Histochem. Cytochem. 16, 507.

— — 1969: Visualization of peroxisomes (microbodies) and mitochondria with diamino-benzidine. J. Histochem. Cytochem. 17, 675—680.

— NOVIKOFF, P. M., 1973: Microperoxisomes. J. Histochem. Cytochem. 21, 963—966.

— SHIN, W. Y., 1964: The endoplasmic reticulum in the Golgi zone and its relations to micro-bodies, Golgi apparatus and autophagic vacuoles in rat liver cells. J. Microsc. 3, 187—206.

— NOVIKOFF, P. M., DAVIS, C., QUINTANA, N., 1972 a: Studies on microperoxisomes. II. A cytochemical method for light and electron microscopy. J. Histochem. Cytochem. 20, 1006—1023.

— — QUINTANA, N., DAVIS, C., 1972 b: Diffusion artifacts in 3,3-diaminobenzidine cyto-chemistry. J. Histochem. Cytochem. 20, 745—749.

— — DAVIS, C., QUINTANA, N., 1973 a: Studies on microperoxisomes. V. Are microper-oxisomes ubiquitous in mammalian cells? J. Histochem. Cytochem. 21, 737—755.

— — QUINTANA, N., DAVIS, C., 1973 b: Studies on microperoxisomes. III. Observations on human and rat hepatocytes. J. Histochem. Cytochem. 21, 540—558.

— — — — 1973 c: Studies on microperoxisomes. IV. Interrelations of microperoxisomes, endoplasmic reticulum, and lipofuscin granules. J. Histochem. Cytochem. 21, 1010—1020.

NOVIKOFF, P. M., NOVIKOFF, A. B., 1972: Peroxisomes in absorptive cells of mammalian small intestine. J. Cell Biol. **53**, 532—560.

— — 1973: The importance of specimen tilting for studying organelle interrelationship in cells. Philips Electronic EM, Bull. January 1973, N. V. Philips Gloeilampenfabrieken, Eindhoven, The Netherlands.

— ROHEIM, P., NOVIKOFF, A. B., 1973: Lipid accumulation induced in hepatocytes by clofibrate in rats fed high sucrose diets and its inhibition by vrotic acid. Fed. Proc. **32**, 837.

OERTEL, D., LINBERG, K. A., CASE, J. F., 1976: Ultrastructure of the larval firefly light organ as related to control of light emission. Cell Tissue Res. **164**, 27—44.

OGATA, M., MIZUGAKI, J., TAKAHARA, S., 1974: Catalase activity in the organs of Japanese acatalasemias. Tohoku J. Exp. Med. **113**, 239—243.

OGURA, Y., 1955: Catalase activity at high concentration of hydrogen peroxide. Arch. Biochem. Biophys. **57**, 288—300.

OLIVER, M. F., 1963: Further observations on the effects of atromid and of ethyl chloro-phenoxyisobutyrate on serum lipid levels. J. Atheroscler. Res. **3**, 427—439.

ONO, T., 1966: Enzyme patterns and malignancy of experimental hepatomas. Gann Monogr. **1**, 189—205.

OPPERDOES, F. R., BORST, P., 1977: Localization of nine glycolytic enzymes in a microbody-like organelle in *Trypanosoma brucei*. The glycosome. FEBS Lett. **80**, 360—364.

— — DE RIJKE, D., 1976: Oligomycin sensitivity of the mitochondrial ATPase as a marker for fly transmissability and the presence of functional kinetoplast DNA in African trypanosomes. Comp. Biochem. Physiol. **55 B**, 25—30.

— — BAKKER, S., LEENE, W., 1977 a: Localization of glycerol-3-phosphate oxidase in the mitochondrion and particulate NAD$^+$-linked glycerol-3-phosphate dehydrogenase in the microbodies of the bloodstream form of *Trypanosoma brucei*. Eur. J. Biochem. **76**, 29—39.

— — SPITS, H., 1977 b: Particle-bound enzyme in the bloodstream form of *Trypanosoma brucei*. Eur. J. Biochem. **76**, 21—28.

ORO, J., RAPPOPORT, D. A., 1959: Formate metabolism by animal tissues. II. The mechanism of formate oxidation. J. Biol. Chem. **234**, 1661—1665.

OSHINO, N., CHANCE, B., 1977: Properties of glutathione release observed during reduction of organic hydroperoxide, demethylation of aminopyrine and oxidation of some substances in perfused rat liver, and their implications for the physiological function of catalase. Biochem. J. **162**, 509—525.

— OMURA, T., 1973: Immunochemical evidence for the participation of cytochrome b_5 in microsomal stearyl-CoA desaturase reaction. Arch. Biochem. Biophys. **157**, 395—404.

— SATO, R., 1971: Stimulation by phenols of the reoxidation of microsomal bound cyto-chrome b_5 and its implication to fatty acid desaturation. J. Biochem. **69**, 169—180.

— IMAI, Y., SATO, R., 1971: A function of cytochrome b_5 in fatty acid desaturation by rat liver microsomes. J. Biochem. **69**, 155—168.

— CHANCE, B., SIES, H., 1973 a: The properties of the secondary catalase-peroxidase complex (compound II) in the hemoglobin-free perfused rat liver. Arch. Biochem. Biophys. **159**, 704—711.

— — — BÜCHER, TH., 1973 b: The role of H_2O_2 generation in perfused rat liver and the reaction of catalase compound I and hydrogen donors. Arch. Biochem. Biophys. **154**, 117—131.

— JAMIESON, D., CHANCE, B., 1975 a: The properties of hydrogen peroxide production under hyperoxic and hypoxic conditions of perfused rat liver. Biochem. J. **146**, 53—65.

— — SUGANO, T., CHANCE, B., 1975 b: Optical measurement of the catalase-hydrogen peroxide intermediate (compound I) in the liver of anaesthetized rats and its implication to hydrogen peroxide production in situ. Biochem. J. **146**, 67—77.

OSMUNDSEN, H., NEAT, C. E., 1979: Regulation of peroxisomal fatty acid oxidation. FEBS Lett. **107**, 81—85.

— NEAT, C. E., NORUM, K. R., 1979: Peroxisomal oxidation of long chain fatty acids. FEBS Lett. **99**, 292—296.

OSUMI, M., 1976: Possible existence of DNA in yeast microbody. J. Electron Microsc. 25, 43—47.

— HASHIMOTO, T., 1978 a: Enhancement of fatty acyl-CoA oxidizing activity in rat liver peroxisomes by di(2-ethylhexyl)phthalate. J. Biochem. 83, 1361—1367.

— — 1978 b: Acyl-CoA oxidase of rat liver: A new enzyme for fatty acid oxidation. Biochem. Biophys. Res. Commun. 83, 479—485.

— — 1979 a: Subcellular distribution of the enzymes of the fatty acyl-CoA β-oxidation system and their induction by di(2-ethylhexyl)phthalate in rat liver. J. Biochem. 85, 131—139.

— — 1979 b: Occurrence of two 3-hydroxyacyl-CoA dehydrogenases in rat liver. Biochim. Biophys. Acta 574, 258—267.

— — 1979 c: Peroxisomal β-oxidation system of rat liver. Copurification of enoyl-CoA hydratase and 3-hydroxyacyl-CoA dehydrogenase. Biochem. Biophys. Res. Commun. 89, 580—584.

— MIWA, N., TERANISHI, Y., TANAKA, A., FUKUI, S., 1974: Ultrastructure of Candida yeasts grown on n-alkanes. Appearance of microbodies and its relationship to high catalase activity. Arch. Microbiol. 99, 181—201.

— FUKUZUMI, F., TERANISHI, Y., TANAKA, A., FUKUI, S., 1975 a: Development of microbodies in Candida tropicalis during incubation in a n-alkane medium. Arch. Mikrobiol. 103, 1—11.

— IMAIZUMI, F., IMAI, M., SATO, H., YAMAGUCHI, H., 1975 b: Isolation and characterization of microbodies from Candida tropicalis PK 233 cells grown on normal alkanes. J. Gen. Appl. Microbiol. 21, 375—387.

— KAZAMA, H., SATO, S., 1978: Microbody-associated DNA in Candida tropicalis pK 223 cells. FEBS Lett. 90, 309—312.

OWEN, G., 1972 a: Lysosomes, peroxisomes and bivalves. Sci. Prog. 60, 299—318.

— 1972 b: Peroxisomes in the digestive diverticula of the bivalve mollusc Nucula sulcata. Z. Zellforsch. 132, 15—24.

PAGET, G., 1963: Experimental studies of the toxicity of bromid with particular reference to fine structural changes in the livers of rodents. J. Atheroscler. Res. 3, 729—736.

PAIGEN, K., 1954: The occurrence of several biochemically distinct types of mitochondria in rat liver. J. Biol. Chem. 206, 945—957.

PAL, S. G., 1971: The fine structure of the digestive tubules of Mya arenaria L. Proc. Malacol. Soc. (Lond.) 39, 303—310.

— 1972: The fine structure of the digestive tubules of Mya arenaria. II. Digestive cells. Proc. Malacol. Soc. (Lond.) 40, 161—170.

PALTAUF, F., MAGNET, K., 1979: Induktion der Fettsäureoxidation in Peroxysomen der Rattenleber durch Magnesium-4-chlorphenoxyisobutyrat. Drug Res. 29, 1132—1134.

PARISH, R. W., 1975 a: The isolation and characterization of peroxisomes (microbodies) from baker's yeast, Saccharomyces cerevisiae. Arch. Microbiol. 105, 187—192.

— 1975 b: Mitochondria and peroxisomes from the cellular slime mould Dictyostelium discoideum. Isolation techniques and urate oxidase association with peroxisomes. Eur. J. Biochem. 58, 523—531.

PARKER, R., KARIYA, T., GRISAR, J., PETROW, V., 1975: 5(Tetradecyloxy)-2-furan carboxylic acid (RMI 14,514) and related hypolipidemic fatty acid-like alkyloxyaryl carboxylic acids. Abstr. Papers Am. Chem. Soc. 170, 25.

— — — — 1977: 5(Tetradecyloxy)-2-furan-carboxylic acid (RMI 14,514) and related hypolipidemic fatty acid-like alkyloxyaryl carboxylic acids. J. Med. Chem. 20, 781—791.

PASSARGE, E., McADAMS, A. J., 1967: Cerebro-hepato-renal syndrome. J. Pediatr. 71, 691—702.

PAUL, P. B., STRAUSS, R. R., JACOBS, A. A., SBARRA, A. J., 1970: Function of H_2O_2, myeloperoxidase and hexose monophosphate shunt enzymes in phagocytizing cells from different species. Infect. Immunol. 1, 338—344.

PAVELKA, M., GOLDENBERG, H., HÜTTINGER, M., KRAMAR, R., 1976: Enzymic and morphological studies on catalase positive particles from brown fat of cold adapted rats. Histochemistry 50, 47—55.

PEARSON, D. J., TUBBS, P. K., 1967: Carnitine and derivatives in rat tissues. Biochem. J. 105, 953—963.

PEETERS-JORIS, C., VANDEVOORDE, A. M., BAUDHUIN, P., 1975: Subcellular localization of superoxide dismutase in rat liver. Biochem. J. 150, 31—39.

PEREIRA, J. N., MEARS, G. A., HOLLAND, G. F., 1975: Studies of the mechanisms of action of tibric acid, a new hypolipidemic agent. Fed. Proc. 34, 789.

PETRIK, P., 1971: Fine structural identification of peroxisomes in mouse and rat bronchiolar and alveolar epithelium. J. Histochem. Cytochem. 19, 339—348.

PFEIFER, U., SCHELLER, H., 1975: A morphometric study of cellular autophagy, including diurnal variations, in kidney tubules of normal rats. J. Cell Biol. 64, 608—621.

— WERDER, E., BERGEEST, H., 1978: Inhibition by insulin of the formation of autophagic vacuoles in rat liver. A morphometric approach to the kinetics of intracellular degradation by autophagy. J. Cell Biol. 78, 152—167.

PHILLIPS, D. R., DULEY, J. A., FENNELL, D. J., HOLMES, R. S., 1976: The self-association of L-α-hydroxyacid oxidase. Biochim. Biophys. Acta 427, 679—687.

PICHERAL, B., 1972: Les tissues élaborateurs d'hormones stéroides chez les urodeles. V. Mise en évidence et localization cytochimique d'une activité de type peroxydasique endogene dans la cellule du tissu glandulaire du testicule de Pleurodeles Waltlii. Comparison avec d'autres types cellulaires. J. Microsc. 13, 247—262.

PIPAN, N., PŠENIČNIK, M., 1975: The development of microperoxisomes in the cells of the proximal tubules of the kidney and epithelium of the small intestine during the embryonic development and postnatal period. Histochemistry 44, 13—21.

PITTS, O. M., PRIEST, D. G., FISH, W. W., 1974: Uricase. Subunit composition and resistance to denaturants. Biochemistry 13, 888—892.

PLATT, D. S., THORP, J. M., 1966: Changes in the weight and composition of the liver in the rat, dog and monkey treated with ethyl-chlorophenoxyisobutyrate. Biochem. Pharmacol. 15, 915—925.

POOLE, B., 1975: Diffusion effects in the metabolism of hydrogen peroxide by rat liver peroxisomes. J. Theor. Biol. 51, 149—167.

— LEIGHTON, F., DE DUVE, C., 1969: The synthesis and turnover of rat liver peroxisomes. II. Turnover of peroxisome proteins. J. Cell Biol. 41, 536—546.

— HIGASHI, T., DE DUVE, C., 1970: The synthesis and turnover of rat liver peroxisomes. III. The size distribution of peroxisomes and the incorporation of new catalase. J. Cell Biol. 45, 408—415.

POSALAKI, Z., BARKA, T., 1968: Alterations of hepatic endoplasmic reticulum in porphyric rats. J. Histochem. Cytochem. 16, 337—345.

POTTER, V. R., ELVEHJEM, C. A., 1936: A modified method for the study of tissue oxidations. J. Biol. Chem. 114, 495—504.

PRAETORIUS, E., 1948: The enzymatic conversion of uric acid. Spectrophotometric analysis. Biochim. Biophys. Acta 2, 602—613.

PRICE, V. E., RECHCIGL, M., JR., HARTLEY, R. W., JR., 1961: Methods for determining the rate of catalase synthesis and destruction in vivo. Nature 189, 62—63.

— STERLING, W. R., TARANTOLA, V. A., HARTLEY, R. W., RECHCIGL, M., 1962: The kinetics of catalase synthesis and destruction in vivo. J. Biol. Chem. 237, 3468—3475.

PŠENIČNIK, M., PIPAN, N., 1977: Nafenopin-induced proliferation of peroxisomes in the small intestine of mice. Virchows Arch. (Zellpathol.) 25 B, 161—169.

RADOMINSKA-PYREK, A., DABROWIECKI, Z., HARROCKS, L. A., 1979: Synthesis and content of ether-linked glycophospholipids in the Harderian gland of rabbits. Biochim. Biophys. Acta 574, 240—257.

RECHCIGL, M., JR., PRICE, V. E., 1968: Studies on the turnover of catalase in vivo. Prog. Exp. Tumor Res. 10, 112—132.

— — MORRIS, H. P., 1962: Studies on the cachexia of tumor-bearing animals. II. Catalase activity in the tissues of hepatoma-bearing animals. Cancer Res. 22, 874—880.

— HRUBAN, Z., MORRIS, H. P., 1969: The roles of synthesis and degradation in the regulation of catalase levels in the neoplastic tissues. Enzymol. Biol. Clin. 10, 161—180.

REDDY, J. K., 1973: Possible properties of microbodies (peroxisomes), microbody proliferation and hypolipidemic drugs. J. Histochem. Cytochem. 21, 967—971.
— 1974: Hepatic microbody proliferation and catalase synthesis induced by methyl clofenapate, a hypolipidemic analog of CPIB. Am. J. Pathol. 75, 103—118.
— KRISHNAKANTHA, T. P., 1975: Hepatic peroxisome proliferation. Induction by two novel compounds structurally unrelated to clofibrate. Science 190, 787—789.
— KUMAR, N. S., 1977: The peroxisome proliferation-associated polypeptide in rat liver. Biochem. Biophys. Res. Commun. 77, 824—829.
— — 1979: Stimulation of catalase synthesis and increase of carnitine acetyltransferase activity in the liver of intact female rats fed clofibrate. J. Biochem. 85, 847—856.
— RAO, M. S., 1977: Malignant tumors in rats fed nafenopin, a hepatic peroxisome proliferator. J. nat. Cancer Inst. 59, 1645—1651.
— SVOBODA, D., 1971 a: Microbodies in experimentally altered cells. VIII. Continuities between microbodies and their possible biologic significance. Lab. Invest. 24, 74—81.
— — 1971 b: Ethyl-α-p-chlorophenoxyisobutyrate induced hepatic microbody proliferation in rat liver and ubiquinone concentration. Experientia 27, 1059.
— — 1971 c: Proliferation of microbodies and synthesis of catalase in rat liver. Induction in tumorbearing host by CPIB. Am. J. Pathol. 63, 99—101.
— — 1972 a: Microbodies (peroxisomes) identification in interstitial cells of the testis. J. Histochem. Cytochem. 20, 140—142.
— — 1972 b: Microbodies (peroxisomes) in the interstitial cells of rodent testes. Lab. Invest. 26, 657—665.
— — 1972 c: Microbodies in Leydig cell tumors of rat testis. J. Histochem. Cytochem. 20, 793—803.
— — 1972 d: Microbodies (peroxisomes) in Leydig cell tumors of rat testis. Fed. Proc. 31, 637.
— — 1973 a: Microbody (peroxisome) matrix: transformation into tubular structures. Virchows Arch. (Zellpathol.) 14 B, 83—92.
— — 1973 b: Further evidence to suggest that microbodies do not exist as individual entities. Am. J. Pathol. 70, 421—438.
— BUNYARATVEJ, S., SVOBODA, D., 1969 a: Microbodies in experimentally altered cells. IV. Acatalasemic (SJ[b]) mice treated with CPIB. J. Cell Biol. 42, 587—596.
— — — 1969 b: Microbodies in experimentally altered cells. V. Histochemical and cytochemical studies on the livers of rats and acatalasemic mice treated with CPIB. Am. J. Pathol. 56, 351—370.
— CHIGA, M., BUNYARATVEJ, S., SVOBODA, D., 1970: Microbodies in experimentally altered cells. VII. CPIB-induced hepatic microbody proliferation in the absence of significant catalase synthesis. J. Cell Biol. 44, 226—234.
— — SVOBODA, D., 1971: Stimulation of liver catalase synthesis in rats by ethyl α-p-chlorophenoxy isobutyrate. Biochem. Biophys. Res. Commun. 43, 318—324.
— SVOBODA, D., AZARNOFF, D., DAWAR, R., 1973 a: Cadmium-induced Leydig cell tumors of rat testis: Morphologic and cytochemical study. J. nat. Cancer Inst. 51, 891—903.
— — — 1973 b: Microbody proliferation in liver induced by nafenopin, a new hypolipidemic drug: Comparison with CPIB. Biochem. Biophys. Res. Commun. 52, 537—543.
— AZARNOFF, D. L., SVOBODA, D. J., PRASAD, J. D., 1974 a: Nafenopin-induced hepatic microbody (peroxisome) proliferation and catalase synthesis in rats and mice. Absence of sex difference in response. J. Cell Biol. 61, 344—358.
— TEWARI, J. P., SVOBODA, D. J., MALHOTRA, S. K., 1974 b: Identification of the hepatic microbody membrane in freeze-fracture replicas. Lab. Invest. 31, 268—275.
— KRISHNAKANTHA, T. P., AZARNOFF, D. L., MOODY, D. E., 1975: 1-Methyl-4-piperidyl-bis (p-chlorophenoxy) acetate: A new hypolipidemic peroxisome proliferator. Res. Commun. Chem. Pathol. Pharmacol. 10, 589—592.
— MOODY, D. E., AZARNOFF, D. L., RAO, M. S., 1976 a: Di(2-ethylhexyl)phthalate: An industrial plasticizer indues hypolipidemia and enhances hepatic catalase and carnitine acetyltransferase activities in rat and mice. Life Sci. 18, 941—945.
— RAO, M. S., MOODY, D. E., 1976 b: Hepatocellular carcinomas in acatalasemic mice treated with nafenopin, a hypolipidemic peroxisome proliferator. Cancer Res. 36, 1211—1217.

REDDY, J. K., RAO, M. S., MOODY, D. E., QURESHI, S. A., 1976 c: Peroxisome development in the regenerating pars recta (P₃ segment) of proximal tubules of the rat kidney. J. Histochem. Cytochem. 24, 1239—1248.

— MOODY, D. E., AZARNOFF, D. L., RAO, M. S., 1977 a: Hepatic catalase is not essential for the hypolipidemic action of peroxisome proliferators. Proc. Soc. Exp. Biol. Med. 154, 483—487.

— — — TOMARELLI, R. M., 1977 b: Hepatic effects of 4-chloro-6-(2,3-xylidino)-pyrimidinyl-thioacetic acid (Wy 14,643) analogs in the mouse. Arch. Int. Pharmacodyn. Ther. 225, 51—57.

— AZARNOFF, D. L., SIRTORI, C. R., 1978: Hepatic peroxisome proliferation: Induction by BR-931, a hypolipidemic analog of WY-14,643. Arch. Int. Pharmacodyn. Ther. 234, 4—15.

REED, P. W., 1969: Glutathione and the hexose monophosphate shunt in phagocytizing and hydrogen peroxide-treated rat leukocytes. J. Biol. Chem. 244, 2459—2464.

REHFELD, D. W., TOLBERT, N. E., 1972: Aminotransferases in peroxisomes from spinach leaves. J. Biol. Chem. 247, 4803—4811.

REHG, J. E., MONTALDI, R. J., SZYMKOWIAK, M. E., 1975: Morphological and histochemical observations on renal microbodies in cats. Vet. Pathol. 12, 186—195.

REID, E., O'NEAL, M., LEVIN, I., 1956: Hormones and liver cytoplasm. 2. Adenosine triphosphatase, glucose-6-phosphatase and xanthine oxidase as affected by hypophysectomy, growth-hormone treatment and adrenalectomy. Biochem. J. 64, 730—734.

REMACLE, J., 1978: Binding of cytochrome b₅ to membranes of isolated subcellular organelles from rat liver. J. Cell Biol. 79, 291—313.

RHODIN, J., 1954: Correlation of ultrastructural organization and function in normal and experimentally changed proximal convoluted tubule cells of the mouse kidney. Doctoral Thesis, Karolinska Institutet, Stockholm: Aktiebolaget Godvil.

— 1958: Electron microscopy of the kidney. Am. J. Med. 24, 661—675.

RICHARDSON, K. E., TOLBERT, N. E., 1961: Oxidation of glyoxylic acid to oxalic acid by glycolic acid oxidase. J. Biol. Chem. 236, 1280—1284.

RIEDE, U. N., ROHR, H. P., 1974: Das Reaktionsmuster der Leberperoxisomen (= Microbodies) im Verlaufe einer Zellschädigung. Beitr. Pathol. 151, 111—133.

RIGATUSO, J. L., LEGG, P. G., WOOD, R. L., 1970: Microbody formation in regenerating rat liver. J. Histochem. Cytochem. 18, 893—900.

ROBBI, M., LAZAROW, P. L., 1978: Synthesis of catalase in two cell-free protein-synthesizing systems and in rat liver. Proc. nat. Acad. Sci. (U.S.A.) 75, 4344—4348.

ROBINSON, J. C., KEAY, L., MOLINARI, R., SIZER, I. W., 1962: L-α-Hydroxyacid oxidases of dog renal cortex. J. Biol. Chem. 237, 2001—2010.

ROBISON, W. G., JR., KUWABARA, T., 1975: Microperoxisomes in retinal pigment epithelium. Invest. Ophthalmol. 14, 866—872.

ROCK, C. O., FITZGERALD, V., SNYDER, F., 1978: Coupling of the biosynthesis of fatty acids and fatty alcohols. Arch. Biochem. Biophys. 186, 77—83.

ROELS, F., 1976: Cytochemical demonstration of extraperoxisomal catalase. I. Sheep liver. J. Histochem. Cytochem. 24, 713—724.

— GOLDFISCHER, S., 1971: Staining of nucleic acids with 3,3'-diaminobenzidine (DAB). J. Histochem. Cytochem. 19, 713—714, Abstract 11.

— WISSE, E., 1973: Distinction cytochimique entre catalase et peroxydases. C. R. Acad. Sci. (Paris) D 276, 391—393.

— SCHILLER, B., GOLDFISCHER, S., 1970: Microbodies (peroxisomes) in the toad, Bufo marinus. A cytochemical study. Z. Zellforsch. 108, 135—149.

— — DE PREST, B., VAN DER MEULEN, J., 1973: Cytochemical discrimination between peroxidases and catalases using diaminobenzidine. In: Electron microscopy and cytochemistry (WISSE, E., DAEMS, W. TH., MOLENAAR, I., VAN DUIJN, P., eds.), pp. 115—118. Amsterdam: North-Holland.

— — — — 1975: Cytochemical discrimination between catalases and peroxidases using diaminobenzidine. Histochemistry 41, 281—312.

— DE COSTER, W., GOLDFISCHER, S., 1977: Cytochemical demonstration of extraperoxisomal catalase. II. Liver of rhesus monkey and guinea pig. J. Histochem. Cytochem. 25, 157—160.

ROGERS, P. J., STEWART, P. R., 1973: Mitochondrial and peroxisomal contribution to the energy metabolism of *Saccharomyces cerevisiae* in continuous culture. J. Gen. Microbiol. 79, 205—217.

ROGGENKAMP, R., SAHM, H., WAGNER, F., 1974: Microbiol assimilation of methanol. Induction and function of catalase in *Candida boidinii*. FEBS Lett. 41, 283—286.

— — HINKELMANN, W., WAGNER, F., 1975: Alcohol oxidase and catalase in peroxisomes of methanol-grown *Candida boidinii*. Eur. J. Biochem. 59, 231—236.

ROHEIM, P. S., BIEMPICA, L., EDELSTEIN, D., ROSOWER, N. S., 1971: Mechanism of fatty liver development and hyperlipemia in rats treated with allylisopropylacetamide. J. Lipid Res. 12, 76—83.

ROHR, H. P., STREBEL, J., BIANCHI, L., 1970: Ultrastrukturell-morphometrische Untersuchungen an der Rattenleberparenchymzelle in der Frühphase der Regeneration nach partieller Hepatektomie. Beitr. Pathol. 141, 52—74.

— WIRZ, A., HENNING, L. L., RIEDE, U. N., BIANCHI, L., 1971: Morphometric analysis of the rat liver cell in the perinatal period. Lab. Invest. 24, 128—140.

ROMER, F., 1974: Ultrastructural changes of the oenocytes of *Gryllus bimaculatus* DEG (Saltatoria, Insecta) during the moulting cycle. Cell Tissue Res. 151, 27—46.

RØRTH, M., JENSEN, P. K., 1967: Determination of catalase activity by means of the Clark oxygen electrode. Biochim. Biophys. Acta 139, 171—173.

ROSE, J. A., HELLMAN, E. S., TSCHUDY, D. P., 1961: Effect of diet on induction of experimental porphyria. Metabolism 10, 514—521.

ROSENTHAL, O., NARASIMHULU, S., 1969: Steroids and terpenoids. In: Methods in enzymology (COLOWICK, S. P., KAPLAN, N. O., eds.), Vol. 15, pp. 596—638. New York: Academic Press.

ROSS, W. T., JR., CARDELL, R. R., JR., 1972: Effects of halothane on the ultrastructure of rat liver cells. Am. J. Anat. 135, 5—21.

ROTHERMAN, J. E., LENARD, J., 1977: Membrane asymmetry. Science 195, 743—753.

ROUILLER, C., BERNHARD, W., 1956: "Microbodies" and the problem of mitochondrial regeneration in liver cells. J. Biophys. Biochem. Cytol., Suppl. 2, 355—359.

ROWE, P. B., WYNGAARDEN, J. B., 1966: The mechanism of dietary alterations in rat hepatic xanthine oxidase levels. J. Biol. Chem. 241, 5571—5576.

RUSSELL, L., BURGUET, S., 1977: Ultrastructure of Leydig cells as revealed by secondary tissue treatment with a ferrocyanide-osmium mixture. Tissue Cell 9, 751—766.

RYLEY, J. F., 1955: Studies on the metabolism of the protozoa. 5. Metabolism of the parasitic flagellate *Trichomonas foetus*. Biochem. J. 59, 361—369.

SACKLER, M. L., 1966: Xanthine oxidase from liver and duodenum of the rat. Histochemical localization and electrophoretic heterogeneity. J. Histochem. Cytochem. 14, 326—333.

SAFRAN, A. P., SCHAFFNER, F., 1967: Chronic passive congestion of the liver in man. Am. J. Pathol. 50, 447—463.

SAITO, T., IWATA, K., OGAWA, K., 1973: Changes of peroxisomes in the regenerating hepatic parenchymal cells in the rat. Acta Histochem. Cytochem. 6, 212—229.

SAKAMOTO, S. I., YAMADA, K., ANZAI, T., WADA, T., 1973: Morphological changes in the liver of rats treated with a new hypolipidemic agent, S-8527. Atherosclerosis 18, 109—116.

SAKAMOTO, T., HIGASHI, T., 1973: Studies on rat liver catalase. VI. Biosynthesis of catalase by free and membrane-bound polysomes. J. Biochem. 73, 1083—1088.

SALOMON, J. C., 1962: Modifications des cellules du parenchyme hepatique du retsous l'effect de la thioacetamide. J. Ultrastruct. Res. 7, 293—307.

SALVATORE, F., ZAPPIA, U., CORTESE, R., 1966: Studies on the determination of L-amino acids in mammalian tissues. Enzymologia 31, 113—127.

SANTILLI, A. A., SCOTESE, A. C., TOMASELLI, R. M., 1974: A potent antihypercholesterolemic agent: [4-chloro-6(2,3-xylidino)-2 pyrimidinylthio]acetic acid (Wy 14,643). Experientia 30, 1110.

SAVOLAINEN, M. J., JAUHONEN, V. P., HASSINEN, I. E., 1977: Effects of clofibrate on ethanol-induced modifications in liver and adipose tissue metabolism. Role of hepatic redox state and hormonal mechanisms. Biochem. Pharmacol. 26, 425—431.

SAWANT, P. L., DESAI, I. D., TAPPEL, A. L., 1964: Factors affecting the lysosomal membrane and availabilities of enzymes. Arch. Biochem. Biophys. 105, 247—253.

SCHAFFNER, F., 1966: Intralobular changes in hepatocytes and the electron microscopic mesenchymal response in acute viral hepatitis. Medicine 45, 547—552.

— JAVITT, N. B., 1966: Morphologic changes in hamster liver during intrahepatic cholestasis induced by taurolithocholate. Lab. Invest. 15, 1783—1792.

SCHMID, R., 1963: Hepatotoxic drugs coupling porphyria in man and animals. Ann. N.Y. Acad. Sci. 104, 1034—1048.

— FIGEN, J. F., SCHWARTZ, S., 1955: Experimental porphyria. IV. Studies of liver catalase and other tissue enzymes in sedormid porphyria. J. Biol. Chem. 217, 263—274.

SCHMUCKER, D., JONES, A., 1975: Hepatic fine structure in young and aging rats treated with oxandrolone: a morphometric study. J. Lipid Res. 16, 143—150.

SCHNEEBERGER, E. E., 1972 a: A comparative cytochemical study of microbodies (peroxisomes) in great alveolar cells of rodents, rabbits and monkey. J. Histochem. Cytochem. 20, 180—191.

— 1972 b: Development of peroxisomes in granular pneumocytes during pre- and postnatal growth. Lab. Invest. 27, 581—589.

SCHNEIDER, W. C., 1948: Intracellular distribution of enzymes. III. The oxidation of octanoic acid by rat liver fractions. J. Biol. Chem. 176, 259—266.

— HOGEBOOM, G. H., 1952: Intracellular distribution of enzymes. IX. Certain purine-metabolizing enzymes. J. Biol. Chem. 195, 161—166.

SCHUMAN, M., MASSEY, V., 1971: Purification and characterization of glycolic acid oxidase from pig liver. Biochim. Biophys. Acta 227, 500—520.

SCOTT, P. J., VISENTIN, L. P., ALLEN, J. M., 1969: The enzymatic characteristics of peroxisomes of amphibian and avian liver and kidney. Ann. N.Y. Acad. Sci. 168, 244—264.

SHAPIRA, E., BEN-YOSEPH, Y., AEBI, H., 1974: Nature of residual erythrocyte catalase activity in Swiss-type acatalasemia. Enzyme 17, 307—318.

SHINDO, Y., HASHIMOTO, T., 1978: Acyl-coenzyme A synthetase and fatty acid oxidation in rat liver peroxisomes. J. Biochem. 84, 1177—1181.

SHIO, H., FARQUHAR, M. G., DE DUVE, C., 1974: Lysosomes of the arterial wall. IV. Cytochemical localization of acid phosphatase and catalase in smooth muscle cells and foam cells from rabbit athermatous aorta. Am. J. Pathol. 76, 1—16.

SHNITKA, T. K., 1966: Comparative ultrastructure of hepatic microbodies in some mammals and birds in relation to species differences in uricase activity. J. Ultrastruct. Res. 16, 598—625.

— TALIBI, G. G., 1971: Cytochemical localization by ferricyanide reduction of α-hydroxy acid oxidase activity in peroxisomes of rat kidney. Histochemistry 27, 137—158.

SIES, H., 1974: Biochemie des Peroxysoms in der Leberzelle. Angew. Chem. 86, 789—801.

— HERZOG, V., MILLER, F., 1972: Electron microscopic and spectrophotometric studies on mitochondrial and peroxisomal reaction with diaminobenzidine in hemoglobin-free perfused rat liver. Proceedings of the Fifth European Congress on Electron Microscopy. University of Manchester, England, p. 274.

SINGH, A., WILLIAMSON, W., GEISSINGER, H. D., BHATNAGAR, R., 1977: Microbodies of porcine hepatocytes. Can. Vet. J. 18, 140—141.

— — — — 1978: Microbodies of the pig liver. A morphologic and morphometric study. Acta Anat. 102, 232—236.

SKREDE, S., HALVORSEN, O., 1979: Increased biosynthesis of CoA in the liver of rats treated with clofibrate. Eur. J. Biochem. 98, 223—229.

SLINDE, E., FLATMARK, T., 1973: Determination of sedimentation coefficients of subcellular particles of rat liver homogenates. A theoretical and experimental approach. Anal. Biochem. 56, 324—340.

SMITH, D. W., OPITZ, J. M., INHORN, S. L., 1964: A syndrome of multiple developmental defects including polycystic kidneys and intrahepatic biliary dysgenesis in two siblings. J. Pediatr. 67, 617—624.

SMITH, R. E., HORWITZ, B. A., 1969: Brown fat and thermogenesis. Physiol. Rev. 49, 330—425.

SNOSWELL, A. M., KOUNDAKJIAN, P. P., 1972: Relationships between carnitine and coenzyme A esters in tissues of normal and alloxan-diabetic sheep. Biochem. J. 127, 133—141.

SOLBERG, H. E., 1971: Carnitine octanoyltransferase. Evidence for a new enzyme in mitochondria. FEBS Lett. **12**, 134—136.

— AAS, M., DAAE, L. N. W., 1972: The activity of different carnitine acyltransferases in the liver of clofibrate-fed rats. Biochim. Biophys. Acta **280**, 434—439.

SPORNITZ, U. M., 1975: Studies on the liver of *Xenopus laevis*. I. The ultrastructure of the parenchymal cell. Anat. Embryol. **146**, 245—264.

SRERE, P. A., BHADURI, A., 1962: Incorporation of radioactive citrate into fatty acids. Biochim. Biophys. Acta **59**, 487—489.

STÄUBLI, W., HESS, R., 1966: Quantitative aspects of hepatomegaly induced by ethylchlorophenoxy-isobutyrate (CPIB). Proceedings of the Sixth International Congress on Electron Microscopy (VEDA, R., ed.), Vol. 2, pp. 625—626. Tokyo: Maruzen Co. Ltd.

— SCHWEIZER, D., SUTER, J., WEIBEL, E. R., 1977: The proliferative response of hepatic peroxisomes in neonatal rats to treatment with SU-13,437 (nafenopin). J. Cell Biol. **74**, 665—689.

STEIGER, R. F., 1973: On the ultrastructure of *Trypanosoma* (*Trypanozoon*) *brucei* in the course of its life cycle and some related aspects. Acta Trop. **30**, 64—168.

STEIN, A. M., SHAVINSKI, E. R., APPLEMAN, D., SHUGARMAN, P. M., 1951: Effect of partial hepatectomy on liver catalase activity in normal and protein-depleted rats. Am. J. Physiol. **167**, 581—585.

STEIN, O., STEIN, Y., 1968: Lipid synthesis, intracellular transport and storage. III. Electron microscopic radioautographic study of rat heart perfused with tritiated oleic acid. J. Cell Biol. **36**, 63—77.

STELLY, N., BALMEFRÉZOL, M., ADOUTTE, A., 1975: Diaminobenzidine reactivity of mitochondria and peroxisomes in *Tetrahymena* and in wild type and cytochrome oxidase-deficient *Paramecium*. J. Histochem. Cytochem. **23**, 686—696.

STENGER, R. J., CONFER, D. B., 1966: Hepatocellular ultrastructure during liver regeneration after subtotal hepatectomy. Exp. Mol. Pathol. **5**, 455—474.

STEPHENS, R. J., BILS, R. F., 1967: Ultrastructural changes in the developing chick liver. 1. General cytology. J. Ultrastruct. Res. **18**, 456—474.

STERNLIEB, I., QUINTANA, N., 1977: The peroxisomes of human hepatocytes. Labor. Invest. **36**, 140—149.

STEVENS, J. B., AUTOR, A. P., 1977: Oxygen induced synthesis of superoxide dismutase and catalase in pulmonary macrophages of neonatal rats. Lab. Invest. **37**, 470—478.

STIRPE, F., DELLA CORTE, E., 1969: The regulation of rat liver xanthine oxidase. Conversion *in vitro* of the enzyme from dehydrogenase (type D) to oxidase. J. Biol. Chem. **244**, 3855—3863.

STRAUS, W., 1971: Inhibition of peroxidase by methanol and by methanol-nitroferricyanide for use in immunoperoxidase procedures. J. Histochem. Cytochem. **19**, 682—688.

— 1972: Phenylhydrazine as inhibitor of horseradish peroxidase for use in immunoperoxidase procedures. J. Histochem. Cytochem. **20**, 949—951.

STREEFKERK, J. G., VAN DER PLOEG, M., 1974: The effect of ethanol on granulocyte and horseradish peroxidase quantitatively studied in a film model system. Histochemistry **40**, 105—111.

STRITTMATTER, P., 1963: Microsomal cytochrome b₅ and cytochrome b₅ reductase. In: The enzymes (BOYER, P. D., LARDY, H., MYRBÄCK, K., eds.), 2nd ed., Vol. 8, pp. 113—145. New York: Academic Press.

— SPATZ, L., CORCORAN, D., ROGERS, M. J., SETLOW, B., REDLINE, R., 1974: Purification and properties of rat liver microsomal stearyl coenzyme A desaturase. Proc. nat. Acad. Sci. (U.S.A.) **71**, 4565—4569.

SUND, H., WEBER, K., MOLEBERT, E., 1967: Dissoziation der Rinderleber-Katalase in ihre Untereinheiten. Eur. J. Biochem. **1**, 400—417.

SUSANI, M., ZIMNIAK, P., FESSL, F., RUIS, H., 1976: Localization of catalase A in vacuoles of *Saccharomyces cerevisiae*: Evidence for the vacuolar nature of isolated "yeast peroxisomes". Hoppe-Seylers Z. Physiol. Chem. **357**, 961—970.

SUZUKI, S., SUGA, T., NIINOBE, S., 1973: Studies on peroxisomes. IV. Intracellular location of NADH₂-glyoxylate reductase in rat liver. J. Biochem. **73**, 1033—1038.

SVEDBERG, T., PEDERSEN, K. O., 1940: The ultracentrifuge, pp. 67—212. Oxford: Clarendon Press.

Svoboda, D., 1976: The response of microperoxisomes in rat small intestinal mucosa to CPIB, a hypolipidemic drug. Biochem. Pharmacol. 25, 2750—2752.
— 1978: Unusual responses of rat hepatic and renal peroxisomes to RMI 14,514, a new hypolipidemic agent. J. Cell Biol. 78, 810—822.
— Azarnoff, D. L., 1966: Response of hepatic microbodies to a hypolipidemic agent, ethyl chlorophenoxyisobutyrate (CPIB). J. Cell Biol. 30, 442—450.
— — 1971: Effects of selected hypolipidemic drugs on cell ultrastructure. Fed. Proc. 30, 841—847.
— Reddy, J., 1972: Microbodies in experimentally altered cells. IX. The fate of microbodies. Am. J. Pathol. 67, 541—554.
— Grady, H., Azarnoff, D., 1967: Microbodies in experimentally altered cells. J. Cell Biol. 35, 127—152.
— Azarnoff, D. L., Reddy, J., 1969: Microbodies in experimentally altered cells. II. The relationship of microbody proliferation to endocrine glands. J. Cell Biol. 40, 734—746.
Szabo, A. S., Avers, J., 1969: Some aspects of regulation of peroxisomes and mitochondria in yeast. Ann. N.Y. Acad. Sci. 168, 302—312.
Szeinberg, A., de Vries, A., Pinkhas, J., Jaldetti, M., Ezra, R., 1963: A dual hereditary red blood cell defect in one family, hypocatalasemia and glucose-6-phosphate dehydrogenase deficiency. Acta Genet. Med. Gemellol. 12, 247—250.
Szmigielski, S., 1972 a: Peroxisomal enzymes in human granulocytes. I. Catalase and peroxidase. Folia Histochem. Cytochem. (Krakow) 10, 47—50.
— 1972 b: Peroxisomal enzymes in human granulocytes. II. D-Aminoacid oxidases and L-α-hydroxyacid oxidases. Folia Histochem. Cytochem. (Krakow) 10, 313—318.

Takahara, S., 1952: Progressive oral gangrene probably due to lack of catalase in the blood (acatalasemia). Lancet 2, 1101—1104.
Tanaka, K., Iida, M., 1977: Microbodies of the hydrocarbon-assimilating yeast Candida rugosa. J. Gen. Appl. Microbiol. 23, 201—206.
Tandler, B., Denning, C. R., Mandel, I. D., Kutscher, A. H., 1969: Ultrastructure of human labial salivary glands. I. Acinar secretory cells. J. Morphol. 127, 383—408.
Tanford, C., Lovrien, R., 1962: Dissociation of catalase into subunits. J. Am. Chem. Soc. 84, 1892—1896.
Tangen, O., Jonsson, J., Orrenius, S., 1973: Isolation of rat liver microsomes by gel filtration. Anal. Biochem. 54, 597—603.
Taylor, R. T., Jenkins, W. T., 1966: Leucine aminotransferase: I. Colorimetric assays. J. Biol. Chem. 241, 4391—4395.
Temple, N. J., Martin, P. A., Connock, M. J., 1979: Intestinal peroxisomes of goldfish (Carassius auratus)-examination for hydrolase, dehydrogenase and carnitine acetyltransferase activities. Comp. Biochem. Physiol. 64 B, 57—63.
Teranishi, Y., Kawamoto, S., Tanaka, A., Osumi, M., Fukui, S., 1974: Induction of catalase activity by hydrocarbons in Candida tropicalis pK 233. Agric. Biol. Chem. 38, 1221—1225.
Teschke, R., Hasumura, Y., Lieber, C. S., 1974: Hepatic microsomal ethanol oxidizing system: Solubilization, isolation and characterization. Arch. Biochem. Biophys. 163, 404—415.
— — — 1975 a: Hepatic microsomal alcohol metabolizing system–affinity for methanol, ethanol, propanol and butanol. J. Biol. Chem. 250, 7397—7404.
— — — 1975 b: Hepatic microsomal alcohol-oxidizing system in normal and acatalasemic mice: Its dissociation from the peroxidatic activity of catalase-H_2O_2. Mol. Pharmacol. 11, 841—849.
— — — 1976: Hepatic ethanol metabolism: Respective roles of alcohol dehydrogenase, the ethanol oxidizing system, and catalase. Arch. Biochem. Biophys. 175, 635—643.
Theorell, H., 1948: Über die Wirkungsweise der Katalasen. Experientia 4, 100—109.
— 1951: The iron containing enzymes. B. Catalases and peroxidases. "Hydroperoxidases". In: The enzymes (Sumner, J. B., Myrbäck, K., eds.), Vol. 2, pp. 397—427. New York: Academic Press.

THINÈS-SEMPOUX, D., BORMANN, A., 1978: Characterization of the fatty acid oxidation by rat liver peroxisomes. Arch. Int. Physiol. Biochim. 86, 456—457.

THOMSON, J. F., KLIPFEL, F. J., 1957: Further studies on cytoplasmic particulates isolated by gradient centrifugation. Arch. Biochem. Biophys. 70, 224—238.

— NANCE, S. L., TOLLAKSEN, S. L., 1974: Preparation of peroxisomes from mouse liver by rate zonal centrifugation. Proc. Soc. Exp. Biol. Med. 145, 1174—1177.

— — — 1978: Spectrophotometric assay of catalase with perborate as substrate. Proc. Soc. Exp. Biol. Med. 157, 33—35.

THORP, J. M., 1970: Hypocholesterolemic and other effects of methyl clofenapate, a novel derivative of clofibrate. In: Atherosclerosis (JONES, R. J., ed.), pp. 541—544. Berlin-Heidelberg-New York: Springer.

— WARING, W. S., 1962: Modification of metabolism and distribution of lipids by ethyl chlorophenoxyisobutyrate. Nature 194, 948—949.

THURMAN, R. G., CHANCE, B., 1969: Inhibition of catalase in perfused rat liver by sodium azide. Ann. N.Y. Acad. Sci. 168, 348—353.

— LEY, H. G., SCHOLZ, R., 1972: Hepatic microsomal ethanol oxidation. Hydrogen peroxide formation and the role of catalase. Eur. J. Biochem. 25, 420—430.

TIEDEMANN, K., 1972: Peroxisomes in cat and sheep mesonephric tubules. Z. Zellforsch. 133, 141—146.

TIMMS, A. R., KELLY, L. A., HOK, S., TRAPOLD, J. H., 1969: Laboratory studies of 1-methyl-1,4-piperidyl-bis(p-chlorophenoxy)acetate (SaH-42,348)—a new hypolipidemic agent. Biochem. Pharmacol. 18, 1861—1871.

TISHER, C. C., BULGER, R. E., TRUMP, B. F., 1966: Human renal ultrastructure. I. Proximal tubule of healthy individuals. Lab. Invest. 15, 1357—1394.

— FINKEL, R. M., ROSEN, S., KENDIG, E. M., 1968: Renal microbodies in the rhesus monkey. Lab. Invest. 19, 1—6.

— ROSEN, S., OSBORNE, G. B., 1969: Ultrastructure of the proximal tubules of the rhesus monkey kidney. Am. J. Pathol. 56, 469—518.

TOKI, K., NAKAMURA, Y., AGATSURUME, K., NATHATANI, H., AONO, S., 1973: Hypolipidemic action of a new aryloxy compound (S-8527) in rats. Atherosclerosis 18, 101—108.

TOLBERT, N. E., 1973: Compartimentation and control in microbodies. Soc. Exp. Biol. 27, 215—239.

— YAMAZAKI, R. K., 1969: Leaf peroxisomes and their relation to photorespiration and photosynthesis. Ann. N.Y. Acad. Sci. 168, 425—441.

— OESER, A., KISAKI, T., HAGEMAN, R. H., YAMAZAKI, R. K., 1968: Peroxisomes from spinach leaves containing enzymes related to glycolate metabolism. J. Biol. Chem. 243, 5179—5184.

TONEGUZZO, F., GOSH, H. P., 1978: *In vitro* synthesis of vesicular stomatitis virus membrane glycoprotein and insertion into membranes. Proc. nat. Acad. Sci. (U.S.A.) 75, 715—719.

TONHAZY, N. E., WHITE, N. G., UMBREIT, W. W., 1950: A rapid method for the estimation of the glutamic aspartic transaminase in tissues and its application to radiation sickness. Arch. Biochem. Biophys. 28, 36—42.

TRELEASE, R. N., BECKER, W. M., BURKE, J. J., 1974: Cytochemical localization of malate synthase in glyoxysomes. J. Cell Biol. 60, 483—495.

TRUMP, B. F., ERICSSON, J. L. E., 1965: The effect of fixative solutions on the ultrastructure of cells and tissues. A comparative analysis with particular attention to the proximal tubule of the rat kidney. Lab. Invest. 14, 1245—1323.

TSUKADA, H., MOCHIZUKI, Y., FUJIWARA, S., 1966: The nucleoids of rat liver cell microbodies. Fine structure and enzymes. J. Cell Biol. 28, 449—460.

— — KONISHI, T., 1968 a: Morphogenesis and development of microbodies of hepatocytes of rats during pre- and postnatal growth. J. Cell Biol. 37, 231—243.

— — — 1968 b: Hepatocyte microbodies in growing rats, partially hepatectomized rats, and tumor-bearing mice. Trans. Soc. Pathol. Jpn. 57, 122.

— KOYAMA, S., GOTOH, M., TADANO, H., 1971: Fine structure of crystalloid nucleoids of compact type in hepatocyte microbodies of guinea pigs, cats, and rabbits. J. Ultrastruct. Res. 36, 159—175.

Tsukada, H., Mochizuki, Y., Gotoh, M., 1975 a: Matrical inclusions induced by clofibrate in in hepatic microbodies of rats fed 2-acetylaminofluorene. J. nat. Cancer Inst. 54, 519—523.
— — Habashi, M., Gotoh, M., Morris, H. P., 1975 b: Response of microbodies in Morris hepatoma 9618 A to clofibrate. J. nat. Cancer Inst. 55, 153—158.
— Gotoh, M., Mochizuki, Y., Furukawa, K., 1979: Changes in peroxisomes in pre-neoplastic liver and hepatoma of mice induced by α-benzene hexachloride. Cancer Res. 39, 1628—1634.
Tuchweber, B., Koursunakis, P., Latour, T., 1976: Drug metabolism and morphologic changes in the liver of nafenopin-treated rats. Arch. Int. Pharmacodyn. Ther. 222, 309—321.

Ushijima, Y., Nakano, M., 1969: Aliphatic L-α-hydroxyacid oxidase from rat liver. Biochim. Biophys. Acta 178, 429—433.
Uzman, L. L., 1956: Histochemical localization of copper with rubeanic acid. Lab. Invest. 5, 299—305.

Van Breeman, V. L., Montgomery, J. D., 1960: Ultrastructure of cytoplasmic "metasomes" found in renal proximal tubular cells. Norelco Rep. 7, 7—18.
Van der Rhee, H. J., de Winter, C. P. M., Daems, W. Th., 1977: Fine structure and peroxidatic activity of rat blood monocytes. Cell Tissue Res. 185, 1—16.
Van Dijken, J. P., Veenhuis, M., Kreger-van Rij, N. J. W., Harder, W., 1975: Micro-bodies in methanol-assimilating yeasts. Arch. Microbiol. 102, 41—44.
Vandor, S. L., Tolbert, N. E., 1970: Glyoxylate metabolism by isolated rat liver per-oxisomes. Biochim. Biophys. Acta 215, 449—455.
Van Gelder, B. F., Slater, E. C., 1962: The extinction coefficient of cytochrome c. Biochim. Biophys. Acta 58, 593—595.
Veenhuis, M., Wendelaar Bonga, S. D., 1977: The cytochemical demonstration of catalase and D-amino acid oxidase in the microbodies of teleost kidney cells. Histochem. J. 9, 171—181.
— van Dijken, J. P., Harder, W., 1976: Cytochemical studies on the localization of methanol oxidase and other oxidases in peroxisomes of methanol from Hansenula poly-morpha. Arch. Microbiol. 111, 123—135.
Vercauteren, R. E., 1962: Oxidoreductase of leukocytes. II. Evidence for particulate bound catalase and peroxidase in leukocyte homogenates. Enzymologia 24, 37—48.
Verheyen, A., Borgers, M., Blaton, H., Sowa, H., 1975: The ultrastructure of human livers after prolonged lidoflavine therapy. Toxicol. Appl. Pharmacol. 34, 224—232.
Versmold, H. T., Bremer, H. J., Herzog, V., Siegel, G., Bassewitz, D. B., Irle, V., Voss, H., Laubel, I., Brauser, B., 1977: A metabolic disorder similar to Zellweger syndrome with hepatic acatalasia and absence of peroxisomes. Eur. J. Pediatr. 124, 261—275.
Vickerman, K., 1965: Polymorphism and mitochondrial activity in sleeping sickness trypano-somes. Nature 208, 762—766.
— 1971: Morphological and physiological considerations of extracellular blood protozoa. In: Ecology and physiology of parasites (Fallis, A. M., ed.), pp. 58—91. Toronto: Uni-versity of Toronto Press.
Virágh, S., Bartók, I., 1966: An electron microscopic study of regeneration of the liver following partial hepatectomy. Am. J. Pathol. 49, 825—839.
Visentin, L. P., Allen, J. M., 1969: Allantoinase: Association with amphibian hepatic peroxisomes. Science 163, 1463—1464.
Vitale, L., Opitz, J. M., Shakidi, N. T., 1969: Congenital and familial iron overload. N. Engl. J. Med. 280, 642—645.

Warburg, O., Christian, W., 1938: Isolierung der prosthetischen Gruppe der d-Aminosäure-oxydase. Biochem. Z. 298, 150—168.
Warton, A., 1975: The ultrastructural localization of microbodies (peroxisomes) in Trypano-soma crucei using 3,3'-diaminobenzidine. West. J. Med. 123, 69 A—70 A.

WARTON, A., MODLINSKA, M., 1976: The ultrastructural localization of peroxisomes in the blood forms of *Trypanosoma cruzei* using 3,3'-diaminobenzidine. Bull. Acad. Pol. Sci. (Biol.) 24, 611—613.

WATTIAUX, R., WIBO, M., BAUDHUIN, P., 1963: Influence of the injection of Triton WR-1339 on the properties of rat-liver lysosomes. In: Lysosomes (Ciba Foundation Symposium), DE REUCK, A. V. S., CAMERON, M. P., eds.), pp. 176—200. London: Churchill.

WATTIAUX-DE CONINCK, S., RUTGEERTS, M. J., WATTIAUX, R., 1965: Lysosomes in rat-kidney tissue. Biochim. Biophys. Acta 105, 446—459.

WEDEL, F. P., BERGER, E. R., 1975: On the quantitative stereomorphology of microbodies in rat hepatocytes. J. Ultrastruct. Res. 51, 153—165.

WEIBEL, E. R., STÄUBLI, W., GUAGI, H. R., HESS, F. A., 1969: Correlated morphometric and biochemical studies on the liver cell. I. Morphometric model, stereolytic methods, and normal morphometric data for rat liver. J. Cell Biol. 42, 68—91.

WERKHEISER, W. C., BARTLEY, W., 1957: The study of steady state concentrations of internal solutes of mitochondria by rapid centrifugal transfer to a fixation medium. Biochem. J. 66, 79—91.

WERNER, D., SCHLEICH, A., KAPOOR, N. K., 1971: Approach to histological detection of metallo-proteins. Evidence for labeling of xanthine oxidase by molybdenum-99. Naturwissenschaften 58, 610—612.

WESTERFIELD, W. W., RICHERT, D. A., RUEGAMER, W. R., 1968: The role of the thyroid hormone in the effect of *p*-chlorophenoxyisobutyrate in rats. Biochem. Pharmacol. 17, 1003—1016.

WHITE, J. E., BRODY, M., 1974: Enzymatic characterization of sucrose-gradient microbodies of dark-grown, greening and constantly light-grown *Euglena gracilis*. FEBS Lett. 40, 325—330.

WIENER, J., LOUD, A. V., KIMBERG, D. V., SPIRO, D., 1968: A quantitative description of cortisone-induced alterations in the ultrastructure of rat liver parenchymal cells. J. Cell Biol. 37, 47—61.

WILLIAMS, N. E., LUFT, J. H., 1968: Use of a nitrogen mustard derivative in fixation for electron microscopy and observations on the ultrastructure of *Tetrahymena*. J. Ultrastruct. Res. 25, 271—292.

WILSON, J. W., LEDUC, E. A., 1963: Mitochondrial changes in the liver of essential fatty acid deficient mice. J. Cell Biol. 16, 281—296.

WIT-PEETERS, E. M., SCHOLTE, H. R., VAN DEN AKKER, F., DE NIE, I., 1971: Intramitochondrial localization of palmitoyl-CoA dehydrogenase, β-hydroxyacyl-CoA dehydrogenase, and enoyl-CoA hydratase in guinea-pig heart. Biochim. Biophys. Acta 231, 23—31.

WOHLRAB, F., 1974: Xanthin-Oxidoreductase-Aktivität in der Aortenwand der Ratte. Acta Histochem. 49, 270—273.

WOOD, R., 1965: The fine structure of hepatic cells in chronic ethionine poisoning and during recovery. Am. J. Pathol. 46, 307—330.

— 1967: Effect of ethionine on the ultrastructure of developing liver cells. J. Ultrastruct. Res. 19, 100—115.

— 1969: Studies on the origin of microbodies in embryonic liver. Anat. Rec. 163, 287 A.

— LEGG, P. G., 1970: Peroxidase activity in rat liver microbodies after aminotriazole inhibition. J. Cell Biol. 45, 576—585.

WYSS, J. R., AEBI, H., 1975: Properties of leukocyte catalase in Swiss type acatalasemia: A comparative study of normals, heterozygotes and homozygotes. Enzyme 20, 257—268.

YASUDA, K., SUZUKI, S., 1965: Immuno-histochemical study on catalase by means of fluorescent antibody method. Acta Anat. Nippon. 38, 19.

YOKOTA, S., 1972: Localization of catalase in mouse liver as detected by fluorescent antibody technique. Med. J. Shinshu Univ. 17, 57—71.

— 1973: Studies on mouse liver urate oxidase. II. Immunochemical and enzymatic distribution of urate oxidase in mouse liver cell fractions. Histochemie 37, 149—159.

— FAHIMI, H. D., 1978: Peroxisome (microbody) membrane—effects of detergents and lipid solvents on its ultrastructure and permeability to catalase. Histochem. J. 10, 469—487.

YOKOTA, S., NAGATA, T., 1974 a: Studies on mouse liver urate oxidase. III. Fine localization of urate oxidase in liver cells revealed by means of ultracryotomy-immunoferritin method. Histochemistry 39, 243—250.

— — 1974 b: Ultrastructural localization of catalase on ultracryotomic sections of mouse liver by ferritin-conjugate antibody technique. Histochemistry 40, 165—174.

— NAKANO, Y., KITAOKA, S., 1978 a: Different effects of some growing conditions on glycolate dehydrogenase in mitochondria and microbodies in *Euglena gracilis*. Agric. Biol. Chem. 42, 115—120.

— — — 1978 b: Metabolism of glycolate in mitochondria of *Euglena gracilis*. Agric. Biol. Chem. 42, 121—129.

Subject Index

Cell Biology Monographs

Price reduction for subscribers: 20%

Springer-Verlag Wien · New York